T0213438

Connections in Discrete Mathematics

Discrete mathematics has been rising in prominence in the past fifty years, both as a tool with practical applications and as a source of new and interesting mathematics. The topics in discrete mathematics have become so well developed that it is easy to forget that common threads connect the different areas, and it is through discovering and using these connections that progress is often made.

For more than fifty years, Ron Graham has been able to illuminate some of these connections and has helped bring the field of discrete mathematics to where it is today. To celebrate his contribution, this volume brings together many of the best researchers working in discrete mathematics, including Fan Chung, Erik Demaine, Persi Diaconis, Peter Frankl, Al Hales, Jeffrey Lagarias, Allen Knutson, Janos Pach, Carl Pomerance, Neil Sloane, and of course Ron Graham himself.

STEVE BUTLER is the Barbara J Janson Professor of Mathematics at Iowa State University. His research interests include spectral graph theory, enumerative combinatorics, mathematics of juggling, and discrete geometry. He is the coeditor of *The Mathematics of Paul Erdős*.

JOSHUA COOPER is Professor of Mathematics at the University of South Carolina. He currently serves on the editorial board of *Involve*. His research interests include spectral hypergraph theory, linear and multilinear algebra, probabilistic combinatorics, quasirandomness, combinatorial number theory, and computational complexity.

GLENN HURLBERT is Professor and Chair of the Department of Mathematics and Applied Mathematics at Virginia Commonwealth University. His research interests include universal cycles, extremal set theory, combinatorial optimization, combinatorial bijections, and mathematical education, and he is recognized as a leader in the field of graph pebbling.

Connections in Discrete Mathematics

A Celebration of the Work of Ron Graham

Edited by

STEVE BUTLER
Iowa State University

JOSHUA COOPER
University of South Carolina

GLENN HURLBERT
Virginia Commonwealth University

CAMBRIDGE
UNIVERSITY PRESS

CAMBRIDGE
UNIVERSITY PRESS

University Printing House, Cambridge CB2 8BS, United Kingdom

One Liberty Plaza, 20th Floor, New York, NY 10006, USA

477 Williamstown Road, Port Melbourne, VIC 3207, Australia

314-321, 3rd Floor, Plot 3, Splendor Forum, Jasola District Centre, New Delhi - 110025, India

79 Anson Road, #06-04/06, Singapore 079906

Cambridge University Press is part of the University of Cambridge.

It furthers the University's mission by disseminating knowledge in the pursuit of education, learning and research at the highest international levels of excellence.

www.cambridge.org
Information on this title: www.cambridge.org/9781107153981
DOI: 9781316650295

© Cambridge University Press 2018

First published 2018

A catalogue record for this publication is available from the British Library

ISBN 978-1-107-15398-1 Hardback
ISBN 978-1-316-60788-6 Paperback

Contents

v

Contributors

Adrian Bolt
Iowa State University, Ames, IA 50011, USA

Joe P. Buhler
Center for Communications Research, San Diego, CA 92121, USA

Steve Butler
Iowa State University, Ames, IA 50011, USA

Fan Chung
University of California at San Diego, La Jolla, CA 92093, USA

Erik D. Demaine
MIT Computer Science and Artificial Intelligence Laboratory, Cambridge, MA 02139, USA

Martin L. Demaine
MIT Computer Science and Artificial Intelligence Laboratory, Cambridge, MA 02139, USA

Persi Diaconis
Departments of Mathematics and Statistics, Stanford University, Stanford, CA 94305, USA

Peter Frankl
Alfréd Rényi Institute of Mathematics, Hungarian Academy of Sciences, H-1053 Budapest, Hungary

Anthony C. Gamst
Center for Communications Research, San Diego, CA 92121, USA

Ron Graham
University of California at San Diego, La Jolla, CA 92093, USA

Alfred W. Hales
Center for Communications Research, San Diego, CA 92121, USA

Neil Hindman
Department of Mathematics, Howard University, Washington, DC 20059,
USA

Espen Hovland
Iowa State University, Ames, IA 50011, USA

Hau-Wen Huang
Department of Mathematics, National Central University, Chung-Li 32001,
Taiwan

Jan Hubička
Computer Science Institute of Charles University (IUUK), Charles University,
11800 Praha, Czech Republic

Allen Knutson
Cornell University, Ithaca, NY 14853, USA

Andrey Kupavskii
Moscow Institute of Physics and Technology, Dolgobrudny, Moscow Region,
141701, Russian Federation; and University of Birmingham, Birmingham,
B15 2TT, UK

Jeffrey C. Lagarias
Department of Mathematics, University of Michigan, Ann Arbor, MI
48109–1043, USA

Karl Levy
Department of Mathematics, Borough of Manhattan Community College
(CUNY), New York, NY 10007, USA

Wen-Ching Winnie Li
Department of Mathematics, Pennsylvania State University, University Park, PA 16802, USA

Rudolph Lorentz
Science Program, Texas A&M University at Qatar, Doha, Qatar

Melvyn B. Nathanson
Department of Mathematics, Lehman College (CUNY), Bronx, NY 10468, USA

Jaroslav Nešetřil
Computer Science Institute of Charles University (IUUK), Charles University, 11800 Praha, Czech Republic

Andrew Odlyzko
School of Mathematics, University of Minnesota, Minneapolis, MN 55455, USA

János Pach
Rényi Institute and EPFL, Station 8, CH-1014 Lausanne, Switzerland

Carl Pomerance
Mathematics Department, Dartmouth College, Hanover, NH 03755, USA

Christian Reiher
Fachbereich Mathematik, Universität Hamburg, 20146 Hamburg, Germany

Vojtěch Rödl
Department of Mathematics and Computer Science, Emory University, Atlanta, GA 30322, USA

Mathias Schacht
Fachbereich Mathematik, Universität Hamburg, 20146 Hamburg, Germany

N. J. A. Sloane
The OEIS Foundation Inc., 11 South Adelaide Ave., Highland Park, NJ 08904, USA

Salvatore Tringali
Institute for Mathematics and Scientific Computing, University of Graz,
NAWI Graz, Heinrichstr. 36, 8010 Graz, Austria

William T. Trotter
School of Mathematics, Georgia Institute of Technology, Atlanta, GA 30332,
USA

Bartosz Walczak
Theoretical Computer Science Department, Faculty of Mathematics and
Computer Science, Jagiellonian University, Kraków 30-348, Poland

Ruidong Wang
Blizzard Entertainment, Irvine, CA 92618, USA

Catherine H. Yan
Department of Mathematics, Texas A&M University, College Station TX
77845, USA

Preface

The last fifty years have seen rapid growth in the rise of discrete mathematics, in areas ranging from the classics of number theory and geometry to the modern tools in computation and algorithms, with hundreds of topics in between. Part of this growth is driven by the increasing availability and importance of computational power, and part is due to the guiding influence and leadership of mathematicians in this field who have helped to encourage generations of mathematicians to pursue research in this area.

Among these mathematicians who have played a leadership role, Ron Graham stands out for his contributions to theory, his visibility to the larger community, his role in mentoring many young mathematicians, and for his longevity.

In 1962, Ron Graham finished his dissertation in combinatorial number theory under the leadership of Derrick Lehmer. He soon found himself at Bell Labs, where he would spend the next thirty-seven years, including as director of the Mathematical Sciences Research Center and as Chief Scientist of AT&T Labs – Research, before his (first) retirement in 1999. During this time he helped bring together many of the best and brightest young minds in discrete mathematics to work in the Labs and would guide their research and prepare them for their later careers. After Bell Labs, Ron went to University of California, San Diego to become the Irwin and Joan Jacobs Chair in Information and Computer Science in the Department of Computer Science and Engineering until his (second) retirement in 2016. He is now emeritus professor at University of California, San Diego and remains mathematically active.

Over the past 50 years, he has had a constant flow of new ideas, with more than 300 papers, a half dozen books, and hundreds of editorial assignments. He has also traveled the world giving talks on mathematics (sometimes demonstrating some of his acrobatic skills such as one-armed handstands and juggling, showing students that mathematicians aren't the stereotypical introverts). Thanks to his friendship with Paul Erdős (30 joint papers!) and his frequent travel, Ron played an important role as a bridge to mathematics in Hungary and several other countries at a time when communication and collaboration were limited. As president of the American Mathematical Society and later the Mathematical Association of America, Ron led the largest mathematical societies in the world. In addition to all of this, Ron Graham has been recognized with numerous awards, accolades, and honorary degrees. It is hard to believe that one individual was able to accomplish so much, but as Ron likes to point out "there are twenty-four hours in the day, and if that's not enough, there are also the nights."

In 2015, Ron Graham turned eighty, and to help mark this occasion a special conference was organized, *Connections in Discrete Mathematics*. This was a chance to bring together many of his friends and colleagues, the best and brightest in discrete mathematics, to celebrate Ron, and also to celebrate discrete mathematics. A major theme of the conference was connections, both the personal connections (as Ron had with so many speakers and participants) as well as the connections between mathematical topics. Both types of connections are what lead to advances in mathematics and open up new ideas for exploration.

This book came out of the conference, with many of the authors having been featured speakers. The chapters here are across the spectrum of discrete mathematics, with topics in number theory, probability, graph theory, Ramsey theory, discrete geometry, algebraic combinatorics, and, of course, juggling. A beautiful mix of topics and also of writing styles, this book has something for everyone.

We thank the many authors for their excellent contributions, including Joe Buhler, Fan Chung, Erik Demaine, Persi Diaconis, Peter Frankl, Al Hales, Jeffrey Lagarias, Allen Knutson, Jarik Nešetřil, Janos Pach, Carl Pomerance, Vojtěch Rödl, Neil Sloane, Tom Trotter, Catherine Yan, and of course Ron Graham. We were happy to see the wealth of ideas that were in these chapters and hope that readers will find something that inspires them.

All three of the editors have been heavily influenced by Ron Graham and his friendship, and like many people in our field, we would not be where we are today without him. Thank you, Ron.

Fig. 0.1. Ron Graham demonstrating a large Rubik's cube to C. K. Cheng, Kevin Milans, and Joel Spencer at the *Connections in Discrete Mathematics* conference in June 2015. Image courtesy of IRMACS. Used with permission.

1

Probabilizing Fibonacci Numbers

Persi Diaconis

Abstract

The question, "What does a typical Fibonacci number look like?" leads to interesting (and impossible) mathematics and many a tale about Ron Graham.

I have talked to Ron Graham every day, more or less, for the past 43 years. Some days we miss a call but some days it's three or four calls; so "once a day" seems about right. There are many reasons: he tells bad jokes, solves my math problems, teaches me things, and is my pal. The following tries to capture a few minutes with Ron.

1.1 A Fibonacci Morning

This term I'm teaching undergraduate number theory at Stanford's Department of Mathematics. It's a course for beginning math majors and the start is pretty dry (I'm in Week 1): every integer is divisible by a prime; if p divides ab then either p divides a or p divides b; unique factorization. I'm using Bill LeVeque's fine *Fundamentals of Number Theory*. It's clear, correct, and cheap (a Dover paperback). He mentions the Fibonacci numbers and I decide to spend some time there to liven things up.

Fibonacci numbers have a "crank math" aspect but they are also serious stuff – from sunflower seeds through Hilbert's tenth problem. My way of understanding anything is to ask, "'What does a typical one look like?" Okay, what does a typical Fibonacci number look like? How many are even? What about the decomposition into prime powers? Are there infinitely many prime Fibonacci numbers? I realize that I don't know and, it turns out, for most of these questions, nobody knows. This is two hours before class and I do what I always do: call Ron.

1

"Hey, do we know what proportion of Fibonacci numbers are even?"

"Sure," he says without missing a beat. "It's $1/3$, and it's easy: let me explain. If you write them out,

$$0, 1, 1, 2, 3, 5, 8, 13, 21, 34, 55, 89, 144, 233, 377, \ldots,$$

you see that every third one is even, and it's easy to see from the recurrence." Similarly, every fourth one is a multiple of three, so $1/4$ are divisible by 3. I note that the period for five is 5, ruining my guess at the pattern. He tells me that after five, the period for the prime p is a divisor of $p \pm 1$; so $1/8$ of all the F_n are divisible by seven. Actually, things are a bit more subtle. For any prime p, the *sequence* of Fibonacci numbers (mod p) is periodic. Let's start them at $F_0 = 0$, when $p = 3$:

$$0, 1, 1, 2, 0, 2, 2, 1, 0, 1, 1, 2, 0, 2, 2, 1, \ldots.$$

The length of the period is called the *Pisano period* (with its own Wikipedia page). The period of three is 8 and there are two zeros, so $1/4$ are divisible by 3; $3/8$ are 1 (mod 3) and $3/8$ are 2 (mod 3). These periods turn out to be pretty chaotic and much is conjectural. The rest of my questions, e.g., is F_n prime infinitely often, are worse: "not in our lifetimes," (Ron says) Erdős said.

Going back to my phone conversation with Ron, he says:

> Here's something your kids can do: You know the Fibonacci numbers grow pretty fast. This means that $\sum_{n=0}^{\infty} 1/F_{2^n}$ converges to its limiting value very fast. It turns out to be a quadratic irrational (!) *and* you can show that if a number has a more rapidly converging rational approximation, it's transcendental. (!)

And then he says, "Here's an easier one for your kids: ask them to add up $F_n/10^n$, $1 \le n < \infty$." Answer: $10/89$ (!)

It turns out that Ron had worked on my questions before. A 1964 paper [7] starts, "Let $S(L_0, L_1) = L_0, L_1, L_2, \ldots$ be the sequence of integers which satisfy the recurrence $L_{n+2} = L_{n+1} + L_n$, $n = 0, 1, 2, \ldots$. It is clear that the values L_0 and L_1 determine $S(L_0, L_1)$, e.g., $L(0, 1)$ is just the sequence of Fibonacci numbers. It is not known whether or not infinitely many primes occur in $S(0, 1)$ ...." He goes on to find an opposite: a pair L_0, L_1 so that *no* primes occur in $S(L_0, L_1)$. His best solution was

$$L_0 = 3316356359982747374472200656430763$$
$$L_1 = 1510028911088401971189590305498785$$

This was subsequently improved by Knuth and then Wilf. The problem itself now has its own Wikipedia page; search for "primefree sequence." The current record is

$$L_0 = 106276436867$$
$$L_1 = 35256392432$$

due to Vesemirsov.

Finding such sequences is related to problems such as, "Is every odd number the sum of a prime plus a power of 2?" The answer is no; indeed Erdös [6] found arithmetic progressions with no numbers of this form. For this, he created the topic/tool of "covering congruences": a sequence $\{a_1 \pmod{n_1}, \ldots, a_k \pmod{n_k}\}$ of finitely many residue classes $\{a_i + n_i x, x \in \mathbb{Z}\}$ whose union covers $\mathbb{Z}$. For example, $\{0 \pmod 2, 0 \pmod 3, 1 \pmod 4, 5 \pmod 6, 7 \pmod{12}\}$ is a covering set where all moduli are distinct. Erdős asked if there were such distinct covering systems where the smallest modulus – two in the example – is arbitrarily large. A variety of number theory hackers found examples. For instance, Nicesend found a set of more than 10^{50} distinct congruences with minimum modulus 40. One of Ron's favorite negative results is a theorem of Hough [9]: there is an absolute upper bound to the minimum modulus of a system of distinct covering congruences. The Wikipedia phrase is "covering sequences."

The preceding paragraphs are amplified from sentences of this same phone conversation. Ron has worked on math problems where Fibonacci facts form a crucial part of the argument "from then to now." For example, in joint work with Fan Chung [2] they solved an old conjecture of D. J. Newman: for a sequence of numbers (mod 1), $x = (x_0, x_1, x_2, \ldots)$, define the strong discrepancy

$$D(x) = \inf_n \inf_m n \|x_n - x_{n+m}\|.$$

They found the following:

Theorem 1.1.

$$\sup_x D(x) \le \frac{1}{\sqrt 5}.$$

The reader who looks will find Fibonacci numbers throughout the proofs; $\sum_{n=0}^{\infty} 1/F_{2n}$ makes an appearance.

As a parting shot in our conversation, Ron moved back to the periods of Fibonacci and Lucus sequences $SL(L_0, L_1)$ in the preceding text. "You know, we had a pretty good trick in our book [4] using Fibonacci periods. You should perform it for your kids." Let me perform it for you. To understand the connection, see Diaconis and Graham [4, p. 187].

The performer draws a 4×4 square on a sheet of paper. A prediction is written on the back (to own up, it's 49).

The patter goes as follows: "They teach kids the craziest things in school nowadays. The other day my daughter came home talking about 'adding mod seven.' That means you add and take away anything over 7, so $5 + 5 = 10 = 3$ (mod 7). Here, let's try it out." Pick any two small numbers; say 5 and 6 are chosen. Write them down in positions $(1, 1)$ and $(1, 2)$. Then sum mod 7 in position $(1, 3)$:

5	6	4			5	6	4	3		5	6	4	3
				$\rightarrow$	0	3	3	6	$\rightarrow$	0	3	3	6
										2	1	3	4
										0	4	4	1

Continue as shown, adding successive pairs row by row until all squares are filled. You can ask spectators to help along the way. At the end, have someone (carefully) add up all 16 numbers in the usual way. The sum will match your prediction, 49. Our write-up gives pointers to the mathematical literature on Fibonacci periods.

In a follow-up call, I mentioned a charming fact pointed out by Susan Holmes. If you want to convert from miles to kilometers (and back) take the next Fibonacci number (or the one before, to go back). Thus 5 miles is close to 8 kilometers, 13 miles is close to 21 kilometers, 144 kilometers is close to 89 miles, and so on. To do general numbers, use Zeckendorf's theorem: any positive integer can be represented as a sum of distinct Fibonacci numbers: uniquely, if you never use two consecutive F_n. So $100 = 89 + 8 + 3$, and 100 miles is about $144 + 13 + 5 = 162$ kilometers. (Really, 100 miles = 160.934 kilometers; it's only an approximation.) The Zeckendorf representation is easy to find, just subtract off the largest possible F_n each time. For much more Fibonacciana, see [8].

1.2 A Second Try

Here is a more successful approach to the question of what a typical F_n looks like. Take one of the many codings of Fibonacci numbers and answer the question there. For example, F_n counts the number of binary strings of length $n - 2$ with no two consecutive ones:

$$F_5 = 5 \longleftrightarrow \{000, 001, 010, 100, 101\}$$

Let $\mathcal{F}_n$ be the Fibonacci strings of length n. So $|\mathcal{F}_n| = F_{n+2}$. This mismatch in notation is unfortunate, but keeping the classical notation for F_n makes the literature easier to use.

For this coding, it is natural to ask, "What does a typical element of $\mathcal{F}_n$ look like?" Throughout, use the uniform distribution

$$P_n(x) = 1/F_{n+2}$$

on $\mathcal{F}_n$. This distribution is well known in statistical physics as the "hardcore model in 1-D." Let $X_i(x)$ be the ith bit of x. Natural questions are:

- What is the chance that $X_i = 1$?
- What is the distribution of $X_1 + \cdots + X_n$?
- How long is the longest zero run of $X_1, X_2, \ldots, X_n$?
- What is the waiting time distribution for the first one?

Indeed, any question asked for coin tossing can be asked for $\mathcal{F}_n$. In [5], a simple, efficient algorithm is given for exact generation of a uniformly chosen element of $\mathcal{F}_n$ (using the Fibonacci number system). Of course, there is literature on the mixing time of the natural Markov chain for generating from the uniform distribution on $\mathcal{F}_n$ – pick a coordinate at random; try to change to its opposite – see [10].

The main results developed (Propositions 1.1 to 1.6): as a process, $X_1, X_2, \ldots, X_n$ is close to a binary Markov chain $\tilde{X}_1, \tilde{X}_2, \ldots, \tilde{X}_n$, where, with

$$\theta = \frac{\sqrt{5} - 1}{2} \doteq 0.6180$$

so $\theta + \theta^2 = 1$,

$$P\{\tilde{X}_1 = 0\} = \theta, \qquad \text{with transition matrix } P = \begin{pmatrix} \theta & \theta^2 \\ 1 & 0 \end{pmatrix}. \qquad (1.1)$$

The closeness is in total variation for $X_1, \ldots, X_k$ with $k = n - f(n)$, $f(n) \to \infty$. This is strong enough to give useful answers to the previous four questions and many others. Let us turn now to mathematics.

Proposition 1.1. *For any i, $1 \le i \le n$,*

$$P_n(X_i = 0) = \frac{F_{i+1}F_{n+2-i}}{F_{n+2}}, \qquad P_n(X_i = 1) = \frac{F_i F_{n+1-i}}{F_{n+2}}.$$

Proof. Sequences with $X_i = 0$ may begin with any Fibonacci sequence of length $i - 1$ (F_{i+1} choices) and end with any Fibonacci sequence of length $n - i$ (F_{n-i+2} choices). Dividing by the total number of Fibonacci sequences of length n (F_{n+2}) gives the first result. The second is similar; a one in position i forces zeros at $i - 1, i + 1$. After this, the start and end are arbitrary Fibonacci sequences.   □

For subsequent use, recall that, with

$$\phi = \frac{1 + \sqrt{5}}{2}, \qquad \psi = \frac{1 - \sqrt{5}}{2} = 1 - \phi = \frac{1}{-\phi} \doteq -0.6180,$$

$$F_n = \frac{\phi^n - (-\phi)^{-n}}{\sqrt{5}} \qquad\qquad (1.2)$$

and F_n is the closest integer to $\phi^n/\sqrt{5}$. Recall $\theta = (\sqrt{5} - 1)/2$. Combining (1.2) and Proposition 1.1, standard asymptotics gives

Proposition 1.2.

(a) $P_n(X_i = 0) = P_n(X_{n-i+1} = 0), \quad 1 \le i \le n.$ *(symmetry)*

(b) $P_n(X_1 = 0) = \theta\left[1 + O(\phi^{-2n})\right].$

(c) $P(X_i = 0) = \dfrac{\theta^2}{\sqrt{5}}\left[1 + O(\phi^{-2i}) + O(\phi^{-2(n-i)})\right].$ □

Remark. Part (c) shows, if i and $n - i$ are large,

$$P(X_i = 0) \sim \frac{\theta^2}{\sqrt{5}}, \qquad P(X_i = 1) \sim 1 - \frac{\theta^2}{\sqrt{5}}.$$

This of course is the stationary distribution of the transition matrix in (1.2). It is useful to collect together properties of the Markov chain.

Proposition 1.3. *For $\theta = (\sqrt{5} - 1)/2$, let a Markov chain $\{\tilde{X}_n\}$ on $\{0, 1\}$ have transition matrix*

$$P = \begin{pmatrix} \theta & \theta^2 \\ 1 & 0 \end{pmatrix},$$

starting distribution $P(\tilde{X}_1 = 0) = \theta$, $P(\tilde{X}_1 = 1) = \theta^2$. Then P has stationary distribution

$$\pi(0) = \frac{\theta^2}{\sqrt{5}}, \qquad \pi(1) = 1 - \frac{\theta^2}{\sqrt{5}},$$

and P is reversible. The eigenvalues are $\beta_0 = 1$, $\beta_1 = \theta - 1$. The right eigenvectors are

$$f_0 = \begin{pmatrix} 1 \\ 1 \end{pmatrix}, \qquad f_1 = \begin{pmatrix} 1 \\ -1/\theta^2 \end{pmatrix}.$$

If the chain is denoted $\tilde{X}_i$, $1 \le i < \infty$, for all n and $e_1, \ldots, e_n \in \{0, 1\}$ an allowable sequence,

$$P(\tilde{X}_1 = e_1, \tilde{X}_2 = e_2, \ldots, \tilde{X}_n = e_n) = \theta^{n+e_n}. \qquad \square$$

The main result of this section gives an explicit bound between the probability distribution $\mu_{n,k}$ of $X_1, \ldots, X_k$ from the Fibonacci chain and $\tilde{\mu}_k$ the probability distribution of the Markov chain $\tilde{X}_1, \ldots, \tilde{X}_k$ as in Proposition 1.3. The total variation distance is

$$\|\mu_{nk} - \tilde{\mu}_k\|_{\mathrm{TV}} = \max_{A \subseteq C_2^k} |\mu_{n,k}(A) - \tilde{\mu}_k(A)|.$$

Proposition 1.4. *With notation as earlier, for all n and $1 \le k \le n$,*

$$\|\mu_{n,k} - \tilde{\mu}_k\| = O(\theta^{2(n-k)}).$$

Proof. As usual,

$$\|\mu_{nk} - \tilde{\mu}_k\|_{\mathrm{TV}} = \frac{1}{2} \sum_{x \in C_2^k} |P(X_1 = x_1, \ldots, X_k = x_k)$$
$$-P(\tilde{X}_1 = x_1, \ldots, \tilde{X}_k = x_k)|.$$

For any $x_1, \ldots, x_k$,

$$P_n(X_1 = x_1, \ldots, X_k = x_k) = \frac{F_{n+2-k-x_n}}{F_{n+2}},$$
$$P\{\tilde{X}_1 = x_1, \ldots, \tilde{X}_n = x_n\} = \theta^{k+x_i}.$$

It follows that

$$\|\mu_{nk} - \tilde{\mu}_k\|_{\mathrm{TV}} = \frac{1}{2} \left\{ \left| \frac{F_{n+2-k}}{F_{n+2}} - \theta^k \right| F_{k+1} + \left| \frac{F_{n+1-k}}{F_{n+2}} - \theta^{k+1} \right| F_k \right\}.$$

The claimed result now follows from (1.2) in a straightforward manner. $\square$

Some of the preceding questions have been previously answered. Let $S_n = X_1 + \cdots + X_n$. Diaconis, Graham, and Holmes [5] prove

Proposition 1.5.

(a) $P_n(S_n = k) = \dfrac{\binom{n+1-k}{k}}{F_{n+2}}, \quad 0 \le k \le \left\lfloor \dfrac{n+1}{2} \right\rfloor.$

(b) $E_n(S_n) = (n+1)\dfrac{\sqrt{5}-1}{2\sqrt{5}} + \dfrac{1}{5\phi} + O(n\phi^{-2n}), \quad \mathrm{Var}(S_n) = \dfrac{n+1}{5\sqrt{5}} + O(1).$

(c) $P_n\left(\dfrac{S_n - E_n}{\sqrt{\mathrm{Var}_n}} \le x\right) \longrightarrow \dfrac{1}{\sqrt{2\pi}}\displaystyle\int_{-\infty}^{x} e^{-t^2/2}\, dt \quad \text{as } n \text{ tends to infinity.} \qquad \square$

The longest zero run can be determined by solving the problem for the Markov chain and then transferring it to $X_1, X_2, \ldots, X_n$ using Proposition 1.4.

Proposition 1.6. *Let M_n be the longest zero run for a uniform element of $\mathcal{F}_n$. Then $M_n/\log_{1/\theta} n \to 1$ in probability.*

Proof. Proposition 1.6 follows by first proving the parallel result for the Markov chain $\tilde{X}_i$ and then tranferring to X_i using Proposition 1.4. Let $l = l(n) = \lfloor \log_{1/\theta} n \rfloor$. From Proposition 1.3, the transition matrix P is explicitly diagonalized as $P = VDV^{-1}$, with V the matrix with column vectors the right eigenvectors of P, and D a diagonal matrix of eigenvalues

$$V = \begin{pmatrix} 1 & 1 \\ 1 & -1/\theta^2 \end{pmatrix}, \quad D = \begin{pmatrix} 1 & 0 \\ 0 & -\theta^2 \end{pmatrix}, \quad V^{-1} = \frac{\theta^2}{1+\theta^2}\begin{pmatrix} 1/\theta^2 & 1 \\ 1 & -1 \end{pmatrix}.$$

Here $\theta + \theta^2 = 1$ and the stationary distribution is $\pi(0) = 1/(1+\theta^2)$, $\pi(1) = \theta^2/(1+\theta^2)$. Thus

$$P_0\{\tilde{X}_i = 0\} = P^i(0,0) = (VD^iV^{-1})_{00} = \pi(0) + O(\theta^{2i}),$$
$$P_1\{\tilde{X}_i = 0\} = \pi(0) + O(\theta^{2i}).$$

Since $P_a\{\tilde{X}_i = \tilde{X}_{i+1} = \cdots = \tilde{X}_{i+l-1} = 1\} = P_a\{\tilde{X}_i = 0\}P^{l-1}(0,0)$, for either starting state a, for any starting distribution σ,

$$P_\sigma(\tilde{X}_i = \cdots = \tilde{X}_{i+l-1} = 1) = \pi(0)P^{l-1}(0,0) + O\left(\theta^{2i}P^{l-1}(0,0)\right). \qquad (1.3)$$

From this,

$$P_\sigma\left\{\tilde{M}_n \ge (1+\epsilon)l\right\} = P_\sigma\left\{\bigcup_{i=0}^{n-l} [\text{0-run from } i \ge (1+\epsilon)l]\right\}$$
$$\le (n-l)\pi(0)P^{\lfloor l(1+\epsilon)\rfloor}(0,0) + O\left(P^{\lfloor l(1+\epsilon)\rfloor}(0,0)\right).$$

From the choice of l, $P^{\lfloor l(1+\epsilon)\rfloor}(0,0) = O(1/n^{1+\epsilon})$, so the right-hand side tends to 0.

To show that $\tilde{M}_n/l \geq (1 - \epsilon)$ with high probability, split $[n]$ into disjoint blocks of length $\lfloor l(1 - \epsilon) \rfloor$. Let $\tilde{Y}_i$ be 1 or 0 as the ith block is all 0's or not. Let $\widetilde{W} = \sum_{i=1}^{n/l(1-\epsilon)} \tilde{Y}_i$. The second moment method will be used to show $P\{\widetilde{W} > 0\} \to 1$. From (1.3), with l replaced by $\lfloor l(1 - \epsilon) \rfloor$,

$$E(\widetilde{W}) = \frac{n}{l(1 - \epsilon)} \pi(0) P^{\lfloor l(1-\epsilon) \rfloor}(0, 0) + O\left(P^{\lfloor l(1-\epsilon) \rfloor}(0, 0)\right) \sim \frac{n^{\epsilon} \pi(0)}{l(1 - \epsilon)}$$

(1.4)

$$\mathrm{Var}(\widetilde{W}) = \sum_i \mathrm{Var}(\tilde{Y}_i) + 2 \sum_{i<j} \mathrm{Cov}(\tilde{Y}_i, \tilde{Y}_j).$$

(1.5)

Since $\tilde{Y}_i$ are binary, the asymptotics of the first sum in (1.5) are as in (1.4). Using the Markov property, $P_\sigma(\tilde{Y}_i = \tilde{Y}_j = 1) = P_\sigma(\tilde{Y}_i = 1)P_0(\tilde{Y}_{j-i} = 1)$. The terms may be bounded using (1.3) and the second sum is of order $n/n^{2(1-\epsilon)}$. It follows that the variance of $\widetilde{W}$ is of the same order as the mean, so a Chebyshev bound shows

$$P\left\{\frac{\tilde{M}_n}{l} > 1 - \epsilon\right\} \longrightarrow 1.$$

This proves Proposition 1.6 with M_n replaced by $\tilde{M}_n$. The transfer of the limit theorem back to M_n is routine from Proposition 1.4.   □

Remark. More refined limiting behavior of M_n will surely be colored by the nonexistence of limiting behavior associated with the maximum of discrete random variables. See [3, 13].

I cannot leave this topic without remarking on some amazing formulas communicated to me by Richard Stanley. Throughout, let $X_i(x)$ be the ith binary digit of a uniformly chosen point in $\mathcal{F}_n$. Define a random variable

$$W_n(x) = \prod_{1 \leq i \leq n} i^{X_i(x)}.$$

Stanley (in personal correspondence) shows

$$E(W_n) = \frac{1}{F_{n+2}} \sum_{\lambda \vdash n+1} f(\lambda), \qquad E(W_n^2) = \frac{1}{F_{n+2}}(n + 1)!.$$

(1.6)

In (1.6), $f(\lambda)$ is the dimension of the irreducible representation of the symmetric group S_n corresponding to the partition λ. It is well known [11, pp. 62–64] that $\sum f(\lambda)$ equals the number of involutions in S_{n+1} and $\sum f^2(\lambda) = (n + 1)!$.

The formulas (1.6) are sufficiently surprising that a numerical check seems called for. Consider $n = 3$; S_4 has 10 involutions and 24 elements. For $W_n(x)$ the product over the empty set is 1:

x	000	100	010	001	101	Sum
$W_3(x)$	1	1	2	3	3	10
$W_3^2(x)$	1	1	4	9	9	24

The asymptotics of $\sum f(\lambda)$ are well known [11, p. 64]. This gives

$$E(W_n) \sim \frac{(n+1)^{(n+1)/2} \exp\left\{-(n+1)/2 + \sqrt{n+1} - 1/2\right\}}{\sqrt{2}\, F_{n+2}},$$

$$\mathrm{Var}(W_n) \sim \frac{(n+1)!}{F_{n+2}} \sim \frac{(n+1)^{(n+1)} e^{-(n+1)}}{\sqrt{2\pi n}\, F_{n+2}}.$$

From this, we see that W_n is concentrated around its mean. Proposition 1.3 and easier calculations show that $L_n = \log(W_n)$ has a limiting normal distribution, so W_n is log normal in the limit.

A somewhat contrived set of steps leading to consideration of W_n may be constructed as follows. Suppose one wanted to consider a random square-free number with factors at most x. One natural way to do this considers the uniform distribution. Another natural approach is to consider $\epsilon_1, \ldots, \epsilon_n$ independent, 0/1 random variables with $P(\epsilon_i = 1) = P(\epsilon_i = 0) = 1/2$ and define

$$\widetilde{W}_n = \prod_{1 \leq i \leq n} p_i^{\epsilon_i}$$

with $2 = p_1 < p_2 < \cdots < p_n$ the distinct primes. An easier (but quite similar) problem considers

$$\widetilde{\widetilde{W}}_n = \prod_{1 \leq i \leq n} i^{\epsilon_i}.$$

This has

$$E\left(\widetilde{\widetilde{W}}_n\right) = \frac{n!}{2^n}, \qquad E\left(\widetilde{\widetilde{W}}_n^2\right) = \frac{(n!)^2}{2^n}.$$

So again $\widetilde{\widetilde{W}}_n$ is concentrated about its mean and $\log \widetilde{\widetilde{W}}_n$ is asymptotically normal. The random variable W_n is the Fibonacci version of $\widetilde{\widetilde{W}}_n$. (Okay, okay; I said it was contrived.)

Stanley's motivation comes from his theory of differential posets. In [14, Problem 8], Stanley constructed a sequence of semisimple algebras A_n of dimension $(n + 1)!$ whose irreducible representations have degree $W_n(x)$ for $x \in \mathcal{F}_n$. Thus the number of irreducible representations is F_{n+2}. The existence of $A_1 \subseteq A_2 \subseteq A_3 \subseteq \ldots$ (with nice restriction properties) is not hard to show by "general nonsense." A useful set of generators and relations was found by

Okada [12]. This has spawned a host of interesting developments that may be found by following the citations to Okada's paper.

The tension between recreational math and "real math" is evident throughout the Fibonacci world. As a parting shot, I offer the following: 144 is a Fibonacci number and it's also a perfect square (uh-oh). Also, 8 is a Fibonacci number that is a cube (uh-oh). Are there any others? No! Bugeaud et al. [1] proved that 1, 8, 144 are the only Fibonacci numbers that are powers. Their proof makes real use of the full machine of modern number theory.

There are other codings of F_n; see https://oeis.org/A000045 at the On-Line Encyclopedia of Integer Sequences. Also, in [15], parts (b), (c), and (d) of Exercise 1.35 are about compositions with specified parts. In [16], part (a) of Exercise 7.66 has a cute proof. There are also Lucas numbers; I don't know any codings for them. Some of these suggest fresh questions. Fortunately, I can call Ron.

Acknowledgments

Thanks to Sourav Chatterjee, Angela Hicks, Susan Holmes, Kannan Soundararajan, Richard Stanley, and most of all to Ron Graham, for help with this work.

References

1. Bugeaud, Y., Mignotte, M., and Siksek, S. Classical and modular approaches to exponential Diophantine equations. I. Fibonacci and Lucas perfect powers. *Ann. Math. (2)* **163** (2006) 969–1018.
2. Chung, F. and Graham, R. On the discrepancy of circular sequences of reals. *J. Number Theory* **164** (2016) 52–65.
3. D'Aristotile, A., Diaconis, P., and Freedman, D. On merging of probabilities. *Sankhyā Ser. A* **50** (1988) 363–380.
4. Diaconis, P., and Graham, R. *Magical Mathematics: The Mathematical Ideas That Animate Great Magic Tricks*. Princeton University Press, Princeton, NJ, 2012.
5. Diaconis, P., Graham, R., and Holmes, S. P. Statistical problems involving permutations with restricted positions. In *State of the Art in Probability and Statistics (Leiden, 1999)*. Vol. 36 of *IMS Lecture Notes Monogr. Ser.* Inst. Math. Statist., Beachwood, OH, 2001, 195–222.
6. Erdös, P. On integers of the form $2^k + p$ and some related problems. *Sum. Bras. Math.* **II** (1950) 113–123.
7. Graham, R. L. A Fibonacci-like sequence of composite numbers. *Math. Mag.* **37** (1964) 322–324.
8. Graham, R. L., Knuth, D. E., and Patashnik, O. *Concrete Mathematics: A Foundation for Computer Science*, 2nd ed. Addison-Wesley Reading, MA, 1994.
9. Hough, B. Solution of the minimum modulus problem for covering systems. *Ann. Math. (2)* **181** (2015) 361–382.

10. Kannan, R., Mahoney, M. W., and Montenegro, R. Rapid mixing of several Markov chains for a hard-core model. In *Algorithms and Computation*. Vol. 2906 of Lecture Notes in Computer Science, Vol. 2906, 663–675. Springer, Berlin, 2003.

11. Knuth, D. E. *The Art of Computer Programming*. Vol. 3, 2nd ed. Addison-Wesley, Reading, MA, 1998.

12. Okada, S. Algebras associated to the Young–Fibonacci lattice. *Trans. Am. Math. Soc.*, **346** (1994) 549–568.

13. Révész, P. Strong theorems on coin tossing. In *Proceedings of the International Congress of Mathematicians (Helsinki, 1978)*. Acad. Sci. Fennica, Helsinki, 749–754, 1980.

14. Stanley, R. P. Differential posets. *J. Amer. Math. Soc.* **1** (1988) 919–961.

15. Stanley, R. P. *Enumerative Combinatorics*. Vol. 1. Cambridge Studies in Advanced Mathematics, Vol. 49, 2nd ed. Cambridge University Press, Cambridge, 2012.

16. Stanley, R. P. *Enumerative Combinatorics*. Vol. 2. Cambridge Studies in Advanced Mathematics, Vol. 62. Cambridge University Press, Cambridge, 1999.

2

On the Number of ON Cells in Cellular Automata

N. J. A. Sloane

Abstract

If a cellular automaton (CA) is started with a single ON cell, how many cells will be ON after n generations? For certain "odd-rule" CAs, including Rule 150, Rule 614, and Fredkin's Replicator, the answer can be found by using the combination of a new transformation of sequences, the run length transform, and some delicate scissor cuts. Several other CAs are also discussed, although the analysis becomes more difficult as the patterns become more intricate.

2.1 Introduction

When confronted with a number sequence, the first thing is to try to conjecture a rule or formula, and then (the hard part) prove that the formula is correct. This chapter had its origin in the study of one such sequence, 1, 8, 8, 24, 8, 64, 24, 112, 8, 64, 64, 192, ... (A160239[1]), although several similar sequences will also be discussed.

These sequences arise from studying how activity spreads in cellular automata (for background see [2, 5, 8, 11, 14, 17, 20, 21, 23, 24, 26]). If we start with a single ON cell, how many cells will be ON after n generations? The preceding sequence arises from the CA known as Fredkin's Replicator [13]. In 2014, Hrothgar sent the author a manuscript [10] studying this CA, and conjectured that the sequence satisfied a certain recurrence. One of the goals of the present chapter is to prove that this conjecture is correct; see (2.31).

In §2.2 we discuss a general class (the "odd-rule" CAs) to which Fredkin's Replicator belongs, and in §2.3 we introduce an operation on number sequences (the "run length transform") that helps in understanding the resulting

---

[1] Six-digit numbers prefixed by A refer to entries in [16].

sequences. Fredkin's Replicator, which is based on the Moore neighborhood, is the subject of §2.4, and §2.5 analyzes another odd-rule CA, based on the von Neumann neighborhood with a center cell. Although these two CAs are similar, different techniques are required for establishing the recurrences. Both proofs involve making scissor cuts to dissect the configuration of ON cells into recognizable pieces.

§2.6 discusses some other CAs in one, two, and three dimensions where it is possible to find a formula, and some for which no formula is known at present. In dimension one, Stephen Wolfram's well-known list [17, 24, 26] of 256 different CAs based on a three-celled neighborhood gives rise to just seven interesting sequences (§2.6.1). Other two-dimensional CAs are discussed in §§2.6.2, 2.6.3, and the three-dimensional analog of Fredkin's Replicator in §2.6.4. The final section (§2.7) gives some additional properties of the run length transform. For many further examples of cellular automata sequences, see [2] and [16] (the index to [16] lists more than 200 such sequences).

2.2 Odd-Rule CAs

We consider cellular automata whose cells form a d-dimensional cubic lattice $\mathbb{Z}^d$, where d is 1, 2, or 3. Each cell is either ON or OFF, and an ON cell with center at the lattice point $u = (u_1, u_2, \ldots, u_d) \in \mathbb{Z}^d$ will be identified with the monomial $x^u = x_1^{u_1} x_2^{u_2} \cdots x_d^{u_d}$, which we regard as an element of the ring of Laurent polynomials $\mathcal{R} = \mathrm{GF}(2)[x_1, x_1^{-1}, \ldots, x_d, x_d^{-1}]$ with mod 2 coefficients. The state of the CA is specified by giving the formal sum S of all its ON cells. As long as only finitely many cells are ON, S is indeed a polynomial in the variables x_i and x_i^{-1}, and is therefore an element of $\mathcal{R}$. We write $u \in S$ to indicate that u is ON, that is, that x^u is a monomial in S.

In most of this chapter we focus on what may be called "odd-rule" CAs. An odd-rule CA is defined by specifying a neighborhood of the cell at the origin, given by an element $F \in \mathcal{R}$ listing the cells in the neighborhood. A typical example is the Moore neighborhood in $\mathbb{Z}^2$, which consists of the eight cells surrounding the cell at the origin in the square grid (see Fig. 2.1(ix)), and is specified by

$$F := \frac{1}{xy} + \frac{1}{y} + \frac{x}{y} + \frac{1}{x} + x + \frac{y}{x} + y + xy$$

$$= \left(\frac{1}{x} + 1 + x\right)\left(\frac{1}{y} + 1 + y\right) - 1 \in \mathcal{R} = \mathrm{GF}(2)[x, x^{-1}, y, y^{-1}]. \quad (2.1)$$

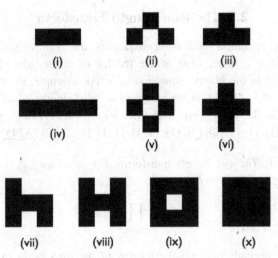

Fig. 2.1. Some two-dimensional neighborhoods. (Parts i and iv are three and five cells wide, respectively).

The neighborhood of an arbitrary cell u is obtained by shifting F so it is centered at u, that is, by the product $x^u F \in \mathcal{R}$. Given F, the corresponding *odd-rule* CA is defined by the rule that the cell at u is ON at generation $n + 1$ if it is the neighbor of an odd number of cells that were ON at generation n, and is otherwise OFF.

Our goal is to find $a_n(F)$, the number of ON cells at the nth generation when the CA is started in generation 0 with a single ON cell at the origin. For odd-rule CAs there is a simple formula. The number of nonzero terms in an element $P \in \mathcal{R}$ will be denoted by $|P|$.

Theorem 2.1. *For an odd-rule CA with neighborhood F, the state at generation n is equal to F^n, and $a_n(F) = |F^n|$.*

Proof. We use induction on n. By definition, the initial state is $1 = F^0$, and $a_0(F) = 1$. The ON cell at the origin turns ON all the cells in F, so the state at generation 1 is F itself, and $a_1(F) = |F|$. Suppose the state at generation n is F^n. An ON cell $w \in F^n$ will affect a cell u if and only if u is in the neighborhood of w, that is, if and only if $u \in wF$. For u to be turned ON, there must be an odd number of cells $w \in F^n$ with $u \in wF$. Since the coefficients in $\mathcal{R}$ are evaluated mod 2, u will be turned ON if and only if $u \in \sum_{w \in F^n} wF = F \sum_{w \in F^n} w = F \cdot F^n = F^{n+1}$. So F^{n+1} is precisely the state at generation $n + 1$, and $a_{n+1}(F) = |F^{n+1}|$. $\qquad\square$

2.3 The Run Length Transform

We define an operation on number sequences, the "run length transform."
For an integer $n \geq 0$, let $\mathcal{L}(n)$ denote the list of the lengths of the maximal runs of 1s in the binary expansion of n. For example, since the binary
expansion of 55 is 110111, containing runs of 1s of lengths 2 and 3, $\mathcal{L}(55) =$
$[2, 3]$. $\mathcal{L}(0)$ is the empty list, and $\mathcal{L}(n)$ for $n = 1, \ldots, 12$ is respectively
$[1], [1], [2], [1], [1, 1], [2], [3], [1], [1, 1], [1, 1], [1, 2], [2]$ (A245562).

Definition 2.1. *The* run length transform *of a sequence* $[S_n, n \geq 0]$ *is the
sequence* $[T_n, n \geq 0]$ *given by*

$$T_n = \prod_{i \in \mathcal{L}(n)} S_i. \tag{2.2}$$

Note that T_n depends only on the lengths of the runs of 1s in the binary
expansion of n, not on the order in which they appear. For example, since
$\mathcal{L}(11) = [1, 2]$ and $\mathcal{L}(13) = [2, 1]$, $T_{11} = T_{13} = S_1 S_2$. Also $T_0 = 1$ (the empty
product), so the value of S_0 is never used, and will usually be taken to be 1. Fur-
ther properties and additional examples of the run length transform are given
in §2.7. See especially Table 2.4, which shows how the transformed sequence
has a natural division into blocks.

Define the *height* ht(P) of an element $P \in \mathcal{R}$ to be the maximal value of $|e_i|$
in any monomial $x_1^{e_1} \cdots x_d^{e_d}$ in P. If ht$(P) = h$, the cells in P are contained in a
d-dimensional cube centered at the origin with edges that are $2h + 1$ cells long.
Note that ht$(PQ) \leq$ ht$(P) +$ ht(Q) and ht$(P^k) = k$ ht(P).

The second property that makes odd-rule CAs easier to analyze than most
is the following.

Theorem 2.2. *If* ht$(F) \leq 1$, *then* $[a_n(F), n \geq 0]$ *is the run length transform of
the subsequence*

$$[a_0(F), a_1(F), a_3(F), a_7(F), a_{15}(F), \ldots, a_{2^k-1}(F), \ldots]. \tag{2.3}$$

Proof. The proof depends on the identity sometimes called the Freshman's
Dream, which in its simplest form states that $(x + y)^2 \equiv x^2 + y^2$ mod 2, and
more generally that for $P(x_1, x_1^{-1}, \ldots) \in \mathcal{R}$,

$$P(x_1, x_1^{-1}, \ldots)^{2^k} = P(x_1^{2^k}, x_1^{-2^k}, \ldots), \tag{2.4}$$

for any integer $k \geq 0$, and in particular that $|P(x_1, x_1^{-1}, \ldots)^{2^k}| = |P|$. Suppose
first that the binary expansion of n contains exactly two runs of 1s, separated

by one or more 0s, say

$$n = \overbrace{111\cdots1}^{m_1}00\cdots0\overbrace{111\cdots1}^{m_3}, \quad \text{where } m_1, m_2, m_3 \geq 1,$$

that is,

$$n = (2^{m_1} - 1)2^{m_2+m_3} + (2^{m_3} - 1),$$

with $\mathcal{L}(n) = [m_1, m_3]$. Then

$$F^n = (F^{2^{m_1}-1})^{2^{m_2+m_3}} F^{2^{m_3}-1} = P^{2^{m_2+m_3}} Q \text{ (say)}, \tag{2.5}$$

where $P := F^{2^{m_1}-1}$, $Q := F^{2^{m_3}-1}$. Equation (2.5) states that F^n is a sum of copies of Q centered at the cells of of $P^{2^{m_2+m_3}}$. By the Freshman's Dream, $P^{2^{m_2+m_3}}$ is a polynomial in the variables $x_i^{\pm 2^{m_2+m_3}}$, so the cells in $P^{2^{m_2+m_3}}$ are separated by at least $2^{m_2+m_3}$. Also, $|P^{2^{m_2+m_3}}| = |P|$. On the other hand, since $\mathrm{ht}(F) \leq 1$, $\mathrm{ht}(Q) \leq 2^{m_3} - 1$, and since

$$2(2^{m_3} - 1) + 1 < 2^{m_3+1} \leq 2^{m_2+m_3},$$

the copies of Q in F^n are disjoint from each other, and so $|F^n| = |P||Q|$, or in other words

$$a_n(F) = a_{2^{m_1}-1}(F)\, a_{2^{m_3}-1}(F) = \prod_{i \in \mathcal{L}(n)} a_{2^i-1}(F).$$

It is straightforward to generalize this argument to the case in which there are more than two runs of 1s in the binary expansion of n, and to establish that for any n,

$$a_n(F) = \prod_{i \in \mathcal{L}(n)} a_{2^i-1}(F), \tag{2.6}$$

thus completing the proof. □

In several interesting cases the subsequence (2.3) satisfies a three-term linear recurrence, in which case there is also a simple recurrence for the run length transform.

Theorem 2.3. *Suppose the sequence $[S_n, n \geq 0]$ is defined by the recurrence $S_{n+1} = c_2 S_n + c_3 S_{n-1}$, $n \geq 1$, with $S_0 = 1$, $S_1 = c_1$. Then its run length transform $[T_n, n \geq 0]$ satisfies the recurrence*

$$T_{2t} = T_t, \quad T_{4t+1} = c_1 T_t, \quad T_{4t+3} = c_2 T_{2t+1} + c_3 T_t, \tag{2.7}$$

for $t > 0$, with $T_0 = 1$.

Proof. $T_{2t} = T_t$ is immediate from the definition of the run length transform, since $\mathcal{L}(2n) = \mathcal{L}(n)$. The binary expansion of $4t + 1$ ends in 01, so $T_{4t+1} = T_t S_1 = c_1 T_t$. If $t = 2^k - 1$ for some $k \geq 1$ then $T_{4t+3} = S_{k+2} = c_2 S_{k+1} + c_3 S_k$, $T_{4t+1} = c_1 S_k$, $T_{2t+1} = S_{k+1}$, implying

$$T_{4t+3} = c_2 T_{2t+1} + c_3 T_t. \tag{2.8}$$

On the other hand, if t has a zero in its binary expansion, say $t = i.2^{k+1} + (2^k - 1), k \geq 0$, then $T_{4t+3} = T_i S_{k+2} = T_i(c_2 S_{k+1} + c_3 S_k)$, $T_{4t+1} = T_i c_1 S_k$, $T_{2t+1} = T_i S_{k+1}$, and again (2.8) follows.   □

2.4 Fredkin's Replicator

The cellular automaton known as *Fredkin's Replicator* [7, 8, 15] is the two-dimensional odd-rule CA defined by the Moore neighborhood F shown in Fig. 2.1(ix) and Eq. (2.1). (This is the eight-neighbor totalistic Rule 52428 in the Wolfram numbering scheme [17, 24, 26].)

We study the evolution of this CA when it is started at generation 0 with a single ON cell at the origin. Generations 0 through 8 are shown in Fig. 2.2. The name of this CA comes from the fact that any configuration of ON cells will be replicated eight times at some later stage. For example, generation 1 is replicated eight times at generation 5. Although distinctive, the name is not especially appropriate, since by (2.4) any odd-rule CA has a similar replication property. Let $a_n(F) = a_n$ denote the number of ON cells at the nth generation. The initial values of a_n are shown in Table 2.1.

Since $\text{ht}(F) = 1$, we know from Theorem 2.2 that $[a_n, n \geq 0]$ is the run length transform of the subsequence $[b_n = a_{2^n-1}, n \geq 0] = 1, 8, 24, 112, 416, 1728, \ldots$ (shown in bold in Table 2.1; it will turn out to be A246030). The main result of this section is the identification of this subsequence.

Theorem 2.4. *The sequence* $[b_n, n \geq 0]$ *satisfies the recurrence*

$$b_{n+1} = 2b_n + 8b_{n-1}, \text{ with } b_0 = 1, b_1 = 8. \tag{2.9}$$

Proof. Let $G_n := F^n$, $H_n := G_{2^n-1} = F^{2^n-1}$. (Figure 2.2 shows $G_0 = H_0$, $G_1 = H_1, G_2, G_3 = H_2, G_4, G_5, G_6, G_7 = H_3$, and G_8.) By definition,

$$H_{n+1} = F^{2^n} H_n, \tag{2.10}$$

and, from Theorem 2.1, $a_n = |G_n|, b_n = |H_n|$.

Since F has diameter 3, the nonzero terms $x^i y^j$ in H_n satisfy

$$-(2^n - 1) \leq i, j \leq 2^n - 1, \tag{2.11}$$

Table 2.1. *Number of ON cells at nth generation of Fredkin's Replicator* (*A160239*).

n	a_n
0	**1**
1	**8**
2–3	8 **24**
4–7	8 64 24 **112**
8–15	8 64 64 192 24 192 112 **416**
16–31	8 64 64 192 64 512 192 896 24 192 192 576 112 896 416 **1728**
32–63	8 64 64 192 64 512 192 896 64 512 512 1536 ...

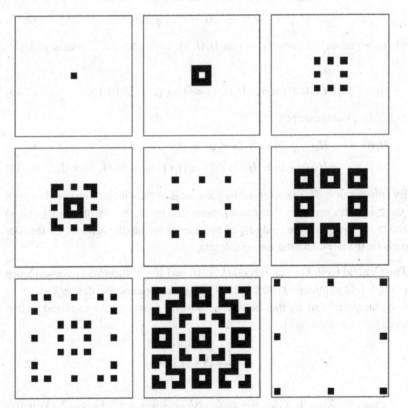

Fig. 2.2. Generations 0 through 8 of the evolution of Fredkin's Replicator. ON cells are black; OFF cells are white.

so we can write

$$H_n = \sum_{i=-(2^n-1)}^{2^n-1} \sum_{j=-(2^n-1)}^{2^n-1} H_n(i, j)x^i y^j, \qquad (2.12)$$

where the coefficient $H_n(i, j) \in GF(2)$ gives the state of the cell (i, j) at generation n.

From (2.10) and (2.4), H_{n+1} is the sum (in $\mathcal{R}$) of eight copies of H_n, translated by 2^n in each of the N, NW, W, SW, S, SE, E, and NE directions. That is, for $n \geq 1$,

$$H_{n+1}(i, j) = H_n(i, j - 2^n) + H_n(i - 2^n, j - 2^n) + H_n(i - 2^n, j)$$
$$+ H_n(i - 2^n, j + 2^n) + H_n(i, j + 2^n) + H_n(i + 2^n, j + 2^n)$$
$$+ H_n(i + 2^n, j) + H_n(i + 2^n, j - 2^n), \qquad (2.13)$$

where we adopt the convention that $H_n(i, j) = 0$ unless i and j satisfy (2.11). Also,

$$H_0(0, 0) = 1, \quad H_0(i, j) = 0 \text{ for } (i, j) \neq (0, 0), \qquad (2.14)$$

and $H_1(i, j) = 0$ except for

$$H_1(0, 1) = H_1(-1, 1) = H_1(-1, 0) = H_1(-1, -1)$$
$$= H_1(0, -1) = H_1(1, -1) = H_1(1, 0) = H_1(1, 1) = 1. \qquad (2.15)$$

By construction, H_n is preserved by the action of the dihedral group of order 8 (the symmetry group of the square), generated by the action of $(x, y) \leftrightarrow (y, x)$ and $(x, y) \leftrightarrow (\frac{1}{x}, y)$. We study H_n by breaking it up into the central cell, the four parts on the axes, and the four quadrants.

The Central Cell. The central cell $H_n(0, 0) = 1$ if $n = 0$, and (as a consequence of the 8-fold symmetry) is 0 for $n > 0$. **The axial parts.** We define X_n $(n \geq 1)$ to be the portion of H_n that lies on the positive x-axis, but normalized so that its center is at the origin:

$$X_n := \frac{1}{x^{2^{n-1}}} \sum_{i=1}^{2^n-1} H_n(i, 0)x^i. \qquad (2.16)$$

For example, $X_1 = 1, X_2 = \frac{1}{x} + x, X_3 = \frac{1}{x^3} + \frac{1}{x} + x + x^3$. From (2.13) it follows by induction that, for $n \geq 2$,

$$X_n = \sum_{i=0}^{2^{n-2}-1} \left(x^{-(2i+1)} + x^{2i+1}\right) = \left(\frac{1}{x} + x\right)^{2^{n-1}-1}. \qquad (2.17)$$

Similarly, the portion of H_n that lies on the negative x-axis, normalized so that its center is at the origin, is

$$\tilde{X}_n := x^{2^{n-1}} \sum_{i=1}^{2^n-1} H_n(-i, 0)x^{-i} = \left(\frac{1}{x} + x\right)^{2^{n-1}-1} = X_n. \qquad (2.18)$$

Likewise, the normalized portions of H_n on the positive and negative y-axes are

$$Y_n = \tilde{Y}_n = \left(\frac{1}{y} + y\right)^{2^{n-1}-1}. \qquad (2.19)$$

The Four Quadrants. Next, define I_n for $n \geq 1$ to consist of the portion of H_n lying in the first quadrant, again normalized so that its center is at the origin:

$$I_n := \frac{1}{(xy)^{2^{n-1}}} \sum_{i=1}^{2^n-1} \sum_{j=1}^{2^n-1} H_n(i, j)x^i y^j. \qquad (2.20)$$

Similarly, we define

$$II_n := \left(\frac{x}{y}\right)^{2^{n-1}} \sum_{i=1}^{2^n-1} \sum_{j=1}^{2^n-1} H_n(-i, j)x^{-i} y^j,$$

$$III_n := (xy)^{2^{n-1}} \sum_{i=1}^{2^n-1} \sum_{j=1}^{2^n-1} H_n(-i, -j)x^{-i} y^{-j},$$

$$IV_n := \left(\frac{y}{x}\right)^{2^{n-1}} \sum_{i=1}^{2^n-1} \sum_{j=1}^{2^n-1} H_n(i, -j)x^i y^{-j}. \qquad (2.21)$$

Assembling the parts, we see that, for $n \geq 1$, $H_n =$

$$\begin{array}{ccc} (y/x)^{2^{n-1}} II_n + & y^{2^{n-1}} Y_n + & (xy)^{2^{n-1}} I_n \\ + \ (1/x)^{2^{n-1}} \tilde{X}_n + & 0 + & x^{2^{n-1}} X_n \\ + \ (xy)^{-2^{n-1}} III_n + & y^{-2^{n-1}} \tilde{Y}_n + & (x/y)^{2^{n-1}} IV_n, \end{array}$$

which we write as a matrix

$$H_n = \begin{bmatrix} II_n & Y_n & I_n \\ \tilde{X}_n & 0 & X_n \\ III_n & \tilde{Y}_n & IV_n \end{bmatrix}, \qquad (2.22)$$

where it is to be understood that the blocks are to be shifted by the appropriate amounts (i.e., the I_n in the top right corner is to be multiplied by $(xy)^{2^{n-1}}$, and so on). By summing the eight translated copies of H_n, as in (2.13), we obtain

H_{n+1}

$$
= \begin{bmatrix}
\text{II}_n & Y_n & \text{I}_n + \text{II}_n & Y_n & \text{I}_n + \text{II}_n & Y_n & \text{I}_n \\
\tilde{X}_n & 0 & X_n + \tilde{X}_n & 0 & X_n + \tilde{X}_n & 0 & X_n \\
\text{II}_n + \text{III}_n & Y_n + \tilde{Y}_n & \text{I}_n + \text{III}_n + \text{IV}_n & \tilde{Y}_n & \text{II}_n + \text{III}_n + \text{IV}_n & Y_n + \tilde{Y}_n & \text{I}_n + \text{IV}_n \\
\tilde{X}_n & 0 & X_n & 0 & \tilde{X}_n & 0 & X_n \\
\text{II}_n + \text{III}_n & Y_n + \tilde{Y}_n & \text{I}_n + \text{II}_n + \text{IV}_n & Y_n & \text{I}_n + \text{II}_n + \text{III}_n & Y_n + \tilde{Y}_n & \text{I}_n + \text{IV}_n \\
\tilde{X}_n & 0 & X_n + \tilde{X}_n & 0 & X_n + \tilde{X}_n & 0 & X_n \\
\text{III}_n & \tilde{Y}_n & \text{III}_n + \text{IV}_n & \tilde{Y}_n & \text{III}_n + \text{IV}_n & \tilde{Y}_n & \text{IV}_n
\end{bmatrix}.
$$

$$(2.23)$$

Using (2.18) and (2.19), we have

$$
\text{II}_{n+1} = \begin{bmatrix}
\text{II}_n & Y_n & \text{I}_n + \text{II}_n \\
X_n & 0 & 0 \\
\text{II}_n + \text{III}_n & 0 & \text{I}_n + \text{III}_n + \text{IV}_n
\end{bmatrix}, \quad
\text{I}_{n+1} = \begin{bmatrix}
\text{I}_n + \text{II}_n & Y_n & \text{I}_n \\
0 & 0 & X_n \\
\text{II}_n + \text{III}_n + \text{IV}_n & 0 & \text{I}_n + \text{IV}_n
\end{bmatrix},
$$

$$(2.24)$$

$$
\text{III}_{n+1} = \begin{bmatrix}
\text{II}_n + \text{III}_n & 0 & \text{I}_n + \text{II}_n + \text{IV}_n \\
X_n & 0 & 0 \\
\text{III}_n & Y_n & \text{III}_n + \text{IV}_n
\end{bmatrix}, \quad
\text{IV}_{n+1} = \begin{bmatrix}
\text{I}_n + \text{II}_n + \text{III}_n & 0 & \text{I}_n + \text{IV}_n \\
0 & 0 & X_n \\
\text{III}_n + \text{IV}_n & Y_n & \text{IV}_n
\end{bmatrix}.
$$

$$(2.25)$$

By adding these four matrices we find that $\text{I}_{n+1} + \text{II}_{n+1} + \text{III}_{n+1} + \text{IV}_{n+1} = 0$ for $n \geq 1$. This identity is also true for $n = 0$, and we conclude that

$$
\text{I}_n + \text{II}_n + \text{III}_n + \text{IV}_n = 0, \quad n \geq 1, \tag{2.26}
$$

and so

$$
\text{II}_{n+1} = \begin{bmatrix}
\text{II}_n & Y_n & \text{I}_n + \text{II}_n \\
X_n & 0 & 0 \\
\text{II}_n + \text{III}_n & 0 & \text{II}_n
\end{bmatrix}, \quad
\text{I}_{n+1} = \begin{bmatrix}
\text{I}_n + \text{II}_n & Y_n & \text{I}_n \\
0 & 0 & X_n \\
\text{I}_n & 0 & \text{I}_n + \text{IV}_n
\end{bmatrix},
$$

$$(2.27)$$

etc., and finally that, for $n \geq 1$,

$$
H_{n+1} = \begin{bmatrix}
\text{II}_n & Y_n & \text{I}_n + \text{II}_n & Y_n & \text{I}_n + \text{II}_n & Y_n & \text{I}_n \\
X_n & 0 & 0 & 0 & 0 & 0 & X_n \\
\text{II}_n + \text{III}_n & 0 & \text{II}_n & Y_n & \text{I}_n & 0 & \text{I}_n + \text{IV}_n \\
X_n & 0 & X_n & 0 & X_n & 0 & X_n \\
\text{II}_n + \text{III}_n & 0 & \text{III}_n & Y_n & \text{IV}_n & 0 & \text{I}_n + \text{IV}_n \\
X_n & 0 & 0 & 0 & 0 & 0 & X_n \\
\text{III}_n & Y_n & \text{III}_n + \text{IV}_n & Y_n & \text{III}_n + \text{IV}_n & Y_n & \text{IV}_n
\end{bmatrix}. \tag{2.28}
$$

In (2.28) we see that H_{n+1} contains a copy of H_n at its center. The four corner blocks together with two copies each of X_n and Y_n form another, "deconstructed," copy of H_n.

Suppose $n \geq 2$, and consider the blocks $[\mathrm{I}_n + \mathrm{II}_n \ Y_n \ \mathrm{I}_n + \mathrm{II}_n \ Y_n]$ in the top row of (2.28). Using (2.27) these blocks can be expanded to give

$$
\begin{bmatrix}
\mathrm{I}_{n-1} & 0 & \mathrm{II}_{n-1} & Y_{n-1} & \mathrm{I}_{n-1} & 0 & \mathrm{II}_{n-1} & Y_{n-1} \\
X_{n-1} & 0 & X_{n-1} & 0 & X_{n-1} & 0 & X_{n-1} & 0 \\
\mathrm{IV}_{n-1} & 0 & \mathrm{III}_{n-1} & Y_{n-1} & \mathrm{IV}_{n-1} & 0 & \mathrm{III}_{n-1} & Y_{n-1}
\end{bmatrix}. \tag{2.29}
$$

The central three columns (columns 3, 4, and 5) give a copy of H_{n-1}, and columns 7, 8, and 1, in that order, give another copy. We get two further copies of H_{n-1} from the analogous blocks in each of the other three sides of (2.28), so in total H_{n+1} contains as many ON cells as are in two copies of H_n plus eight copies of H_{n-2}. This implies (2.9) for $n \geq 2$. Equation (2.9) is certainly true for $n = 1$, so this completes the proof of the theorem. □

The characteristic polynomial of (2.9) is $x^2 - 2x - 8 = (x + 2)(x - 4)$, and it follows that

$$
b_n = \frac{5 \cdot 4^n + (-2)^{n+1}}{3}, n \geq 0. \tag{2.30}
$$

In geometric terms, the proof of Theorem 2.4 shows that H_{n+1} can be dissected into pieces that can be reassembled to give two copies of H_n and eight copies of H_{n-1}. Figure 2.3 shows this dissection in the case of H_4. The central dashed square (red in the online version) encloses a copy of H_3. The smaller dashed squares on the four sides (blue) enclose copies of H_2. The four pairs of vertical parallel lines (green) enclose copies of Y_3, and the four pairs of horizontal parallel lines (also green) enclose copies of X_3. The four corners (outside the parallel lines) are, reading counterclockwise from the top right corner, respectively I_3, II_3, III_3, and IV_3, and combine with two copies each of X_3 and Y_3 to give the second copy of H_3. Along the top edge, the figure is divided into seven pieces, as in the top row of (2.28). Taking the fifth, sixth, and third pieces in that order gives another copy of H_2, and three further copies are obtained from the other edges of the figure.

From Theorem 2.3 and (2.9), we see that a_n is given by the recurrence

$$
a_{2t} = a_t, \quad a_{4t+1} = 8a_t, \quad a_{4t+3} = 2a_{2t+1} + 8a_t, \tag{2.31}
$$

for $t > 0$, with $a_0 = 1$, as conjectured by Hrothgar [10].

Fig. 2.3. Dissection of H_4 into pieces that can be reassembled to give two copies of H_3 and eight copies of H_2, illustrating (2.9) in the case $n = 3$.

2.5 The Centered von Neumann Neighborhood

In this section we analyze the two-dimensional odd-rule CA defined by the five-celled neighborhood

$$F = \frac{1}{x} + 1 + x + \frac{1}{y} + y \in \mathcal{R} = \mathrm{GF}(2)[x, x^{-1}, y, y^{-1}] \qquad (2.32)$$

shown in Fig. 2.1(vi), and consisting of the von Neumann neighborhood together with its center. (This is the five-neighbor totalistic Rule 614.) We use the same notation as in the previous section, except that now F is defined by (2.32) instead of (2.1).

The initial values of $[a_n, n \geq 0]$ are

$$1, \mathbf{5}, 5, \mathbf{17}, 5, 25, 17, \mathbf{61}, 5, 25, 25, 85, 17, 85, 61, \mathbf{217}, \ldots$$

(A072272), which we know from Theorem 2.2 is the run length transform of the subsequence $[b_n, n \geq 0] = 1, 5, 17, 61, 217, 773, 2753, \ldots$, shown in bold

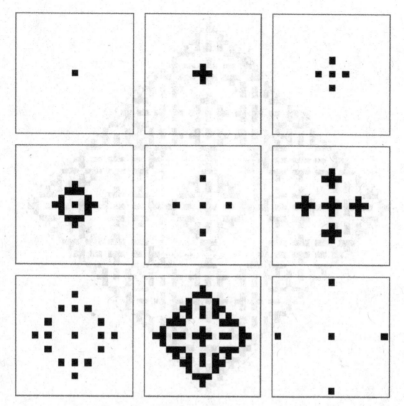

Fig. 2.4. Generations 0 through 8 of the odd-rule CA defined by the centered von Neumann neighborhood. Generations 0, 1, 3, 7 show H_0, H_1, H_2, H_3 respectively.

(A007483). Generations 0 through 8 are shown in Fig. 2.4, and Figs. 2.5 and 2.6 show generation 31.

Theorem 2.5. *The sequence $[b_n, n \geq 0]$ satisfies the recurrence*

$$b_{n+1} = 3b_n + 2b_{n-1}, \text{ with } b_0 = 1, b_1 = 5. \tag{2.33}$$

Proof. As in the proof of Theorem 2.4, $H_n = F^{2^n-1}$, $b_n = |H_n|$, and again H_n is preserved under the action of a dihedral group of order 8. The first step in the proof is to show that for $n \geq 2$, H_n can be dissected into a central copy of H_{n-2} and four disjoint pentagonal or "haystack"-shaped regions (see Fig. 2.5 for the dissection of H_5). The four haystacks in H_n are denoted by N_n, W_n, S_n, and E_n, according to the direction in which they point (a precise definition is given in the text that follows). They are equivalent under the action of the dihedral

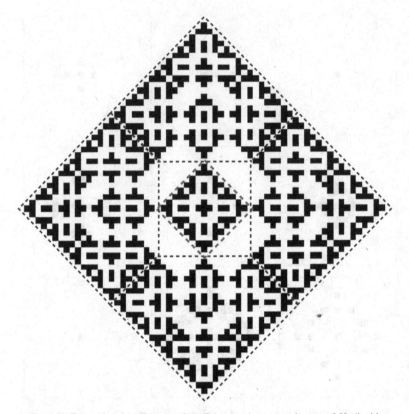

Fig. 2.5. Generation 31 (H_5), showing dissection into central copy of H_3 (inside red dashed line) and four "haystacks" N_5, W_5, S_5, E_5 (each enclosed by blue dashed line).

group. Algebraically, we will show that

$$H_n = H_{n-2} + y^{2^{n-1}} N_n + x^{-2^{n-1}} W_n + y^{-2^{n-1}} S_n + x^{2^{n-1}} E_n, \qquad (2.34)$$

where the five polynomials on the right are disjoint (i.e., have no monomials in common with each other).

Once we have established (2.34), the second step in the proof will be to show that each haystack can be dissected into five smaller haystacks (see Fig. 2.6 for the dissection of N_5). In particular, we will show that for $n \geq 3$,

$$N_n = y^{2^{n-2}} N_{n-1} + x^{-2^{n-2}} W_{n-1} + x^{2^{n-2}} E_{n-1} + N_{n-2} + y^{-2^{n-3}} S_{n-2}, \qquad (2.35)$$

where again the polynomials on the right are disjoint. Let $v_n := |N_n| = |W_n| = |S_n| = |E_n|$. Then (2.35) implies $v_n = 3v_{n-1} + 2v_{n-2}$. From (2.34) we have

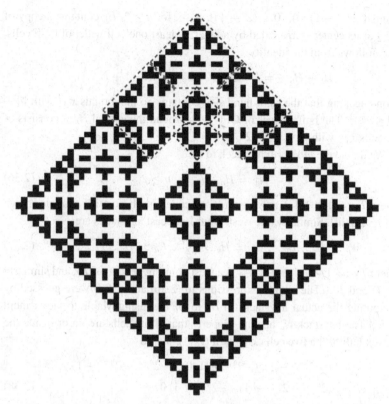

Fig. 2.6. Another view of generation 31 (H_5) showing dissection of haystack N_5 into three smaller haystacks N_4, W_4, E_4 (each enclosed in green dashed line) and two still smaller haystacks N_3 and S_3 (inside blue dashed lines).

$b_n = b_{n-2} + 4v_n$, so

$$b_{n+1} - 3b_n - 2b_{n-1} = b_{n-1} - 3b_{n-2} - 2b_{n-3} = \cdots$$
$$= \text{ either } b_3 - 3b_2 - 2b_1 \text{ or } b_2 - 3b_1 - 2b_0,$$

and each of the last two expressions evaluates to zero. This will complete the proof of (2.33). It is worth remarking that these dissections are of a different nature from the dissection in the previous section. There it was necessary to make some nonobvious cuts through the contiguous blocks of ON cells, as shown by the parallel (green) lines in the corners of Fig. 2.3. In contrast, in the present proof, the dissections are carried out by "tearing" along the obvious "perforations," rather like tearing apart a block of postage stamps.

Now to the details. It follows from the definition (see the sequence of successive states in Fig. 2.4) that H_n is a diamond-shaped configuration with extreme points $(\pm(2^n - 1), 0)$, $(0, \pm(2^n - 1))$. Also, for $n \geq 2$, H_n contains a copy of H_{n-2} at its center, surrounded by a layer, at least one cell wide, of OFF cells. This follows from the identity

$$H_n - H_{n-2} = F^{2^n - 1} - F^{2^{n-2} - 1} = H_{n-2}(1 + H_2^{2^{n-2}}),$$

upon checking that the right-hand side contains no monomials $x^i y^j$ with $|i| + |j| \leq 2^{n-2}$. The buffer layer of OFF cells around the central H_{n-2} consists of the cells $x^i y^j$ with $|i| + |j| = 2^{n-2}$.

We define the nth north haystack to be

$$N_n := H_{n-2} T_N^{2^{n-2}}, \quad n \geq 2, \tag{2.36}$$

where $T_N := 1/x + 1 + x + y$ is the four-celled north-pointing triangle shown in Fig. 2.1(iii). Similarly, the west, south, and east haystacks are

$$W_n := H_{n-2} T_W^{2^{n-2}}, \quad S_n := H_{n-2} T_{S2}^{2^{n-2}}, \quad E_n := H_{n-2} T_E^{2^{n-2}}, \tag{2.37}$$

where $T_W := 1/x + 1/y + 1 + y$ is a west pointing version of T_N, and similarly for T_S and T_E. (The simple expressions in (2.36) and (2.37) were guessed by computing the actual haystacks in H_3 to H_6 and using Maple to factor them in $\mathcal{R}$.) The haystack N_n has the property that all its cells are on or inside the convex hull of the five cells $x^i y^j$ with (i, j) equal to

$$(0, 2^{n-1} - 1), (-2^{n-1} + 1, 0), (-2^{n-2}, -2^{n-2} + 1),$$
$$(2^{n-2} - 2^{n-2} + 1), (-2^{n-1} + 1, 0). \tag{2.38}$$

To see this, consider what happens to N_n one generation later: it becomes

$$FN_n = (FT_N)^{2^{n-2}}, \tag{2.39}$$

using the Freshman's Dream (2.4), where FT_n is the six-celled configuration

$$y^2 + \frac{1}{x^2} + \frac{1}{xy} + \frac{1}{y} + \frac{x}{y} + x^2.$$

From (2.39), FN_n is this configuration with the cells moved 2^{n-2} steps apart, and the cells (2.38) lie just inside it.

Note that the point $(0, 0)$ is in the interior of N_n at the intersection of the vertical line through the apex and the horizontal line joining the most western and eastern points. The powers of x and y in (2.34) and (2.35) are needed in order to translate the haystacks into their correct positions. Since we now know the boundaries of all the terms on the right side of (2.34), we can check that these five polynomials are indeed disjoint. We must still check that (2.34) is an identity. Using the Freshman's Dream, this reduces to checking the identity

$$H_2 = 1 + y^2 T_N + x^{-2} T_W + y^{-2} T_S + x^2 T_E,$$

which is true. This completes the proof that (2.34) is a proper dissection of H_n. The correctness of the dissection (2.35) is verified in a similar way; we omit the details.   □

2.6 Other Cellular Automata

2.6.1 The 256 One-Dimensional Rules

There are 256 possible CAs based on the one-dimensional three-celled neighborhood shown in Fig. 2.1(i). These are the CAs labeled Rule 0 through Rule 255 in the Wolfram numbering scheme [17, 24, 26]. As usual, we assume the automaton is started with a single ON cell, and let a_n denote the number of ON cells after n generations. Illustrations of the initial generations of all 256 CAs are shown on pp. 54–56 of [26]. Many of these sequences were analyzed in [23]; see also [22]. If we eliminate those in which some a_n is infinite, or the sequence $[a_n, n \geq 0]$ is trivial (essentially linear), or is a duplicate of one of the others, we are left with just seven sequences:

- Rule 18 (or Rule 90; Rule 182 is very similar): $a_n = 2^{wt(n)}$ (Gould's sequence A001316), where wt(n) is the number of 1s in the binary expansion of n. This is the run length transform of the powers of 2.
- Rule 22: $a_n = 2^{wt(n)}$ if n even, $3 \cdot 2^{wt(n)-1}$ if n odd (A071044).
- Rule 30: The behavior appears chaotic, even when started with a single ON cell [22, 26]. The sequence, A070952, has roughly linear growth, but it seems likely that there is no simpler way to obtain it than by its definition.
- Rule 62: $a_{n+7} = a_{n+4} + a_{n+3} - a_n$ with initial terms $1, 3, 3, 6, 5, 8, 9$ (A071047).
- Rule 110: Although the initial behavior is chaotic, it is an astonishing fact, pointed out by Wolfram [26, p. 39], that after about 3,000 terms all the irregularities disappear. By using the Salvy–Zimmermann gfun package in Maple [18], we find that the sequence, A071049, satisfies a linear recurrence of order 469: for $n \geq 2,854$,

$$a_{n+469} = -a_{n+453} + a_{n+256} + a_{n+240} + a_{n+229} + a_{n+213} - a_{n+16} - a_n.$$
$$(2.40)$$

This recurrence is far nicer than it initially appears: the coefficients are palindromic, and its characteristic polynomial is the product of twenty-five irreducible factors.
- Rule 126: $a_n = 2^{wt(n)+1}$, except we must subtract 1 if $n = 2^k - 1$ for some k (A071051).

Table 2.2. *Odd-rule CAs defined by height-1 neighborhoods F in Fig. 2.1.*

F	$a_n(F)$	$b_n(F)$	g.f.	Notes
(i)	A071053	A001045	$\frac{1+2x}{1-x-2x^2}$	Rule 150, §2.6.1
(ii)	A048883	A000244	$\frac{1}{1-3x}$	$a_n = 3^{\text{wt}(n)}$
(iii)	A253064	A087206	$\frac{1+2x}{1-2x-4x^2}$	$b_n = 2^n \text{Fib}_{n+2}$
(v)	A102376	A000302	$\frac{1}{1-4x}$	$a_n = 4^{\text{wt}(n)}$
(vi)	A072272	A007483	$\frac{1+2x}{1-3x-2x^2}$	§2.5
(vii)	A253069	A253070	(2.41)	
(viii)	A246039	A246038	$\frac{(1+2x)(1+2x+4x^2)}{1-3x-8x^3-8x^4}$	
(ix)	A160239	A246030	$\frac{1+6x}{1-2x-8x^2}$	Fredkin Replicator, §2.4
(x)	A246035	A139818	$\frac{1+6x-8x^2}{(1-x)(1+2x)(1-4x)}$	Squares of entries from Fig. 2.1i

- Rule 150: This is the odd-rule CA defined by the three-celled neighborhood. The sequence $[a_n, n \geq 0]$ (A071053) was analyzed by Wolfram [23] (see also [9], [19]). In the notation of this chapter, $[a_n, n \geq 0]$ is the run length transform of the Jacobsthal sequence A001045. Theorems 2.1 and 2.2 were suggested by reading Sillke and Postl's analysis [19].

2.6.2 Other Odd-Rule CAs

In this section we discuss the odd-rule CAs defined by the height-one neighborhoods in Fig. 2.1. (The height-two neighborhood of Fig. 2.1(iv) is discussed in the last section of the chapter.) Table 2.2 summarizes the results. The first column specifies the neighborhood F in Fig. 2.1, $a_n(F)$ is the number of ON cells at generation n, $b_n(F)$ denotes the sequence of which $a_n(F)$ is the run length transform, and the fourth column gives a generating function (g.f.) for $b_n(f)$. For Fig. 2.1(iii), Fib_{n+2} denotes a Fibonacci number. The g.f. for (vii), found by Doron Zeilberger [3], is

$$\frac{(1+2x)\left(1+x-x^2+x^3+2x^5\right)}{1-3x-3x^2+x^3+6x^4-10x^5+8x^6-8x^7}. \tag{2.41}$$

An expanded version of this table, analyzing all the sequences arising from odd-rule CAs defined by height-one neighborhoods on the square grid, will be published elsewhere [4].

Table 2.3. *Number of ON cells at nth generation of CA defined by Rule 750.*

n	a_n							
1	1							
2–3	5	9						
4–7	21	25	37	57				
8–15	85	89	101	121	149	169	213	281
16–31	341	345	357	377	405	425	469	...

2.6.3 Further Two-Dimensional CAs

If we drop the "odd-rule" definition, the number of CAs grows astronomically—there are 2^{512} based on the Moore neighborhood alone. Pages 171–175 of [26] show many examples of the subset of "totalistic" rules, in which the next state of a cell depends only on its present state and the total number of ON cells surrounding it. All of these are potential sources of sequences. In a few cases it is possible to analyze the sequence, but usually it seems that no formula or recurrence exists. In this section we give three examples: one that can be analyzed, one that might be analyzable with further research, and one (typical of the majority) where the state diagrams are aesthetically appealing but finding a formula seems hopeless. All three are totalistic rules, the first two being based on the von Neumann neighborhood (Fig. 2.1(v)), and the third on the Moore neighborhood (Fig. 2.1(ix)).

The first example is the Rule 750 automaton, in which an OFF cell turns ON if an odd number of its four neighbors are ON, and once a cell is ON it stays ON [26, p. 925]. This CA is a hybrid of the "odd-rule" CAs studied in the preceding text and the "once a cell is ON it stays ON" rules studied in [2]. Here it is convenient to call the initial ON cell generation 1 (rather than 0). The numbers of ON cells in the first few generations (A169707) are given in Table 2.3.

The evolution of this CA is similar to several that were studied in [2]: at generation 2^k, for $k \geq 2$, the structure is enclosed in a diamond-shaped region, which is saturated in the sense that no additional interior cells can ever be turned ON, and contains $a_{2^k} = (4^{k+1} - 1)/3$ ON cells. Then in generations $2^k + 1$ to $2^{k+1} - 1$, the structure grows outward from the four vertices of the diamond, and the first half of the growth that follows generation 2^k is the same as the growth that followed generation 2^{k-1}. Figure 2.7 shows generation $20 = 2^4 + 4$, where we can see that 16 cells have grown out of each vertex. Pictures of generations $16 = 2^3 + 4$ and $36 = 2^5 + 4$ show exactly the same

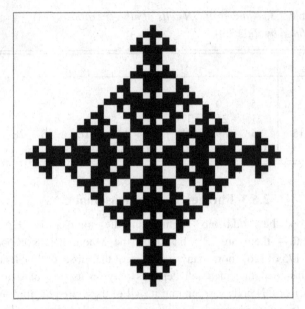

Fig. 2.7. Generation 20 of CA defined by Rule 750, showing
$341 + 4(1 + 3 + 5 + 7) = 405$ ON cells, illustrating (2.43).

growth from the vertices (although with different numbers of ON cells in the central diamond).

The successive numbers of ON cells added to a vertex in the generations from 2^k to $2^{k+1} - 1$ are $0, 1, 3, 5, 7, 5, 11, 17, 15, 5, \ldots$, which form the initial terms $v_0, v_1, v_2, \ldots$ of a sequence ([A151548](#)) encountered in [2]. The v_i have generating function

$$\frac{x}{1+x} + 4x^2 \prod_{r=1}^{\infty} (1 + x^{2^r - 1} + 2x^{2^r}). \tag{2.42}$$

Then, for $k \geq 0$ and $0 \leq m < 2^k$, we have

$$a_{2^k + m} = \frac{4^{k+1} - 1}{3} + 4 \sum_{i=0}^{m} v_i. \tag{2.43}$$

Bearing in mind the warning in the first sentence of this chapter, we must admit that we have not written out a complete proof that (2.43) is correct. However, there should be no difficulty in filling in the details: as the automaton evolves from generation 2^k to 2^{k+1}, the structure has a natural dissection into polygonal pieces.

The second example is more speculative: this is the Rule 493 automaton [26, p. 173], A246333. The binary expansions of 493 and 750 differ in just four places, so it is not surprising that this is similar to the previous example. Now an ON cell stays ON unless exactly zero or four of its neighbors are ON, in which case it turns OFF, and an OFF cell turns ON unless exactly two of its neighbors are ON. Assuming here that we start with a single ON cell at generation 0, in the even-numbered generations the number of ON cells is finite (A246334):

$$1, 5, 17, 29, 61, 73, 109, 157, 229, 241, 277, 329, 429, 477, 573, 633, 861, \ldots,$$

whereas in the odd-numbered generations the number of OFF cells is finite (A246335):

$$1, 5, 9, 21, 25, 37, 57, 85, 89, 101, 121, 165, 169, 213, 217, 317, 321, 333, \ldots.$$

The reason for hoping this automaton might be analyzable is that the latter sequence agrees with the sequence in Table 2.3 up though the eleventh term, 121, after which the sequences diverge. Even the respective states are the same up through the sixth term, 37, although to see this one has to work with the negatives—in the photographer's sense, interchanging black and white cells—and then rotating the result by 45 degrees. This needs further investigation.

The third example in the eight-neighbor Rule 780 (A246310), in which a cell turns ON if one or four of its neighbors in ON, and otherwise turns OFF. Although the initial generations are simple enough, already by generation 15 (Fig. 2.8) the structure is extremely complicated. Is there a recurrence? Is the five-neighbor analog (A253086) any easier to understand?

2.6.4 The Three-Dimensional analog of Fredkin's Replicator

The three-dimensional Moore neighborhood, that is, the $3 \times 3 \times 3$ cube without its center cell, gives rise to the sequence 1, 26, 26, 124, 26, 676, 124, 1400, ... (A246031), which by Theorem 2.2 is the run length transform of the subsequence

$$1, 26, 124, 1400, 10000, 89504, 707008, 5924480, 47900416, 393069824,$$
$$3189761536, 25963397888, \ldots \tag{2.44}$$

(A246032), computed by Roman Pearce and Michael Monagan. Doron Zeilberger [3] has found a generating function, a rational function with numerator of degree 10 and denominator of degree 11, as well as a proof that it is correct.

Fig. 2.8. Generation 15 of CA defined by eight-neighbor Rule 780.

2.7 Further Remarks about Run Length Transforms

Block Structure. It is a surprising fact that the growth sequences of many CAs have a natural division into blocks of successive lengths $(1,)1, 2, 4, 8, 16, 32, \ldots$. This is true even for some CAs that are defined on lattices other than $\mathbb{Z}^d$ [2]. Some of these examples are explained by the fact that the run length transform always has this property—the division into blocks of the run length transform $[T_n, n \geq 0]$ of an arbitrary sequence $S = [1, A, B, C, D, \ldots]$ is shown in Table 2.4. Table 2.1 gives a concrete example. The first half of each row is given by A times the beginning of the $[T_n]$ sequence itself.

Further Examples. We briefly mention four additional examples of run length transforms. The run length transform of $0, 1, 2, 3, 4, \ldots$ is $1, 1, 1, 2, 1, 1, 2, 3, 1, \ldots$ (A227349), which gives the product of the lengths of runs of 1s in the binary representation of n. The primes, prefixed by 1, give $1, 2, 2, 3, 2, 4, 3, 5, 2, \ldots$ (A246029). The squares give $1, 1, 1, 4, 1, 1, 4, 9, \ldots$ (A246595). The powers of 2 give $1, 2, 2, 4, 2, 4, 4, 8, 2, \ldots, 2^{\text{wt}(n)}$ (A001316, already mentioned in §2.6.1).

Table 2.4. *The run length transform $[T_n, n \geq 0]$ of a sequence
$S = [1, A, B, C, D, \ldots]$, showing the division into blocks of sizes
$1, 1, 2, 4, 8, 16, \ldots$*

n	T_n
0	1
1	A
2–3	A B
4–7	A A^2 B C
8–15	A A^2 A^2 AB B AB C D
16–31	A A^2 A^2 AB A^2 A^3 AB AC B AB AB B^2 C AC D E
32–63	A A^2 A^2 AB A^2 A^3 AB AC ...

Graphs. The graphs of run length transforms are usually highly irregular, as one expects from Table 2.4. The partial sums of these sequences are naturally smoother, and generally have a family resemblance, with a bumpy appearance somewhat similar to what is seen in the Takagi curve [1, 12]. The partial sums of the four examples in the previous paragraph are A253083, A253081, A253082, A006046, respectively, and the partial sums of the sequences arising from Fredkin's Replicator, the sequence in §2.5, and the Rule 150 sequence in §2.6 are respectively A245542, A253908, A134659.[2] The latter sequence is discussed in [6], and it would be interesting to see if the methods of that paper can be applied to the other six sequences. Also, is there any direct connection between the limiting form of these graphs for large n and the Takagi curve?

The Generalized Run Length Transform. There are analogs of Theorem 2.2 that apply to larger neighborhoods, although they are more complicated and not as useful. The following is a version that applies when the neighborhood F has height at most 2. Whereas in Theorem 2.2, $a_n(F)$ was expressed as a product of terms from the subsequence $a_m(F)$ where the binary expansion of m contained no zeros, now we need the values $a_m(F)$ where m is any number whose binary expansion begins and ends with 1 and does not contain any pair of adjacent zeros. These are the numbers (A247648):

$$1, 3, 5, 7, 11, 13, 15, 21, 23, 27, 29, 31, 43, 45, 47, 53, 55, 59, 61, 63, \ldots.$$

$$(2.45)$$

---

[2] The "graph" button in [16] makes it easy to compare these graphs. However, it is not clear how the growth rate of the original sequence affects the "bumpiness" of the partial sums.

Table 2.5. *Number of ON cells at nth generation of of odd-rule one-dimensional CA defined by a five-celled neighborhood (A247649).*

n	a_n							
0	1							
1	5							
2–3	5	7						
4–7	5	17	7	19				
8–15	5	25	17	19	7	31	19	25
16–31	5	25	25	35	17	61	19	71 ...

Suppose for simplicity that the binary expansion of n has the form

$$n = \overbrace{**\cdots*}^{m_1}\overbrace{00\cdots0}^{m_2}\overbrace{**\cdots*}^{m_3}, \quad \text{where } m_1, m_3 \geq 1, m_2 \geq 2,$$

and the asterisks indicate strings of 0s and 1s that begin and end with 1s and do not contain any pair of adjacent zeros. If the first such string represents N_1 and the second N_2, then $a_n(F) = a_{N_1}(F)a_{N_2}(F)$. There is an analogous expression in the general case, expressing $a_n(F)$ as a product of terms $a_m(F)$ where m belongs to (2.45).

To illustrate, suppose F is the five-celled one-dimensional neighborhood shown in Fig. 2.1(iv), with height 2. The initial values of $a_n(F)$ are given in Table 2.5, with $a_m(F)$ for m in (2.45) shown in bold. For example, the binary expansion of 167 is 10100111, so the generalized run length transform tells us that $a_{167}(F) = a_5(F)a_7(F) = 17 \cdot 19 = 323$. It follows from the generalized run length transform property that in each row of the table, the first one-eighth of the terms coincide with five times the beginning of the sequence itself.

The bold-faced entries in Table 2.5 form A253085, and we end with one last question: is there an independent characterization of this sequence? An affirmative answer might make the generalized run length transform a lot more interesting. Much remains to be done in this subject!

Postscript, March 2015

After seeing an initial version of this article, Doron Zeilberger observed that it is possible to use Theorems 2.1 and 2.2 to automate calculation of sequences giving the number of ON cells in odd-rule CAs, and in the case of height-one neighborhoods, to find and rigorously prove the correctness of generating functions for the sequences of which they are the run length transforms. This provides an alternative, computer proof of Theorems 2.4 and 2.5. Details will appear elsewhere [3, 4].

Acknowledgments

Theorems 2.1 and 2.2 were suggested by reading Torsten Sillke and Helmut Postl's paper [19]. Thanks to Hrothgar for sending a copy of [10]. Figures 2.2–2.8 were produced with the help of the CellularAutomaton command in Mathematica® [25]. Kellen Myers showed me how to make an animated gif with Mathematica. Thanks to Roman Pearce and Michael Monagan for computing the initial terms of sequence (2.44). Stephen Wolfram, Todd Rowland, Hrothgar, and Omar Pol provided helpful comments on the manuscript.

References

1. P. C. Allaart and K. Kawamura. The Takagi function: A survey. *Real Analysis Exchange*, **37** (2011/12), 1–54; http://arxiv.org/abs/1110.1691.
2. D. Applegate, O. E. Pol, and N. J. A. Sloane. The toothpick sequence and other sequences from cellular automata. *Congress. Numerant.* **206** (2010), 157–191; http://arxiv.org/abs/1004.3036.
3. S. B. Ekhad, N. J. A. Sloane, and D. Zeilberger. A meta-algorithm for creating fast algorithms for counting ON cells in odd-rule cellular automata, 2015. http://arxiv.org/abs/1503.01796.
4. S. B. Ekhad, N. J. A. Sloane, and D. Zeilberger. "Odd-rule" cellular automata on the square grid, 2015. http://arxiv.org/abs/1503.04249.
5. D. Eppstein. Growth and decay in life-like cellular automata, 2009. http://arxiv.org/abs/0911.2890.
6. S. Finch, P. Sebah, and Z.-Q. Bai. Odd entries in Pascal's trinomial triangle, 2008. http://arxiv.org/abs/0802.2654.
7. E. Fredkin. Digital mechanics, an informational process based on reversible universal cellular automata. In *Cellular Automata, Theory and Experiment*, ed. H. Gutowitz. MIT Press, Cambridge, MA, 1990, pp. 254–270.
8. E. Fredkin. *Digital Mechanics* (Working Draft), 2000. http://64.78.31.152/wp-content/uploads/2012/08/digital_mechanics_book.pdf.
9. H. Havermann, H. Havermann, B. Cloitre, R. Stephan, et al. Entry A071053 in [16], 2002–2015.
10. Hrothgar. Notes on a replicating automaton. Unpublished manuscript, July 2014.
11. J. Kari. Theory of cellular automata: A survey. *Theoret. Comput. Sci.* **334** (2005), 3–33.
12. J. C. Lagarias. The Takagi function and its properties. In *Functions in Number Theory and Their Probabilistic Aspects*, ed. K. Matsumoto, W. Layman, Hrothgar, O. E. Pol, et al. RIMS Lecture Notes, Vol. **B34**, Res. Inst. Math. Sci., Kyoto, 2012, pp. 153–189. http://arxiv.org/abs/1112.4205.

13. J. Layman, Hrothgar, O. E. Pol, et al. Entry A160239 in [16], 2009–2015.
14. O. Martin, A. M. Odlyzko, and S. Wolfram. Algebraic properties of cellular automata. *Commun. Math. Phys.* **93** (1984), 219–258.
15. S. Mitra and S. Kumar. Fractal replication in time-manipulated one-dimensional cellular automata. *Complex Systems* **16** (2006), 191–207.
16. The OEIS Foundation Inc. *The On-Line Encyclopedia of Integer Sequences*, 1996–present. https://oeis.org.
17. N. H. Packard and S. Wolfram. Two-dimensional cellular automata. *J. Statist. Phys.* **38** (1985), 901–946.
18. B. Salvy and P. Zimmermann. GFUN: A Maple package for the manipulation of generating and holonomic functions in one variable. *ACM Trans. Math. Software*, **20** (1994), 163–177.
19. T. Sillke and H. Postl. Odd trinomials: $t(n) = (1 + x + x^2)^n$, 2004. www.mathematik.uni-bielefeld.de/~sillke/PUZZLES/trinomials.
20. D. Singmaster. On the cellular automaton of Ulam and Warburton. *M500 Magazine of the Open University*, No. **195** (December 2003), 2–7; https://oeis.org/A079314/a079314.pdf.
21. S. M. Ulam. On some mathematical problems connected with patterns of growth of figures. In *Mathematical Problems in the Biological Sciences*, ed. R. E. Bellman, Proc. Sympos. Applied Math., Vol. **14**, Amer. Math. Soc., 1962, 215–224.
22. E. W. Weisstein. MathWorld, Entries for Rules 30, 90, 110, 150, 182, etc. http://mathworld.wolfram.com/, 2004–present.
23. S. Wolfram. Statistical mechanics of cellular automata. *Rev. Mod. Phys.* **55** (1983), 601–644.
24. S. Wolfram. Universality and complexity in cellular automata (*Cellular Automata, Los Alamos, 1983*). *Physica D* **10** (1984), 1–35.
25. S. Wolfram. *The Mathematica Book*. Cambridge University Press and Wolfram Research, Inc., New York 2000.
26. S. Wolfram. *A New Kind of Science*. Wolfram Media, Champaign, IL, 2002.

3

Search for Ultraflat Polynomials with Plus and Minus One Coefficients

Andrew Odlyzko

Abstract

It is not known whether there exist polynomials with plus and minus one coefficients that are almost constant on the unit circle (called ultraflat). Extensive computations described in this chapter strongly suggest such polynomials do not exist, and lead to conjectures about the precise degree to which flatness can be approached according to various criteria. The evidence shows surprisingly rapid convergence to limiting behavior. Connections to problems about the Golay merit factor, Barker sequences, Golay–Rudin–Shapiro polynomials, and others are discussed. Some results are presented on extensions where the coefficients are allowed to be roots of unity of orders larger than 2. It is pointed out that one conjecture of Littlewood about polynomials with plus and minus one coefficients is true, while another is very likely to be false, as it is inconsistent with another Littlewood conjecture that is supported by the data.

3.1 Introduction

There are many very appealing and easy to state problems about the behavior of polynomials with restricted coefficients that have proved very hard. One that was raised in pure mathematics context by Erdős [11] and later extended and popularized by Littlewood in several of his papers, such as [22], and in his book [23], concerns the degree to which the absolute value of a polynomial can be almost constant when the argument runs over the unit circle. This problem also arose in several engineering problems, cf. [32, 35]. If the coefficients are not constrained, this is of course trivial. But what if the coefficients are all forced to be of absolute value 1, or, even more restrictively, equal ± 1?

We define

$$\mathbb{U}_n = \left\{ F(z) = \sum_{k=0}^{n} a_k z^k, \; a_k = \pm 1 \right\} \tag{3.1}$$

and similarly $\mathbb{V}_n$, where we allow $a_k \in \mathbb{C}$, $|a_k| = 1$. A simple computation (trivial case of Parseval's identity) shows that for $F(z) \in \mathbb{V}_n$ (and therefore also for $F(z) \in \mathbb{U}_n$),

$$\|F\|_2^2 = \frac{1}{2\pi} \int_0^{2\pi} |F(e^{i\theta})|^2 d\theta = \frac{1}{2\pi} \sum_{j,k=0}^{n} \int_0^{2\pi} a_j \bar{a}_k e^{i(j-k)\theta} d\theta$$

$$= \sum_{k=0}^{n} |a_k|^2 = n + 1. \tag{3.2}$$

Hence if $F(z) \in \mathbb{V}_n$ is ultraflat, its absolute value must be close to $\sqrt{n+1}$. This motivates the definition, for $F(z) \in \mathbb{V}_n$,

$$M(F) = \max_{|z|=1} \frac{|F(z)|}{\sqrt{n+1}}, \quad m(F) = \min_{|z|=1} \frac{|F(z)|}{\sqrt{n+1}}, \quad W(F) = M(F) - m(F). \tag{3.3}$$

All graphs in this chapter are of $(\text{Re}(F(e^{i\theta}), \text{Im}(F(e^{i\theta}))/\sqrt{n+1}$ for $0 \le \theta \le 2\pi$ to obtain comparisons that are independent of the degree of $F(z)$.

$W(F)$ is the minimal width of an annulus centered at the origin that contains the graph of $F(z)/\sqrt{n+1}$. Ultraflat polynomials $F(z)$ of high degree would have $M(F) \approx m(F) \approx 1$ and $W(F) \approx 0$.

For random $F(z)$ taken from $\mathbb{V}_n$ or $\mathbb{U}_n$,

$$M(F) \sim \sqrt{\log n} \tag{3.4}$$

as $n \to \infty$ with probability tending to 1. An upper bound of this form was obtained by Salem and Zygmund, and the asymptotic form for a special case by Halasz [17] and in the general form by Gersho, Gopinath, and Odlyzko [13]. (The last result allows for the a_k to be drawn from very general distributions, with the main requirement being that the a_k be independent. The a_k do not even have to be identically distributed. That paper also shows that for most $F(z)$, large values are taken on in at least $\log n$ regions.) By similar methods one can show that $m(F) \to 0$ for most $F(z)$. Thus ultraflat polynomials in either $\mathbb{U}_n$ or $\mathbb{V}_n$, if they exist, are relatively rare, which proves the first part of Littlewood's conjecture (C_3), [23], p. 29. However, the second part of that conjecture, which predicts the number of such polynomials is extremely small, namely $O(n^3)$, is almost surely false. If, for a large n, there exists even one polynomial with $M(F)$ bounded above, and $m(F)$ bounded away from zero (in both cases with bounds independent of n), as is predicted by Littlewood's conjecture (C_1), and as is strongly suggested by the results of this chapter, then Section 3.3 shows there have to be many such.

For a long time the prevailing opinion seemed to be that there were no ultraflat polynomials in $\mathbb{V}_n$. It came as a surprise to many, therefore, when in 1980 Kahane [19], building on earlier work of Körner [20] and other investigators, showed this was not correct, and that for any $\epsilon > 0$, for sufficiently large n there are $F(z) \in \mathbb{V}_n$ with $W(F) < \epsilon$. Kahane's method is not constructive, as it uses a randomization procedure to guarantee that $|a_k| = 1$. Kahane's construction was recently improved by Bombieri and Bourgain [2], who obtained smaller error terms, and, even more important, obtained explicit constructions, thus eliminating the nonconstructive element.

This chapter investigates the problem that the Kahane and also the Bombieri and Bourgain papers left open, namely what happens if we insist the coefficients be ± 1. Let

$$M_n = \min_{F \in \mathbb{U}_n} M(F), \tag{3.5}$$

$$m_n = \max_{F \in \mathbb{U}_n} m(F), \tag{3.6}$$

$$W_n = \min_{F \in \mathbb{U}_n} W(F). \tag{3.7}$$

Very little is known theoretically, although some very recent work [9] that is yet to be verified claims to prove that ultraflat polynomials do not exist in $\mathbb{U}_n$.

For $n = 2^k - 1$, it is known that $M_n \leq \sqrt{2}$, since the Golay–Rudin–Shapiro (GRS) polynomials achieve this bound; cf. [3]. These polynomials are usually referred to in the literature as the Rudin–Shapiro polynomials [30, 33]. However, they were discovered independently by Golay [14], so it is appropriate to attach his name to them. For references to some of the large literature on these remarkable polynomials, see, for example, [3, 6]. It can be shown, using GRS polynomials, that M_n is bounded as n ranges over all positive integers. But that is just about all that is known rigorously. On the lower side, it is not even known whether $\limsup_{n\to\infty} m_n > 0$. (See Section 3.5.)

This chapter is based on exhaustive computational searches for extremal polynomials in $\mathbb{U}_n$. The first part examined all $F(z) \in \mathbb{U}_n$ for $n \leq 52$. These computations extend unpublished computations of the author in the late 1980s and smaller scale searches of Robinson [28]. These searches led to the

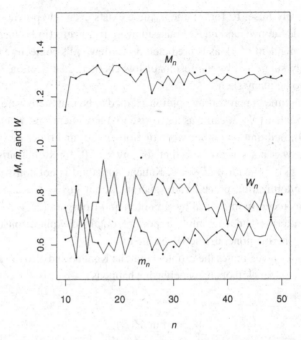

Fig. 3.1. Values of M_n, m_n and W_n for $10 \leq n \leq 50$. Scatterplot gives values of $M_n^*, m_n^*,$ and W_n^* for even n in that range.

conjecture that each of the following limits exists:

$$\lim_{n \to \infty} M_n = M, \tag{3.8}$$

$$\lim_{n \to \infty} m_n = m, \tag{3.9}$$

$$\lim_{n \to \infty} W_n = W. \tag{3.10}$$

The computations reported here suggest that $M \approx 1.27, m \approx 0.64$, and $W \approx 0.79$. These conjectures imply that ultraflat polynomials in $\mathbb{U}_n$ do not exist, but that GRS polynomials are far from optimal in terms of never being large. The conjecture that W exists and that $W < 1$ implies there exist constants $0 < c_1 < c_2$ so that for all high degrees N there are polynomials $F(z) \in \mathbb{U}_n$ with $c_1 < m(F) < M(F) < c_2$, which is Littlewood's conjecture (C_1), [23], p. 29.

Figure 3.1 shows a graph of the values of M_n, m_n, and W_n for $10 \leq n \leq 50$. The numerical values for all $n \leq 52$ and the polynomials that achieve them (as well as a large collection of other polynomials that come close to the record values) are available on the author's home page. Perhaps some patterns will be found in their coefficients that will help in explicit constructions of high-degree polynomials that are close to ultraflat.

It should be noted that the optimal polynomials are not isolated, as in most cases there are many others that have similar values; see Section 3.3.

Figure 3.1 shows M_n converging rapidly, and m_n and W_n considerably more slowly. However, given the low degrees involved, even the rate of convergence for m_n and W_n is rather remarkable; see Section 3.3.

The dots in Fig. 3.1 represent the optimal values when we restrict consideration to the very important class of skew-symmetric polynomials. When $F(z) \in \mathbb{U}_n$, $M(F)$ and $m(F)$ are the same for $F(z)$ and for

$$z^n F\left(\frac{1}{z}\right), \quad -F(z), \quad F(-z). \tag{3.11}$$

Therefore $F(z) \in \mathbb{U}_n$ can be grouped naturally into octuplets, which makes the search easier. Sometimes these octuplets collapse. The symmetric case, $F(z) = z^n F(\frac{1}{z})$, leads to very poor results for our problem. (This is easily observed with even low-degree polynomials, and there are some rigorous results that show, for example, that symmetric polynomials cannot be ultraflat. For the latest in this area, see [10].) However, when n is even, we usually obtain excellent outcomes from the skew-symmetric polynomials, namely those with

$$F(z) = \pm z^n F\left(\frac{-1}{z}\right)$$

(where the sign has to be $(-1)^{n/2}$). Those, as was first noticed by Golay in a slightly different context, see Section 3.2, contain the optimal polynomials in a majority of known cases. This is visible in Fig. 3.1 when the dot is right on the line. Even when a non–skew-symmetric polynomial is better than any skew-symmetric ones, visible in Fig. 3.1 when the dot is not right on the corresponding curve, the difference is usually slight. The largest exception to this claim that has been found so far is for W_{24}, which stands out in Fig. 3.1, with $W_{24} = 0.8344$, whereas the best result obtainable from skew-symmetric polynomials requires an annulus of width 0.9528.

Figure 3.1 naturally suggests the conjecture that if we define M_n^*, m_n^*, and W_n^* in analogy to M_n, m_n, and W_n, but limiting consideration to skew-symmetric polynomials of degree n, then

$$\lim_{\substack{n \to \infty \\ n \text{ even}}} M_n^* = \lim_{n \to \infty} M_n = M,$$

$$\lim_{\substack{n \to \infty \\ n \text{ even}}} m_n^* = \lim_{n \to \infty} m_n = m,$$

$$\lim_{\substack{n \to \infty \\ n \text{ even}}} W_n^* = \lim_{n \to \infty} W_n = W.$$

Since skew-symmetric polynomials have only $n/2 + 1$ free coefficients, as opposed to $n + 1$ for general ones, searches can be carried out about twice as

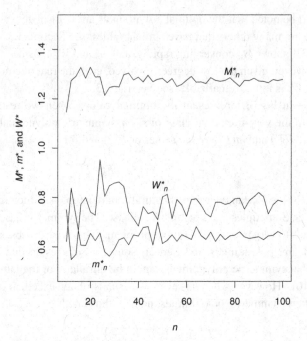

Fig. 3.2. Extreme values for skew-symmetric polynomials, of M_n^*, m_n^*, and W_n^* for even n, $10 \leq n \leq 100$.

far, and so far have been taken up to $n = 104$. The results for $10 \leq n \leq 100$ are shown in Fig. 3.2, and the numerical values for all even $n \leq 104$ are in the online tables, together with the corresponding polynomials, and nearly optimal polynomials. As with general $F(z) \in \mathbb{U}_n$, convergence is fastest for M_n^*, but now m_n^* also appears to converge rapidly, and it is only W_n^* that oscillates to a substantial extent.

The conjectured values for the limits, $M \approx 1.27$, $m \approx 0.64$, and $W \approx 0.79$ were derived from the computed values of M_n^*, m_n^*, and W_n^*.

3.2 Golay Merit Factor and Barker Polynomials

Ultraflat polynomials have close connections to the much larger field of discrete sequences and their correlation properties. (For some general information and references, see, for example, [3]. For much more detail and extensive references, including numerous applications, see [12, 16].) Here we just cite some of the basic results about the Golay merit factor and Barker polynomials. Much more can be found in the references just cited, as well as on the web

pages of the Centre for Experimental and Constructive Mathematics at Simon
Fraser University, and the home pages for Michael Mossinghoff and Tamás
Erdélyi, for example.

For $a_0, \ldots, a_n = \pm 1$, and $0 \le k \le n$, let

$$c_k = \sum_{j=0}^{n-k} a_j a_{j+k}, \tag{3.12}$$

and for $k < 0$ let $c_k = c_{-k}$. A Barker sequence $a_0, \ldots, a_n$ is defined by the
property that the nontrivial c_k are as small as possible, namely $c_k = 0, \pm 1$
for $1 \le k \le n$. The only nontrivial Barker sequences that are known have
$n = 2, 3, 4, 6, 10$, and 12, and are of great utility in communications and radar
applications. It is conjectured that there are no more, and it is known [21] that
there are no other ones with lengths $n < 4 \cdot 10^{33}$.

To any sequence $a_0, \ldots, a_n = \pm 1$ we can associate the polynomial
$F(z) \in \mathbb{U}_n$ with the a_k as coefficients. For this sequence and its polynomial,
we define the Golay merit factor

$$G(F) = \frac{(n+1)^2}{2 \sum_{k=1}^{n} c_k^2}. \tag{3.13}$$

A Barker sequence has the denominator $\sim n$, and so $G(F) \sim n$. The largest
known $G(F)$ comes from the Barker sequence with $n = 12$, and equals
$169/12 = 14.08\ldots$. It is conjectured that this is the largest merit factor among
all sequences of all lengths. The second largest is 12.1, coming from the Barker
sequence with $n = 10$, and no other merit factors exceeding 10 are known.
Golay's conjecture [15] that the highest merit factors should approach 12.32
asymptotically is generally not accepted as likely to be correct. For the latest
computations of highest merit factors for all sequences with $n \le 65$ and all
skew-symmetric sequences with $n \le 116$, see [26].

If merit factors are bounded, then the c_k are in absolute value on the order
of $\sqrt{n}$ on average. That is what random choices of a_k produce.

If the a_k are chosen at random, then for large n, for most sequences
$G(F) \sim 1$. GRS polynomials have $G(F) \sim 3$, and the best currently known
constructions give $G(F) \sim 6.34$ for large n [18]. However, exhaustive computations for lengths up to 60, and heuristic searches for greater n, frequently
find sequences with $G(F) > 9$.

Getting back to polynomials, if $F(z) \in \mathbb{U}_n$, $F(z) = \sum_{k=0}^{n} a_k z^k$, then (all on
$|z| = 1$)

$$F(z)\bar{F}(z) = F(z)F\left(\frac{1}{z}\right) = \sum_{k=-n}^{n} c_k z^k,$$

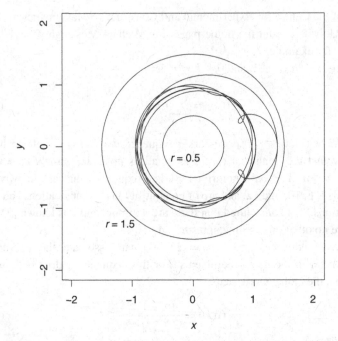

Fig. 3.3. Real and imaginary parts of the Barker polynomial of degree 12 (scaled by $\sqrt{13}$) as the argument runs over the unit circle, and circles of radii 0.5 and 1.5.

so

$$\|F(z)\|_4^4 = \|F(z)F\left(\frac{1}{z}\right)\|_2^2 = \sum_{k=-n}^{n} c_k^2.$$

Hence

$$G(F) = \frac{c_0^2}{\sum_{k\neq 0} c_k^2} = \frac{(n+1)^2}{\|F\|_4^4 - \|F\|_2^4}. \tag{3.14}$$

Ultraflat polynomials $F(z)$ would have $\|F\|_4 \sim \|F\|_2 \sim (1 + o(1))\sqrt{n+1}$ and so would give $G(F) \to \infty$. The conjecture that $G(F)$ is bounded therefore implies there are no ultraflat polynomials of high degrees. (For more on connections between sequences and flat polynomials, see also [3, 4].)

Equation (3.14) shows that there is a relation between high merit factors and flatness. In some cases, the correspondence is very close. The Barker polynomial of degree 10 has the smallest $M(F)$ ($= 1.1464$) of all polynomials that have been tested, and the Barker polynomial of degree 12 (whose behavior on the unit circle is displayed in Fig. 3.3) has the largest $m(F)$ ($= 0.8375$) and the smallest $W(F)$ ($= 0.5493$) that have been found.

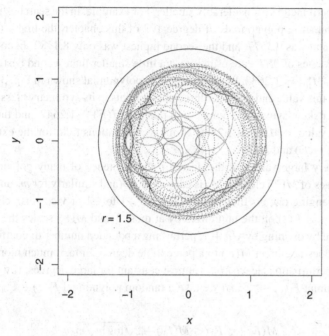

Fig. 3.4. Real and imaginary parts of the extremal skew-symmetric polynomial of degree 102 that achieves $M(F) = M_{102}^* = 1.2633\ldots$, at 10,000 uniformly spread points over the unit circle, scaled by $\sqrt{103}$, and a circle of radius 1.5.

The correlation between high Golay merit factors and flatness is not perfect, as can be seen by the example of the skew-symmetric polynomials of degree 102. Figure 3.4 shows the behavior on the unit circle of the polynomial in this set which has the smallest $M(F)$ (= 1.2633) among all skew-symmetric polynomials of degrees $84 \leq n \leq 104$. It has $G(F) = 5.7973$ (and $m(F) = 0.0985$). However, there is a skew-symmetric polynomial $F(z)$ of degree 102 that achieves $G(F) = 9.5577$ (the highest value that has been found in searches just for high merit factors among skew-symmetric sequences of that length), but it has $M(F) = 1.3689$, $m(F) = 0.5667$. This polynomial does not produce the best values for either $m(F)$ or $W(F)$ for degree 102.

3.3 Lack of Isolation of Extreme Flat Polynomials

In searches for high merit factors, it has been noted that for a given length, the sequence with the highest merit factor is usually not just unique (aside from the obvious symmetries given by (3.11)), but is also isolated, in that the second largest merit factor is considerably smaller. For example, in the searches for flat skew-symmetric polynomials of degree 102 of this chapter, the highest merit factor found was 9.5577, and the second highest was only 8.1482. In contrast, the best values of $M(F)$ are usually just a little smaller than second best. As as example, $M_{102}^* = 1.2633$, but aside from the polynomial shown in Fig. 3.4 that achieves this value and the other three polynomials in its symmetry class, there is another skew-symmetric polynomial that has $M(F) = 1.2647$, and the 10th smallest value of $M(F)$ is 1.2876. Similar observations hold for the extremal values of $m(F)$ and $W(F)$.

For very large values of the degree n the existence of many polynomials with values of $M(F)$ close to M_n is easy to show (and similarly for m_n and W_n). Since changing one coefficient in $F(z)$ from $+1$ to -1 or vice versa changes the values of $F(z)$ on the unit circle by at most 2, and $M(F)$ scales the maximal value by dividing by $\sqrt{n+1}$, perturbing a bounded number of coefficients of $F(z)$ does not affect $M(F)$ to a perceptible degree. In fact, much more substantial perturbations leave $M(F)$ almost constant for large degrees. If we have a polynomial $F_1(z) \in \mathbb{U}_n$ and we take a random polynomial $F_2(z) \in \mathbb{U}_m$, then, by (3.4),

$$M(F_1 + z^n F_2) \le M(F_1) + 2\sqrt{\log m}\sqrt{m/n} \tag{3.15}$$

for most choices of $F_2(z)$. Hence if $m = o(n/(\log n))$ as $n \to \infty$, we obtain close to 2^m polynomials $F(z) \in \mathbb{U}_{n+m}$ that have $M(F)$ just about the same as $M(F_1)$. So we should expect M_n to vary smoothly with n. A modification of the argument that led to (3.4) can be used to show that one can also alter many coefficients of a given $F(z) \in \mathbb{U}_m$ without affecting $M(F)$ significantly, so that there will be many polynomials in $\mathbb{U}_n$ with $M(F)$ close to M_n.

One can obtain even stronger results by invoking the work of Spencer [34], who showed that there are exponentially many $F(z) \in \mathbb{U}_m$ with $M(F) < C$ for large constants C. This shows that in the foregoing construction one can take $m = o(n)$, and not just $m = o(n/(\log n))$. Thus non-isolation of extremal polynomials is to be expected for high degrees. But the same argument also shows that sequences that achieve maximal merit factors cannot be isolated for large lengths. So why the difference in behavior for modest lengths? If we consider the effect that the change of a single coefficient can make for n on the order of 100, intuitively we might expect more isolation for extremal polynomials, and much more variation in M_n and M_n^* as n varies than is visible in Figs. 3.1 and 3.2. Even the values of W_n and W_n^*, which show more variation, are surprisingly smooth.

Yet another puzzle is why M_n and M_n^* converge so much faster than W_n and W_n^*.

3.4 More General Coefficients

Kahane showed that ultraflat polynomials do exist with $a_k \in \mathbf{C}$, $|a_k| = 1$. The computations of this chapter strongly suggest such polynomials don't exist if we require $a_k = \pm 1$, but that there do exist constants $0 < \delta < C$ (even with $\delta = 0.5$ and $C = 1.5$) so that for all large n, there exist $F(z) \in \mathbb{U}_n$ with

$$\delta < |F(z)|/\sqrt{n+1} < C \qquad (3.16)$$

for z on the unit circle. Beck [1] has shown, through a nonconstructive argument, that there do exist polynomials satisfying (3.16) for some positive δ, C when the a_k are required to satisfy $a_k^{400} = 1$. (Higher orders of roots of unity lead to similar results.) So it is natural to conjecture that for each integer $r \geq 2$, if we require the a_k to be rth roots of unity, the limits corresponding to M and m will exist, and will go to 1 as $r \to \infty$.

Some small-scale computations for $r = 3, 4$ do support the conjecture about existence of the limits. They also show that for $r = 3$, it is harder to approach flatness than it is for the $r = 2$ (± 1) case, and that $r = 4$ does not produce polynomials much flatter than $r = 2$, at least for small degrees.

3.5 Polynomials That Are Never Too Small

GRS polynomials show that M_n is bounded. However, it is not known whether m_n is bounded away from zero, even for a sparse sequence of degrees n. The best known results come from a recursive construction of Carroll, Eustice, and Figiel [7]. If $F(z) \in \mathbb{U}_n$, and $m(F) \geq \alpha$, then

$$F(z^{n+1})F(z) \in \mathbb{U}_{(n+1)^2 - 1} \qquad (3.17)$$

and for $|z| = 1$,

$$|F(z^{n+1})F(z)| \geq \alpha^2. \qquad (3.18)$$

Repeating this construction, we obtain a sequence of polynomials $G(z)$ of degrees $r = (n+1)^k - 1$ with $m(G) \geq \alpha^k$, which gives

$$m_r \geq (r+1)^{-\beta}, \qquad (3.19)$$

where

$$\beta = \frac{1}{2} - (\log \alpha)/(\log(n+1)). \qquad (3.20)$$

The smallest value of β that has been found among all the polynomials examined so far comes from the Barker polynomial of degree 12, which has

$m(F) = 0.8375$, $\alpha = 3.0196$ and gives $\beta = 0.069$. (This example was already featured in [7].) It is disappointing that even the skew-symmetric polynomials with $m(F) = m_n^*$ for $n = 100, 102$, and 104 do not provide a better bound. If the conjecture about the limit of m_n being about 0.64 is valid, and the limit is approached as smoothly as suggested by Fig. 3.1, it will require an example with the degree n on the order of 500 to improve on the Barker example of degree 12. This provides yet another demonstration of the uniqueness and nice behavior of Barker sequences.

Carroll, Eustice, and Figiel [7] have shown, using an interpolation procedure, that lower bounds similar to (3.19) hold for all large degrees r, not just for $r = 13^k - 1$.

3.6 Uniform Distribution Conjectures

B. Saffari and H. Montgomery developed conjectures about uniform distribution of GRS polynomials; cf. [25]. These have been proved recently, see [8, 29], so we now know that as the degrees of GRS polynomials $F(z)$ grow, the values of $F(z)/\sqrt{n+1}$ as z runs over the unit circle approach the uniform distribution in the disk of radius $\sqrt{2}$.

For the polynomials $F(z)$ that are conjectured here to exist with $M(F) < 1.3$, say, the distribution of values cannot be uniform in this sense, as the L_2-norm of $F(z)/\sqrt{n+1}$ would be < 1. If we look at the distribution of values of the skew-symmetric polynomial of degree $n = 102$ that has the smallest maximal value, pictured in Fig. 3.4, we see there the expected concentration close to the unit circle.

If $F(z)/\sqrt{n+1}$ were to have its values approximately uniformly distributed in an annulus with radii $r < R$, the L_2-norm would be

$$(R^2 + r^2)/2$$

so for it to equal 1, we would need

$$r = \sqrt{2 - R^2}. \tag{3.21}$$

Figure 3.5 shows the behavior on the unit circle of the skew-symmetric polynomial $F(z)$ of degree 94 that has $W(F) = 0.733\ldots$ and thus fits in the smallest annulus of all skew-symmetric polynomials of degrees $72 \le n \le 104$. Whether that approximates a uniform distribution is difficult to say. This polynomial has $M(F) = 1.3162\ldots$ and $m(F) = 0.5830\ldots$, so does not fit the formula (3.21) too well, since substituting $R = 1.3162$ in that formula

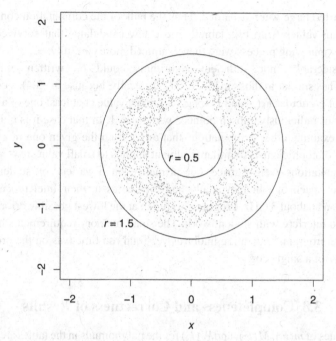

Fig. 3.5. Real and imaginary parts of the extremal skew-symmetric polynomial of degree 94 that achieves $W(F) = W_{94}$ at 2,000 points uniformly spread over the unit circle, scaled by $\sqrt{95}$, and circles of radii 0.5 and 1.5.

produces $r = 0.5173\ldots$ Thus we do not have much evidence to suggest whether uniform distribution will prevail in the limit.

3.7 Algorithms

Computations were carried out with a simple program that examined all candidates (after taking advantage of the symmetries of (3.11)). If we write $F(z) = F_1(z) + F_2(z)$, where

$$F_1(z) = \sum_{k=0}^{15} a_k z^k,$$

say, then the values of $F_1(z)$ for all possible choices of $F_1(z)$ were precomputed at a small set of points on the upper half of the unit circle (typically 32 values). A candidate for $F_2(z)$ was evaluated at those points (with cosine evaluations replaced by table lookups, since only a small number of arguments were relevant), and choices of $F_1(z)$ that produced values of $F(z)$ that were either too small or too large were discarded. Thus the bulk of the computation consisted of adding values from two tables. Those few candidates that survived this simple winnowing process were then examined more carefully.

Considerably more efficient algorithms could be written, utilizing approaches similar to those in [5, 26, 27]. Typically, because of (3.4), a combination of particular $F_1(z)$ and $F_2(z)$ gives a large value at at least one of a small number of judiciously chosen points z with $|z| = 1$. In that case it is unnecessary to examine other combinations that differ from the given one in a small number of coefficients. A similar argument applies to small values.

Computations were carried out primarily on a variety of student lab machines, each of which typically had four cores in their Intel processors, which ran at about 3 GHz. Programs were run at the lowest possible priority, so as not to interfere with student work. The small memory requirements helped keep the programs' operation unobtrusive. Total run time was on the order of 30 years on a single core.

3.8 Completeness and Correctness of Results

The values of $m(F)$, $M(F)$, and $W(F)$ for the polynomials in the tables are trustworthy. They were computed for the candidates identified by the main program with an inefficient but straightforward program. It used the trivial bounds on the first and second derivatives of $F(z)$ to find the extremal values.

What is not completely certain is whether all extremal polynomials were found. Typically around 100 cores were doing the searches, sending the promising candidates to a file over a local area network. This took several months in all, and there were some network hitches that triggered warnings that led to repeats of some computations. There is a chance that some network or storage system abnormalities may not have been detected, so that some good polynomials may have been missed. This probability is slight, though, since extremal polynomials are so rare.

3.9 Conclusions

The computations of this chapter support the conjecture that ultraflat polynomials with ± 1 coefficients do not exist. However, it seems extremely likely that the Golay–Rudin–Shapiro polynomials are far from best possible in terms of never being large.

It is to be hoped that the extremal polynomials produced by this project will be helpful in finding some patterns that will lead to rigorous constructions.

The approach of the measures M_n, m_n, and W_n to their asymptotic values is surprisingly rapid. It was also unexpected that there would be as many polynomials close to the extremal ones.

Acknowledgments

Thanks are due to Enrico Bombieri, Charles Jackson, Stephan Mertens, Hugh Montgomery, and Michael Mossinghoff for their comments and helpful information.

The author acknowledges the Minnesota Supercomputing Institute (MSI) at the University of Minnesota for providing resources that contributed to the research results reported within this chapter. The hospitality of the Institute for Mathematics and its Applications (IMA), also at the University of Minnesota, during part of the time this research was carried out, is also greatly appreciated.

References

1. J. Beck. Flat polynomials on the unit circle – Note on a problem of Littlewood. *Bull. London Math. Soc.* **23** (1991), 269–277.
2. E. Bombieri and J. Bourgain. On Kahane's ultraflat polynomials. *J. Eur. Math. Soc.* **11** (2009), 627–703.
3. P. Borwein. *Computational Excursions in Analysis and Number Theory.* Springer Science+Business Media, New York, 2002.
4. P. Borwein and M. Mossinghoff. Barker sequences and flat polynomials. pp. 71–88 in *Number Theory and Polynomials,* London Math. Soc. Lecture Note Ser., Vol. 352, Cambridge University Press, Cambridge, 2008. Available at ⟨http://academics.davidson.edu/math/mossinghoff/BarkerSeqsFlatPolys_BorMoss.pdf⟩.
5. B. Bošković, F. Brglez, and J. Brest. Low-autocorrelation binary sequences: On improved merit factors and runtime predictions to achieve them. *Appl. Soft Comput.* **56** (July 2017), 262–285.
6. J. Brillhart and P. Morton. A case study in mathematical research: The Golay–Rudin–Shapiro sequence. *Amer. Math. Monthly* **103** (Dec. 1996), 854–869.
7. F. W. Carroll, D. Eustice, and T. Figiel. The minimum modulus of polynomials with coefficients of modulus one. *J. London Math. Soc.,* Ser. 2, **16** (1977), 76–82.
8. S. B. Ekhad and D. Zeilberger. Integrals involving Rudin-Shapiro polynomials and a sketch of a proof of Saffari's conjecture. 2016 preprint, available at ⟨https://arxiv.org/abs/1605.06679⟩.
9. E. H. el Abdalaoui. On the Erdős flat polynomials problem, Chowla conjecture and Riemann Hypothesis. Preprint, version 2, 11 January 2017, available at ⟨https://arxiv.org/abs/1609.03435⟩.
10. T. Erdélyi. On the flatness of conjugate reciprocal unimodular polynomials. *J. Math. Anal. Appl.* **432**, no. 2 (2015), 699–714.
11. P. Erdős. Some unsolved problems. *Michigan Math. J.* **4** (1957), 291–300.

12. P. Fan and M. Darnell. *Sequence Design for Communications Applications*. Research Studies Press, 1996.

13. A. Gersho, B. Gopinath, and A. Odlyzko. Coefficient inaccuracy in transversal filtering. *Bell System Tech. J.* **58** (1979), 2301–2316.

14. M. J. E. Golay. Static multislit spectrometry and its application to the panoramic display of infrared spectra. *J. Opt. Soc. Amer.* **41** (1951), 468–472.

15. M. J. E. Golay. The merit factor of long low autocorrelation binary sequences. *IEEE Trans. Information Theory* **IT-28**, no. 3 (1982), 543–549.

16. S. W. Golomb and G. Gong. *Signal Design for Good Correlation for Wireless Communication, Cryptography, and Radar*. Cambridge University Press, Cambridge, 2005.

17. G. Halasz. On a result of Salem and Zygmund concerning random polynomials. *Studia Scient. Math. Hungar.* **8** (1973), 369–377.

18. J. Jedwab, D. J. Katz, and K.-W. Schmidt. Advances in the merit factor problem for binary sequences. *J. Combinatorial Theory, Ser. A* **120** (2012), 882–906.

19. J.-P. Kahane. Sur les polynômes à coefficients unimodulaires. *Bull. London Math. Soc.* **12** (1980), 321–342.

20. T. W. Körner. On a polynomial of Byrnes. *Bull. London Math. Soc.* **12** (1980), 219–224.

21. K. H. Leung and B. Schmidt. The anti-field-descent method. *J. Combinat. Theory, Ser. A* **139** (April 2016), 87–131.

22. J. E. Littlewood. On polynomials $\sum \pm z^m$, $\sum e^{\alpha_m i}$, $z = e^{\theta i}$. *J. London Math. Soc.* **41** (1966), 367–376.

23. J. E. Littlewood. *Some Problems in Real and Complex Analysis*. Heath, Lexington, MA, 1968.

24. I. D. Mercer. *Autocorrelation and Flatness of Height One Polynomials*. Ph.D. thesis, Simon Fraser University, 2005. Available at ⟨http://people.math.sfu.ca/~idmercer/thesis.pdf⟩.

25. H. L. Montgomery. Littlewood polynomials. In *Analytic Number Theory: Modular Forms and q-Hypergeometric Series*, G. Andrews and F. Garvan, eds., Springer Science+Business Media, forthcoming.

26. T. Packebusch and S. Mertens. Low autocorrelation binary sequences. *J. Physics A: Mathematical and Theoretical* **49** (2016), 165001.

27. S. D. Prestwich. Improved branch-and-bound for low autocorrelation binary sequences. available at ⟨http://arxiv.org/abs/1305.6187⟩.

28. L. Robinson. *Polynomials with Plus or Minus One Coefficients: Growth Properties on the Unit Circle*. M.S. thesis, Simon Fraser University, 1997. Available at ⟨http://summit.sfu.ca/system/files/iritems1/7393/b18765270.pdf⟩.

29. B. Rodgers. On the distribution of Rudin-Shapiro polynomials. 2016 preprint, available at ⟨http://arxiv.org/abs/1606.01637⟩.

30. W. Rudin. Some theorems on Fourier coefficients. *Proc. Amer. Math. Soc.* **10** (1959), 855–859.

31. B. Saffari. Some polynomial extremal problems which emerged in the twentieth century. pp. 201–233 in J. S. Byrnes, ed., *Twentieth Century Harmonic Analysis – A Celebration*. Kluwer, Dordrecht, 2001.

32. M. R. Schroeder. *Number Theory in Science and Communication: With Applications in Cryptography, Physics, Biology, Digital Information, and Computing,* Springer-verlag, New York, 1984.
33. H. S. Shapiro, *Extremal Problems for Polynomials and Power Series,* M.S. thesis, MIT, 1951.
34. J. Spencer. Six standard deviations suffice. *Trans. Amer. Math. Soc.,* **289**, no. 2 (June 1985), 679–706.
35. N. Xiang and G. M. Sessler, eds. *Acoustics, Information, and Communication: Memorial Volume in Honor of Manfred R. Schroeder.* Springer Science+Business Media, New York, 2014.

4

Generalized Gončarov Polynomials

Rudolph Lorentz, Salvatore Tringali, and Catherine H. Yan

Abstract

We introduce the sequence of generalized Gončarov polynomials, which is a basis for the solutions to the Gončarov interpolation problem with respect to a delta operator. Explicitly, a generalized Gončarov basis is a sequence $(t_n(x))_{n \geq 0}$ of polynomials defined by the biorthogonality relation $\varepsilon_{z_i}(\eth^i(t_n(x))) = n! \, \delta_{i,n}$ for all $i, n \in \mathbb{N}$, where $\eth$ is a delta operator, $\mathcal{Z} = (z_i)_{i \geq 0}$, a sequence of scalars, and ε_{z_i} the evaluation at z_i. We present algebraic and analytic properties of generalized Gončarov polynomials and show that such polynomial sequences provide a natural algebraic tool for enumerating combinatorial structures with a linear constraint on their order statistics.

4.1 Introduction

This chapter is a work combining three areas: interpolation theory, finite operator calculus, and combinatorial enumeration. Lying in the center is a sequence of polynomials, the generalized Gončarov polynomials, that arose from the Gončarov Interpolation problem in numerical analysis.

The classical Gončarov Interpolation problem is a special case of Hermite-like interpolation. It asks for a polynomial $f(x)$ of degree n such that the ith derivative of $f(x)$ at a given point a_i has value b_i, for $i = 0, 1, \ldots, n$. The problem was introduced by Gončarov [4, 5] and Whittaker [26], and the solution is obtained by taking linear combinations of the (classical) Gončarov polynomials, or the Abel–Gončarov polynomials, which have been studied extensively by analysts, owing to their considerable significance in the interpolation theory of smooth and analytic functions, see, for instance, [3, 4, 6, 13] and references therein.

Surprisingly, Gončarov polynomials also play an important role in combinatorics because of their close relations to parking functions. A parking function is a sequence $(a_1, a_2, \ldots, a_n)$ of positive integers such that for every $i = 1, 2, \ldots, n$, there are at least i terms that are less than or equal to i. Parking functions are one of the most fundamental objects in combinatorics and are related to many different structures, for example, labeled trees and graphs, linear probing in computer algorithms, hyperplane arrangements, noncrossing partitions, and diagonal harmonics and representation theory, to name a few. See, for example, [28] for a comprehensive survey on parking functions. It is shown by Kung and Yan [12] that Gončarov polynomials are the natural basis of polynomials for working with parking functions, and the enumeration of parking functions and their generalizations can be obtained using Gončarov polynomials. Khare, Lorentz, and Yan [9] investigated a multivariate Gončarov interpolation problem and defined sequences of multivariate Gončarov polynomials, which are solutions to the interpolation problem and enumerate k-tuples of integer sequences whose order statistics are bounded by certain weight functions along lattice paths in $\mathbb{N}^k$.

In [21], Rota, Kahaner, and Odlyzko introduced a unified theory of special polynomials by exploiting to the hilt the duality between x and d/dx. The main technique is a rigorous version of symbolic calculus, also called *finite operator calculus*, since it has an emphasis on operator methods. This algebraic theory is particularly useful in dealing with polynomial sequences of binomial type, which occur in a large variety of combinatorial problems when one wants to enumerate objects that can be pieced together out of small, disjoint objects. Each polynomial sequence of binomial type can be characterized by a linear operator called a *delta operator*, which possesses many properties of the differential operator. A few basic principles of delta operators lead to a series of expansion and isomorphism theorems on families of special polynomials, which in turn lead to new identities and solutions to combinatorial problems.

Inspired by the rich theory on delta operators, we extend the Gončarov interpolation problem by replacing the differential operator with a delta operator and consider the following interpolation.

Generalized Gončarov Interpolation. Given two sequences $z_0, z_1, \ldots, z_n$ and $b_0, b_1, \ldots, b_n$ of real or complex numbers and a delta operator $\mathfrak{d}$, find a (complex) polynomial $p(x)$ of degree n such that

$$\varepsilon_{z_i}\mathfrak{d}^i(p(x)) = b_i \qquad \text{for each } i = 0, 1, \ldots, n. \tag{4.1}$$

The solution of this problem gives the generalized Gončarov polynomial. When $\mathfrak{d} = D$, we recover the classical Gončarov polynomials. Generalized

Gončarov polynomials enjoy many nice algebraic properties and carry a combinatorial interpretation that combines the ideas of binomial enumeration and order statistics. Roughly speaking, if each combinatorial object is associated with a sequence of numbers and we rearrange those numbers in nondecreasing order, then the generalized Gončarov polynomial counts those objects of which the nondecreasing rearrangements are bounded by a predetermined sequence. Such structures give a new generalization of the classical parking functions.

The main objective of this chapter is to present the algebraic and combinatorial properties of generalized Gončarov polynomials. The chapter is organized as follows. In Section 4.2 we recall the basic theory of delta operators, polynomial sequences of binomial type, and polynomial sequences biorthogonal to a sequence of linear operators. Using this theory we introduce the sequence of generalized Gončarov polynomials $t_n(x; \eth, \mathcal{Z})$ associated with a delta operator $\eth$ and a grid $\mathcal{Z}$. In the subsequent Sections 4.3–4.5, we discuss the algebraic properties, present a combinatorial formula for $t_n(x; \eth, \mathcal{Z})$, and describe the combinatorial interpretation by counting reluctant functions, a kind of combinatorial structure arising from the study of binomial enumeration, with constraints on the order statistics. Many examples are given in Section 4.6. We finish the chapter with some further remarks in the last section.

4.2 Delta Operators and Generalized Gončarov Polynomials

4.2.1 Delta Operators and Basic Polynomials

We start by recalling the basic theory of delta operators and their associated sequence of basic polynomials, as developed by Mullin and Rota [21].

Consider the vector space $\mathbb{K}[x]$ of all polynomials in the variable x over a field $\mathbb{K}$ of characteristic zero. For each $a \in \mathbb{K}$, let E_a denote the shift operator $\mathbb{K}[x] \to \mathbb{K}[x] : f(x) \mapsto f(x + a)$. A linear operator $\mathfrak{s} : \mathbb{K}[x] \to \mathbb{K}[x]$ is called *shift-invariant* if $\mathfrak{s}E_a = E_a\mathfrak{s}$ for all $a \in \mathbb{K}$, where the multiplication of operators is the composition of operators.

Definition 4.1. *A* delta operator $\eth$ *is a shift-invariant operator satisfying* $\eth(x) = a$ *for some nonzero constant a.*

Delta operators possess many of the properties of the differentiation operator D. For example, $\deg(\eth(f)) = \deg(f) - 1$ for any $f \in \mathbb{K}[x]$ and $\eth(a) = 0$ for every constant a.

We say that a shift-invariant operator $\mathfrak{s}$ is *invertible* if $\mathfrak{s}(1) \neq 0$. Note that delta operators are not invertible.

Definition 4.2. *Let* $\eth$ *be a delta operator. A polynomial sequence* $(p_n(x))_{n\geq 0}$ *is called the* sequence of basic polynomials, *or the* basic sequence, *of* $\eth$ *if*

(1) $p_0(x) = 1$;
(2) $p_n(0) = 0$ *whenever* $n \geq 1$;
(3) $\eth(p_n(x)) = np_{n-1}(x)$.

Every delta operator has a unique sequence of basic polynomials, which is a sequence of binomial type, i.e., satisfies

$$p_n(x + y) = \sum_{k\geq 0} \binom{n}{k} p_k(x)p_{n-k}(x) \qquad \text{for all } n. \qquad (4.2)$$

Conversely, every sequence of polynomials of binomial type is the basic sequence for some delta operator.

Let $\mathfrak{s}$ be a shift-invariant operator, and $\eth$ a delta operator with basic sequence $p_n(x)$. Then $\mathfrak{s}$ can be expanded as a formal power series of $\eth$, as

$$\mathfrak{s} = \sum_{k\geq 0} \frac{a_k}{k!}\eth^k, \qquad (4.3)$$

with $a_k = \varepsilon_0(\mathfrak{s}(p_k(x)))$. We will say that the formal power series $f(t) = \sum_{k\geq 0} \frac{a_k}{k!}t^k$ is the $\eth$-indicator of $\mathfrak{s}$. In fact, there exists an isomorphism from the ring $\mathbb{K}[\![t]\!]$ of formal power series in the variable t over $\mathbb{K}$ onto the ring Σ of shift-invariant operators, which carries

$$f(t) = \sum_{k\geq 0} \frac{a_k}{k!}t^k \qquad \text{into} \qquad \sum_{k\geq 0} \frac{a_k}{k!}\eth^k. \qquad (4.4)$$

Under this isomorphism, a shift-invariant operator $\mathfrak{s}$ is invertible if and only if its $\eth$-indicator $f(t)$ satisfies $f(0) \neq 0$, and $\mathfrak{s}$ is a delta operator if and only if $f(0) = 0$ and $f'(0) \neq 0$.

Another result we will need is the generating function for the sequence of basic polynomials $\{p_n(x)\}$ associated to a delta operator $\eth$. Let $q(t)$ be the D-indicator of $\eth$, i.e., $q(t)$ is a formal power series satisfying $\eth = q(D)$. Let $q^{-1}(t)$ be the compositional inverse of $q(t)$. Then

$$\sum_{n\geq 0} \frac{p_n(x)}{n!}t^n = e^{xq^{-1}(t)}. \qquad (4.5)$$

4.2.2 Biorthogonal Sequences

Generalized Gončarov polynomials are defined by a biorthogonality condition posted in the Gončarov interpolation problem. Many properties of these polynomials follow from a general theory of *sequences of polynomials biorthogonal to a sequence of linear functionals.* The idea behind this theory is well-known (for examples, see [1, 2]), and an explicit description for the differential operator D is given in [12, section 2]. Here we briefly describe this theory with a general delta operator $\eth$. The proofs are analogous to the ones in [12] and hence omitted.

Let $\eth$ be a delta operator with the basic sequence $\mathcal{P} = (p_n(x))_{n \geq 0}$. Let Φ_i, $i = 0, 1, 2, \ldots$, be a sequence of shift-invariant operators of the form $\sum_{j \geq 0} a_j^{(i)} \eth^{i+j}$, where $(a_j^{(i)}) \in \mathbb{K}$ and $a_0^{(i)} \neq 0$. Then we have:

Theorem 4.1. (1) *There exists a unique sequence $\mathcal{F} = (f_n(x))_{n \geq 0}$ of polynomials such that $f_n(x)$ is of degree n and*

$$\varepsilon_0(\Phi_i(f_n(x))) = n!\delta_{i,n} \qquad \text{for all } i, n \in \mathbb{N}$$

In addition, for every n we have

$$f_n(x) = \frac{n!}{a_0^{(0)} \cdots a_0^{(n)}} \det{}_{\mathbb{K}[x]}(\Lambda^{(n)}),$$

with $\Lambda^{(n)}$ the $(n+1)$-by-$(n+1)$ matrix whose (i, j)-entry, for $0 \leq i, j \leq n$, is given by

$$\lambda_{i,j}^{(n)} := \begin{cases} a_{j-i}^{(i)}, & \text{if } 0 \leq i \leq \min(j, n-1) \\ \frac{1}{j!} p_j(x), & \text{if } i = n \\ 0, & \text{otherwise.} \end{cases}$$

(2) *The sequence $\mathcal{F}$ defined in the foregoing forms a basis of $\mathbb{K}[x]$. For every $f(x) \in \mathbb{K}[x]$ it holds*

$$f(x) = \sum_{n \geq 0} \frac{\varepsilon_0(\Phi_i(f))}{n!} f_n(x) = \sum_{n=0}^{\deg(f)} \frac{\varepsilon_0(\Phi_i(f))}{n!} f_n(x). \qquad (4.6)$$

4.2.3 Generalized Gončarov Polynomials

Let $\mathcal{Z} = (z_i)_{i \geq 0}$ be a fixed sequence with values in $\mathbb{K}$; in this context, we may refer to $\mathcal{Z}$ as an (interpolation) $\mathbb{K}$-grid (or simply a grid), and to the scalars $z_i \in \mathbb{K}$ as (interpolation) nodes. For every $a \in \mathbb{K}$, E_a is an invertible shift-invariant operator and hence can be expressed as $E_a = f_a(\eth)$ where $f_a(t) \in \mathbb{K}[[t]]$ with $f_a(0) \neq 0$. It follows from Theorem 4.1 that there is a unique sequence of polynomials $\mathcal{T} = (t_n(x))_{n \geq 0}$ biorthogonal to the sequence of operators $\{\Phi_i = E_{z_i}\eth^i : i \geq 0\}$. More precisely, $t_n(x)$ satisfies

$$\varepsilon_{z_i}\eth^i(t_n(x)) = n!\delta_{i,n}. \qquad (4.7)$$

Definition 4.3. *We call the polynomial sequence* $\mathcal{T} = (t_n(x))_{n \geq 0}$ *determined by* (4.7) the sequence of generalized Gončarov polynomials, *or the* generalized Gončarov basis, *associated with the pair* $(\eth, \mathcal{Z})$, *and* $t_n(x)$ *the nth generalized Gončarov polynomial relative to the same pair. Accordingly,* (4.7) *will be referred to as the biorthogonality property of generalized Gončarov bases.*

By Theorem 4.1(2), for any polynomial $f(x) \in \mathbb{K}[x]$, we have the expansion formula

$$f(x) = \sum_{i \geq 0} \frac{\varepsilon_{z_i}\left(\eth^i(f)\right)}{i!} t_i(x) = \sum_{i=0}^{\deg(f)} \frac{\varepsilon_{z_i}\left(\eth^i(f)\right)}{i!} t_i(x). \tag{4.8}$$

In particular, the solution of the generalized Gončarov interpolation (4.1) described in Section 4.1 is given by the polynomial

$$p(x) = \sum_{i=0}^{n} \frac{b_i}{i!} t_i(x).$$

In some cases, to emphasize the dependence of $\mathcal{T}$ on $\eth$ and $\mathcal{Z}$, we write $t_n(x)$ as $t_n(x; \eth, \mathcal{Z})$. When the delta operator is the differentiation D, we get the classical Gončarov polynomials, which were studied in [12]. We reserve the symbols $g_n(x)$ and $g_n(x; \mathcal{Z})$, respectively, for the classical Gončarov polynomials to avoid confusion when we compare the results of generalized Gončarov polynomials with those of the classical case. Another special case that has been considered is the difference Gončarov polynomials [11], for which $\eth$ is the backward difference operator $\Delta_{0,-1} = I - E_{-1}$. The present paper is the first one describing the theory of biorthogonal polynomials with an arbitrary delta operator, and hence connecting the theory of interpolation to finite operator calculus. In the next sections we will describe the algebraic properties of the generalized Gončarov polynomials, and reveal a deeper connection between binomial enumerations and structure of order statistics.

We remark that every generalized Gončarov basis is a sequence of biorthogonal polynomials, but the converse is not true in general.

4.3 Algebraic Properties of Generalized Gončarov Polynomials

Let $\mathcal{Z} = (z_i)_{i \geq 0}$ be a fixed $\mathbb{K}$-grid. We denote by $\mathcal{Z}^{(j)}$ the grid whose ith term is the $(i + j)$th term of $\mathcal{Z}$, and call $\mathcal{Z}^{(j)}$ the jth shift of the grid $\mathcal{Z}$. The zero grid, herein denoted by $\mathcal{O}$, is the one with $z_i = 0$ for all i.

Proposition 4.1. *If $\mathcal{T} = (t_n(x))_{n\geq 0}$ is the generalized Gončarov basis associated with the pair $(\eth, \mathcal{Z})$, then $t_0(x) = 1$ and $t_n(z_0) = 0$ for all $n \geq 1$.*

Proof. This follows from the biorthogonality property (4.7) with $i = 0$.   □

The next proposition is a generalization of the differential relations satisfied by the classical Gončarov polynomials [12, p. 23].

Proposition 4.2. *Let $\mathcal{T} = (t_n(x))_{n\geq 0}$ be the generalized Gončarov basis associated with the pair $(\eth, \mathcal{Z})$. Fix $j \in \mathbf{N}$ and define for each $n \in \mathbb{N}$ the polynomial $t_n^{(j)}(x)$ by letting*

$$t_n^{(j)}(x) := \frac{1}{(n+j)_{(j)}} \eth^j \left(t_{n+j}(x) \right), \tag{4.9}$$

where $n_{(j)} = n(n-1)\cdots(n-j+1)$ is the jth lower factorial function. Then, $(t_n^{(j)}(x))_{n\geq 0}$ is the generalized Gončarov basis associated with the pair $(\eth, \mathcal{Z}^{(j)})$. In particular, we have

$$\eth^j \left(t_n(x) \right) = n_{(j)} t_{n-j}^{(j)}(x). \tag{4.10}$$

Proof. First, notice that $t_n^{(j)}(x)$ is a polynomial of degree n, since $t_{n+j}(x)$ is a polynomial of degree $n + j$ and delta operators reduce degrees by 1.

Next, pick $i, n \in \mathbb{N}$ with $i \leq n$, and let $z_i^{(j)}$ denote the ith node of the grid $\mathcal{Z}^{(j)}$. We just need to verify that

$$\varepsilon_{z_i^{(j)}}(\eth^i(t_n^{(j)}(x))) = n! \delta_{i,n}.$$

Since $z_i^{(j)} = z_{i+j}$ and $\delta_{i,n} = \delta_{i+j,n+j}$, the preceding equation is equivalent to

$$\frac{1}{(n+j)_{(j)}} \varepsilon_{z_{i+j}} (\eth^{i+j}(t_{n+j}(x))) = n! \delta_{i+j,n+j},$$

which follows from the equations $(n + j)_{(j)} n! = (n + j)!$ and $\varepsilon_{z_l} \left(\eth^l \left(t_k(x) \right) \right) = k! \delta_{l,k}$ for all $l, k \in \mathbb{N}$. The last statement is obtained by replacing $n + j$ by n in (4.9).   □

Following Proposition 4.2 we see that sequences of polynomials of binomial type are a special case of generalized Gončarov polynomials.

Proposition 4.3. *The basic sequence of the delta operator $\eth$ is the generalized Gončarov basis associated with the pair $(\eth, \mathcal{O})$.*

Proof. Let $(p_n(x))_{n\geq 0}$ be the basic sequence of the delta operator $\eth$. Then iterating the equation $\eth(p_n) = np_{n-1}$ yields $\eth^i(p_n(x)) = n_{(i)} p_{n-i}(x)$, which, when evaluated at $x = 0$, is $n! p_0(x) = n!$ if $n = i$, and $n_{(i)} p_{n-i}(0) = 0$ if $i \neq n$.   □

Corollary 4.1. *Let* $\mathcal{P} = (p_n(x))_{n \geq 0}$ *be a sequence of polynomials with* $\deg(p_i) = i$ *for all* i. *Then,* $\mathcal{P}$ *is of binomial type if and only if* $\mathcal{P}$ *is the generalized Gončarov basis associated with the pair* $(\eth, \mathcal{O})$ *for a suitable choice of* $\eth$.

Proof. The necessity follows from Proposition 4.3 and Theorem 1(b) of [21], which states that any sequence of polynomials of binomial type is a basic sequence for some delta operator.

Conversely, let $\mathcal{P} = (p_n(x))_{n \geq 0}$ be the generalized Gončarov basis associated with the pair $(\eth, \mathcal{O})$. If $(p_n^{(1)}(x))_{n \geq 0}$ denotes the generalized Gončarov basis associated with the pair $(\eth, \mathcal{O}^{(1)})$, then by Proposition 4.2, $\eth(p_n) = n p_{n-1}^{(1)}$ for all $n \geq 1$, which in turn implies $\eth(p_n) = n p_{n-1}$, since $\mathcal{O}^{(1)} = \mathcal{O}$. This, together with Proposition 4.1, concludes the proof. $\qquad\square$

Next we investigate the behavior of a generalized Gončarov basis with respect to a transformation of the interpolation grid. Proposition 4.4 extends the shift-invariance property for classical Gončarov polynomials and difference Gončarov polynomials.

Proposition 4.4. *Let* $\mathcal{W} = (w_i)_{i \geq 0}$ *be a translation of the grid* $\mathcal{Z}$ *by* $\xi \in \mathbb{K}$, *i.e.,* $w_i = z_i + \xi$ *for all* i. *Assume that* $\mathcal{T} = (t_n(x))_{n \geq 0}$ *and* $\mathcal{H} = (h_n(x))_{n \geq 0}$ *are the generalized Gončarov bases associated with the pairs* $(\eth, \mathcal{Z})$ *and* $(\eth, \mathcal{W})$, *respectively. Then,* $h_n(x + \xi) = t_n(x)$ *for all* n.

Proof. Clearly $h_n(x + \xi) = E_\xi(h_n(x))$ is a polynomial of degree n, so by the uniqueness of the Gončarov basis associated with the pair $(\eth, \mathcal{Z})$, it suffices to prove that

$$\varepsilon_{z_i}\left(\eth^i\left(E_\xi(h_n(x))\right)\right) = n! \delta_{i,n}.$$

Note that $\eth^i E_\xi = E_\xi \eth^i$ because any two shift-invariant operators commute. Therefore

$$\varepsilon_{z_i}\left(\eth^i\left(E_\xi(h_n(x))\right)\right) = \varepsilon_{z_i}\left(E_\xi\left(\eth^i(h_n(x))\right)\right) = \varepsilon_{z_i + \xi}\left(\eth^i(h_n(x))\right)$$
$$= \varepsilon_{w_i}\left(\eth^i(h_n(x))\right) = n! \delta_{i,n}. \qquad\square$$

Proposition 4.5. *Fix* $\xi \in \mathbb{K}$ *and let* $\mathcal{H} = (h_n(x))_{n \geq 0}$ *be the generalized Gončarov basis associated with the pair* $(\eth, \mathcal{W})$, *where* $\mathcal{W} = (w_i)_{i \geq 0}$ *is the grid given by* $w_i = z_i + i\xi$ *for all* $i \geq 0$. *Then,* $\mathcal{H}$ *are also the generalized Gončarov basis associated with the pair* $(E_\xi \eth, \mathcal{Z})$.

Proof. One checks that

$$\varepsilon_{z_i}\left((E_\xi \eth)^i(h_n(x))\right) = n! \delta_{i,n}.$$

To this end, first observe that $(E_\xi \partial)^i = E_\xi^i \partial^i = E_{i\xi} \partial^i$, since any two shift-invariant operators commute and $E_a E_b = E_{a+b}$ for all $a, b \in \mathbb{K}$. It follows that

$$\varepsilon_{z_i} \left((E_\xi \partial)^i (h_n(x))\right) = \varepsilon_{z_i} \left(E_{i\xi} \left(\partial^i (h_n(x))\right)\right) = \varepsilon_{z_i + i\xi} \left(\partial^i (h_n(x))\right)$$
$$= \varepsilon_{w_i} \left(\partial^i (h_n(x))\right) = n! \delta_{i,n}.$$

The last equation is true because $\mathcal{H}$ is the generalized Gončarov basis associated with $(\partial, \mathcal{W})$. $\qquad \square$

Proposition 4.6 gives a relation between the generalized Gončarov polynomials and the basic polynomials of the same delta operator. It provides a linear recurrence that can be used to compute efficiently the explicit formulas for the generalized Gončarov basis if the basic sequence is known, for example, as in the classical case where the basic sequence is $(x^n)_{n \geq 1}$, or in the case of difference Gončarov polynomials where the n-term of the basic sequence is the upper factorial $x(x+1)(x+2) \cdots (x+n-1)$, see [11].

Proposition 4.6. *Let* $\mathcal{T} = (t_n(x))_{n \geq 0}$ *be the generalized Gončarov basis associated with the pair* $(\partial, \mathcal{Z})$, *and let* $(p_n(x))_{n \geq 0}$ *be the sequence of basic polynomials of the delta operator* ∂. *Then, for all* $n \in \mathbb{N}$ *it holds*

$$p_n(x) = \sum_{i=0}^{n} \binom{n}{i} p_{n-i}(z_i) t_i(x),$$

and hence

$$t_n(x) = p_n(x) - \sum_{i=0}^{n-1} \binom{n}{i} p_{n-i}(z_i) t_i(x). \qquad (4.11)$$

Proof. Let $n \in \mathbb{N}$. Substituting $f(x)$ with $p_n(x)$ in (4.8) we obtain

$$p_n(x) = \sum_{i=0}^{n} \frac{\varepsilon_{z_i}(\partial^i(p_n))}{i!} t_i(x) = \sum_{i=0}^{n} \frac{(n)_{(i)} p_{n-i}(z_i)}{i!} t_i(x) = \sum_{i=0}^{n} \binom{n}{i} p_{n-i}(z_i) t_i(x),$$

where we use the fact that $\partial^i(p_n(x)) = (n)_{(i)} p_{n-i}(x)$ for all $i = 0, 1, \ldots, n$. $\qquad \square$

Proposition 4.7 generalizes the binomial expansion for classical Gončarov polynomials.

Proposition 4.7. *Let* $(t_n^{(j)}(x))_{n \geq 0}$ *be the generalized Gončarov basis associated with the pair* $(\partial, \mathcal{Z}^{(j)})$, *and let* $(p_n(x))_{n \geq 0}$ *be the sequence of basic polynomials of the delta operator* ∂. *Then, for all* $\xi \in \mathbb{K}$ *and* $n \in \mathbb{N}$ *we have the following* *"binomial identity"*:

$$t_n(x + \xi) = t_n^{(0)}(x + \xi) = \sum_{i=0}^{n} \binom{n}{i} t_{n-i}^{(i)}(\xi) p_i(x). \tag{4.12}$$

In particular, letting $\xi = 0$ we have

$$t_n(x) = t_n^{(0)}(x) = \sum_{i=0}^{n} \binom{n}{i} t_{n-i}^{(i)}(0) p_i(x). \tag{4.13}$$

Proof. Fix $\xi \in \mathbb{K}$ and $n \in \mathbb{N}$. Since $(p_i(x))_{0 \le i \le n}$ is a basis of the linear subspace of $\mathbb{K}[x]$ of polynomials of degree $\le n$, there exist $c_0, \ldots, c_n \in K$ such that $E_\xi(t_n(x)) = \sum_{i=0}^{n} c_i p_i(x)$, where $t_n := t_n^{(0)}$.

Pick an integer $j \in [0, n]$. One computes

$$\varepsilon_\xi \left(\eth^j(t_n(x)) \right) = \varepsilon_0 \left(E_\xi \left(\eth^j(t_n(x)) \right) \right) = \varepsilon_0 \left(\eth^j(E_\xi(t_n(x))) \right)$$

$$= \sum_{i=0}^{n} c_i \varepsilon_0 \left(\eth^j(p_i(x)) \right). \tag{4.14}$$

From Proposition 4.2 we have $\eth^j(t_n(x)) = (n)_{(j)} t_{n-j}^{(j)}(x)$, so that $\varepsilon_\xi \left(\eth^j(t_n(x)) \right) = (n)_{(j)} t_{n-j}^{(j)}(\xi)$. On the other hand, the sequence $(p_n(x))_{n \ge 0}$, being the basic polynomials of $\eth$, satisfies $\varepsilon_0 \left(\eth^j(p_i) \right) = i! \delta_{i,j}$ for all $i \in \mathbb{N}$. Combining the foregoing we obtain from (4.14) that

$$c_j = \frac{(n)_{(j)}}{j!} t_{n-j}^{(j)}(\xi) = \binom{n}{j} t_{n-j}^{(j)}(\xi),$$

which proves (4.12). $\qquad\square$

Corollary 4.2. *Assume $\mathcal{Z}$ is a constant grid, namely $z_i = z_j$ for all i, j. Let $(t_n(x))_{n \ge 0}$ be the generalized Gončarov basis associated with the pair $(\eth, \mathcal{Z})$, and $(p_n(x))_{n \ge 0}$ the sequence of basic polynomials of the delta operator $\eth$. Then, for all $\xi \in K$ and $n \in \mathbb{N}$ we have*

$$t_n(x + \xi) = \sum_{i=0}^{n} \binom{n}{i} t_{n-i}(\xi) p_i(x) \quad \text{and} \quad t_n(x) = \sum_{i=0}^{n} \binom{n}{i} t_{n-i}(0) p_i(x). \tag{4.15}$$

Proof. Immediate by Proposition 4.7 and the fact that $\mathcal{Z} = \mathcal{Z}^{(j)}$ for all j. $\qquad\square$

The next proposition gives an extension of the integral formula for classical Gončarov polynomials, see [12, p. 23].

Proposition 4.8. *Let $\eth$ be a delta operator. Then $\eth$ has a right inverse, i.e., there exists a linear operator $\eth^{-1} : \mathbb{K}[x] \to \mathbb{K}[x]$ such that $\eth(\eth^{-1}(f)) = f$ for all $f(x) \in \mathbb{K}[x]$, $\deg(\eth^{-1}(f)) = 1 + \deg(f)$ for $f \neq 0$, and $\eth^{-1}(f)(0) = 0$.*

Proof. Let $\mathcal{P} = (p_n(x))_{n \geq 0}$ denote the sequence of basic polynomials of the delta operator $\eth$. Then we can define an operator $\eth^{-1} : \mathbb{K}[x] \to \mathbb{K}[x]$ as follows: Given a polynomial $f(x) \in \mathbb{K}[x]$ of degree n, let $a_0, \ldots, a_n \in K$ be such that $f(x) = \sum_{i=0}^{n} a_i p_i(x)$, set

$$\eth^{-1}(f(x)) := \sum_{i=0}^{n} \frac{a_i}{i+1} p_{i+1}(x).$$

It is seen that $\eth^{-1}$ is a linear operator on $\mathbb{K}[x]$, and since $\eth(p_n(x)) = n p_{n-1}(x)$ and $p_n(0) = 0$ for all $n \geq 1$, we have as well that $\eth(\eth^{-1}(f(x))) = f(x)$ and $\eth^{-1}(f)(0) = 0$ for every $f(x) \in \mathbb{K}[x]$. The rest is trivial, when considering that $p_n(x)$ is, for each $n \geq 1$, a polynomial of degree n.   □

Thus, we have the following generalization of the integral formula for classical Gončarov polynomials, see [12, p. 23].

Proposition 4.9. *Let $\eth$ be a delta operator and $\eth^{-1}$ its right inverse (which exists by Proposition 4.8), and let $(t_n(x))_{n \geq 0}$ be the generalized Gončarov basis associated with the pair $(\eth, \mathcal{Z})$. Then, for every $n, k \in \mathbb{N}$ with $k \leq n$, it holds that $t_n(x) = (n)_{(k)} \cdot \mathcal{I}_k(t_{n-k}^{(k)}(x))$, where $(t_i^{(n)}(x))_{n \geq 0}$ is the generalized Gončarov basis associated with the pair $(\eth, \mathcal{Z}^{(k)})$ and $\mathcal{I}_k$ the linear operator $\prod_{i=0}^{k-1}(1 - \varepsilon_{z_i})\eth^{-1}$.*

Proof. Fix $n, k \in \mathbb{N}$ with $0 \leq k \leq n$. If $k = 0$, the claim is trivial, because $\mathcal{I}_0$ is the identity operator. Now, suppose we have already confirmed that the statement is true for $0 \leq k < n$; then, by induction, we are just left to show that it continues to be true for $k + 1$.

For this, it is enough to prove that $t_{n-k}^{(k)}(x) = (n - k) \cdot (1 - \varepsilon_{z_k})$ $\left(\eth^{-1}\left(t_{n-k-1}^{(k+1)}(x)\right)\right)$. It follows from the facts that (1) both $t_{n-k}^{(k)}(x)$ and $(1 - \varepsilon_{z_k})$ $\left(\eth^{-1}\left(t_{n-k-1}^{(k+1)}(x)\right)\right)$ are zero when evaluated at z_k, and (2) we have $\eth\left(t_{n-k}^{(k)}(x)\right) = (n - k)t_{n-k-1}^{(k+1)}(x)$ by Proposition 4.2, and

$$\eth\left((1 - \varepsilon_{z_k})\left(\eth^{-1}\left(t_{n-k-1}^{(k+1)}(x)\right)\right)\right) = \eth\left(\eth^{-1}\left(t_{n-k-1}^{(k+1)}(x)\right)\right)$$
$$-\eth\left(\varepsilon_{z_k}\left(\eth^{-1}\left(t_{n-k-1}^{(k+1)}(x)\right)\right)\right) = t_{n-k-1}^{(k+1)}(x),$$

where we used Proposition 4.8 and the fact that $\varepsilon_{z_k}\left(\eth^{-1}\left(t_{n-k-1}^{(k+1)}(x)\right)\right)$ is a constant (and $\eth$ applied to a constant is zero).   □

In addition, we have the following generalization of the "perturbation formulas" obtained in [12, p. 24] and [11, p. 5].

Proposition 4.10. *Let $\eth$ be a delta operator, and let $\mathcal{Z} = (z_i)_{i\geq 0}$ and $\mathcal{Z}' = (z_i')_{i\geq 0}$ be $\mathbb{K}$-grids such that $z_k \neq z_k'$ for a given $k \in \mathbb{N}$ and $z_i = z_i'$ for $i \neq k$. Then we have, for $n > k$, that*

$$t_n(x; \eth, \mathcal{Z}') = t_n(x; \eth, \mathcal{Z}) - \binom{n}{k} t_{n-k}(z_k'; \eth, \mathcal{Z}^{(k)}) t_k(x; \eth, \mathcal{Z}), \qquad (4.16)$$

while $t_n(x; \eth, \mathcal{Z}') = t_n(x; \eth, \mathcal{Z})$ for $n \leq k$.

Proof. Let $n \in \mathbb{N}$ and denote by $f_n(x)$ the polynomial on the right-hand side of (4.16). The claim is straightforward if $n \leq k$, essentially because we get by Theorem 4.1(1) that $t_n(x; \eth, \mathcal{Z})$ and $t_n(x; \eth, \mathcal{Z}')$ depend only on the first n nodes of $\mathcal{Z}$ and $\mathcal{Z}'$, respectively. Accordingly, assume in what follows that $n > k$ and fix $i \in \mathbb{N}$. By the unicity of the generalized Gončarov basis associated to the pair $(\eth, \mathcal{Z}')$, we just have to prove that $\varepsilon_{z_i'}(\eth^i f_n(x)) = n! \delta_{i,n}$. This is immediate if $i > n$, since then $\eth^i(f_n(x)) = 0$, and for $i \leq n$ it is a consequence of Proposition 4.2 (we omit further details). $\qquad \square$

We conclude the present section by proving that generalized Gončarov bases obey an Appell relation, which extends an analogous result from [12, Section 3].

Proposition 4.11. *Let $(t_n(x))_{n\geq 0}$ be the generalized Gončarov basis associated with the pair $(\eth, \mathcal{Z})$. In addition, denote by $d(t)$ the compositional inverse of the D-indicator of $\eth$ in $\mathbb{K}[\![t]\!]$. Then the following identity holds:*

$$e^{xd(t)} = \sum_{n\geq 0} \frac{1}{n!} t_n(x) e^{z_n d(t)} t^n. \qquad (4.17)$$

In particular, if $\eth = D$ then $e^{xt} = \sum_{n\geq 0} \frac{1}{n!} t_n(x) e^{z_n t} t^n$.

Proof. Let $(p_n(x))_{n\geq 0}$ the sequence of basic polynomials of $\eth$. By Proposition 4.6

$$p_n(x) = \sum_{i=0}^{n} \binom{n}{i} p_{n-i}(z_i) t_i(x) \qquad \text{for all } n \in \mathbb{N},$$

whence we get that

$$\sum_{n\geq 0} \frac{p_n(x)}{n!} t^n = \sum_{n\geq 0} \sum_{i=0}^{n} \frac{1}{i!(n-i)!} p_{n-i}(z_i) t_i(x) t^n = \sum_{i\geq 0} \left(\frac{1}{i!} t_i(x) t^i \sum_{j\geq 0} \frac{p_j(z_i)}{j!} t^j \right).$$

$$(4.18)$$

On the other hand, Eq. (4.5) gives

$$e^{xd(t)} = \sum_{n\geq 0} \frac{p_n(x)}{n!} t^n,$$

which, together with (4.18), implies (4.17). The rest is trivial, when considering that the D-indicator of D over $\mathbb{K}[\![t]\!]$ is just t. □

Using Proposition 4.11, we obtain the following characterization of sequences of binomial type, which is complementary to Corollary 4.1.

Proposition 4.12. *Let $\mathcal{T} = (t_n(x))_{n\geq 0}$ be the generalized Gončarov basis associated with the pair $(\eth, \mathcal{Z})$. Then $\mathcal{T}$ is of binomial type if and only if $\mathcal{Z}$ is an arithmetic grid of initial term 0.*

Proof. Suppose first that $\mathcal{T}$ is a sequence of binomial type. We get by [21, Section 3, Corollary 3] that there is a formal power series $f \in \mathbb{K}[\![t]\!]$ with $f(0) = 0$ and $f'(0) \neq 0$ such that

$$e^{xf(t)} = \sum_{n=0}^{\infty} \frac{t_n(x)}{n!} t^n. \tag{4.19}$$

On the other hand, we have from Proposition 4.11 that

$$e^{xd(t)} = \sum_{n\geq 0} \frac{1}{n!} t_n(x) e^{z_n d(t)} t^n, \tag{4.20}$$

where $d(t)$ is the compositional inverse of the D-indicator of the delta operator $\eth$ in $\mathbb{K}[\![t]\!]$.

Let $h \in \mathbb{K}[\![t]\!]$ be the compositional inverse of f, which exists by the assumption that $f(0) = 0$ and $f'(0) \neq 0$ [23]. Then $h(f(t)) = f(h(t)) = t$. Using the change of variable $y \mapsto h(d(t))$ in (4.19) yields that

$$e^{xd(t)} = \sum_{n=0}^{\infty} \frac{1}{n!} t_n(x) (h(d(t)))^n.$$

Comparing this with (4.20) implies, for all $n \in \mathbb{N}$, that

$$(h(d(y)))^n = e^{z_n d(y)} y^n,$$

which holds as an identity between formal power series in $\mathbb{K}[\![t]\!]$, and is in turn possible, by the further change of variables $t \mapsto d^{-1}(t)$, if and only if

$$(h(t))^n = e^{z_n t} (d^{-1}(t))^n. \tag{4.21}$$

Combining with $(h(t))^{n+1} = e^{z_{n+1}t}(d^{-1}(t))^{n+1}$ for all $n \in \mathbb{N}$, we find that

$$h(t) = \frac{(h(t))^{n+1}}{(h(t))^n} = e^{(z_{n+1}-z_n)t}d^{-1}(t).$$

It follows that $z_{n+1} - z_n$ is a constant independent of n, viz. there exists $b \in K$ such that $z_{n+1} - z_n = b$ for all $n \in \mathbb{N}$. Then $z_n = z_0 + nb$ for all $n \in \mathbb{N}$. But evaluating (4.21) at $n = 0$ gives $1 = e^{z_0 t}$, which implies $z_0 = 0$. Thus $\mathcal{Z}$ is an arithmetic grid of initial term 0.

As for the converse, assume now that $\mathcal{Z}$ is an arithmetic grid of common difference $b \in \mathbb{K}$ and initial term 0. Then $\mathcal{T}$ is, by Proposition 4.5, the generalized Gončarov basis associated with the pair $(E_b \eth, \mathcal{O})$, and hence by Corollary 4.1 is a sequence of binomial type. $\square$

4.4 A Combinatorial Formula for Generalized Gončarov Polynomials

Let $\eth$ be a delta operator and $\mathcal{Z}$ a $\mathbb{K}$-grid. Assume $\mathcal{T} = (t_n(x))_{n\geq 0}$ is the generalized Gončarov basis associated with the pair $(\eth, \mathcal{Z})$. The main purpose of this section is to provide a combinatorial interpretation of the coefficients of $t_n(x)$. By (4.13) it is sufficient to consider only the constant terms. We will give an explicit combinatorial formula of $t_n(0)$ as a summation of ordered partitions.

Given a finite set S with n elements, an *ordered partition*, or *preferential arrangement*, of S is an ordered list $(B_1, \ldots, B_k)$ of disjoint nonempty subsets of S such that $B_1 \cup \cdots \cup B_k = S$.

If $\rho = (B_1, \ldots, B_k)$ is an ordered partition of S, then we set $|\rho| = k$. For every $i = 0, 1, \ldots, k$ we let $b_i := b_i(\rho) := |B_i|$ and $s_i := s_i(\rho) := \sum_{j=1}^{i} b_j$. In particular, $s_0(\rho) = 0$.

Let $\mathcal{R}[n]$ be the set of all ordered partitions of the set $[n] := \{1, 2, \ldots, n\}$. It is shown in [12, Theorem 4.2] that the constant coefficient for $g_n(x; \mathcal{Z})$, the classical Gončarov polynomial associated to $(D, \mathcal{Z})$, can be expressed as

$$g_n(0; z_0, \ldots, z_{n-1}) = \sum_{\rho}(-1)^{|\rho|}\prod_{i=0}^{k-1} z_{s_i}^{b_{i+1}} = \sum_{\rho \in \mathcal{R}[n]}(-1)^{|\rho|}z_0^{b_1}\cdots z_{s_{k-1}}^{b_k}. \quad (4.22)$$

A similar formula holds for the generalized Gončarov polynomials associated to the pair $(\eth, \mathcal{Z})$.

Theorem 4.2. *Let $(t_n(x))_{n\geq 0}$ be the generalized Gončarov basis associated with the pair $(\eth, \mathcal{Z})$, and $(p_n(x))_{n\geq 0}$ be the sequence of basic polynomials of $\eth$.*

Then for $n \geq 1$,

$$t_n(0) = \sum_{\rho \in \mathcal{R}[n]} (-1)^{|\rho|} \prod_{i=0}^{k-1} p_{b_{i+1}}(z_{s_i}) = \sum_{\rho \in \mathcal{R}[n]} (-1)^{|\rho|} p_{b_1}(z_0) \cdots p_{b_k}(z_{s_{k-1}}). \quad (4.23)$$

Proof. Using Proposition 4.6 and noting that $p_n(0) = 0$ for $n \geq 1$, we have

$$t_n(0) = - \sum_{i=0}^{n-1} \binom{n}{i} p_{n-i}(z_i) t_i(0). \quad (4.24)$$

Denote by $\mathcal{T}(n)$, for $n \geq 1$, the right-hand side of (4.23), and let $\mathcal{T}(0) = 1$, which agrees with $t_0(0)$. We show by induction that $(\mathcal{T}(n))_{n \geq 0}$ satisfies the same recurrence as (4.24), i.e.,

$$\mathcal{T}(n) = - \sum_{i=0}^{n-1} \binom{n}{i} p_{n-i}(z_i) \mathcal{T}(i). \quad (4.25)$$

To see this, we divide $\mathcal{R}[n]$ into disjoint subsets $\mathcal{R}[n, i]$, where

$$\mathcal{R}[n, i] := \{(B_1, \ldots, B_k) \in \mathcal{R}[n] : |B_k| = n - i\} \qquad \text{for } i = 0, 1, \ldots, n - 1.$$

Given $\rho \in \mathcal{R}[n, i]$ with a fixed last block B_k, we can write ρ as the concatenation of B_k and an ordered partition of a set with i elements. So we get from the inductive hypothesis that

$$\sum_{\rho' \in \mathcal{R}[i]} (-1)^{|\rho'|} \prod_{i=0}^{k-2} p_{b_{i+1}}(z_{s_i}) = \mathcal{T}(i).$$

Since there are $\binom{n}{i}$ ways to choose the elements of B_k, it follows that the total contribution of ordered partitions in $\mathcal{R}[n, i]$ to $\mathcal{T}(n)$ is

$$-\binom{n}{i} p_{n-i}(z_i) \mathcal{T}(i).$$

Then, summing over $i = 0, 1, \ldots, n - 1$ proves the desired recurrence (4.25). $\qquad \qquad \square$

One way of obtaining $t_n(x; \eth, \mathcal{Z})$ is to use the shift-invariance property (Proposition 4.4) as to write $t_n(x; \eth, \mathcal{Z}) = t_n(0; \eth, \mathcal{Z} - x)$. Another way is to use (4.13):

$$t_n(x; \eth, \mathcal{Z}) = \sum_{i=0}^{n} \binom{n}{i} t_{n-i}(0; \eth, \mathcal{Z}^{(i)}) p_i(x),$$

and apply Theorem 4.2 to each $t_{n-i}(0; \eth, \mathcal{Z}^{(i)})$. Comparing this with the analogous equation for classical Gončarov polynomials:

$$g_n(x; \mathcal{Z}) = \sum_{i=0}^{n} \binom{n}{i} g_{n-i}(0; z_i, \ldots, z_{n-1}) x^i,$$

we notice that $t_n(x; \eth, \mathcal{Z})$ can be obtained from $g_n(x; \mathcal{Z})$ by replacing x^i with $p_i(x)$ and z_k^i by $p_i(z_k)$. For example, we have the following formulas, which the reader may want to compare with the ones for $g_n(x; \mathcal{Z})$ in [12, p. 23]:

$t_0(x; \eth, \mathcal{Z}) = 1,$

$t_1(x; \eth, \mathcal{Z}) = p_1(x) - p_1(z_0),$

$t_2(x; \eth, \mathcal{Z}) = p_2(x) - 2p_1(z_1)p_1(x) + 2p_1(z_0)p_1(z_1) - p_2(z_0),$

$t_3(x; \eth, \mathcal{Z}) = p_3(x) - 3p_1(z_2)p_2(x) + (6p_1(z_1)p_1(z_2) - 3p_2(z_1))p_1(x)$

$\qquad - p_3(z_0) + 3p_2(z_0)p_1(z_2) - 6p_1(z_0)p_1(z_1)p_1(z_2) + 3p_1(z_0)p_2(z_1).$

For a generic delta operator $\eth$ and an arbitrary grid $\mathcal{Z}$, the generalized Gončarov polynomials do not usually have a simple closed formula. However, an interesting exception to this "rule" occurs when $\mathcal{Z}$ is an arithmetic progression with $z_i = a + bi$, in which case we refer to the corresponding generalized Gončarov polynomials as $\eth$-Abel polynomials. In fact, $\eth$-Abel polynomials can be expressed in terms of the basic polynomials of $\eth$, as implied by the following:

Theorem 4.3. *Let $\eth$ be a delta operator with basic sequence $(p_n(x))_{n \geq 0}$, and let $\mathcal{Z}$ be the arithmetic grid $(a + bi)_{i \geq 0}$, where $a, b \in \mathbb{K}$. Then the $\eth$-Abel polynomial $t_n(x; \eth, \mathcal{Z})$ can be obtained by*

$$t_n(x; \eth, \mathcal{Z}) = \frac{(x-a)p_n(x-a-nb)}{x-a-nb}. \tag{4.26}$$

Proof. By the shift-invariance formula we have $t_n(x; \eth, \mathcal{Z}) = t_n(x - a; \eth, (bi)_{i \geq 0})$. Using Proposition 4.5, the generalized Gončarov polynomials associated to $(\eth, \mathcal{W})$ with $w_i = bi$ are also the generalized Gončarov polynomials associated to $(E_b \eth, \mathcal{O})$. Let $q_n(x)$ be the basic sequence of $E_b \eth$. Hence we have $t_n(x - a; \eth, (bi)_{i \geq 0}) = q_n(x - a)$.

Now, we have from [17, Theorem 4(3)], along with the fact that shift-invariant operators commute with each other, that

$$q_n(x) = x(E_b \eth)^{-n}(x^n) = xE_{-nb}\left(\eth^{-n}(x^n)\right). \tag{4.27}$$

On the other hand, a further application of [17, Theorem 4(3)] yields that $p_n(x) = x\eth^{-n}(x^n)$, and hence $\eth^{-1}(x^n) = p_n(x)/x$. Together with (4.27), this in turn implies that

$$q_n(x) = xE_{-nb}\left(\frac{p_n(x)}{x}\right) = \frac{xp_n(x - nb)}{x - nb}.$$

So putting it all together, (4.26) follows immediately.   □

We note that Niederhausen has also obtained formula (4.26) in [18], but with other means; he calls the procedure $\eth \mapsto E_a\eth$, for a fixed $a \in \mathbb{K}$, the abelization of the delta operator $\eth$.

Bivariate extensions of $\eth$-Abel polynomials, which are solutions to a multivariate Gončarov Interpolation problem with respect to an affine grid, are further studied and characterized in [15] for $\eth = D$, and in [14] for general delta operators.

4.5 Reluctant Functions and Order Statistics

In the classical paper "Finite Operator Calculus," Rota, Kahaner, and Odlyzko presented a unified theory of special polynomials via operator methods. One open question arising from this algebraic theory is to find the "statistical, probabilistic and combinatorial interpretations of the identities" (of the polynomials), see Problem 5 of Section 14, [21]. For polynomial sequences of binomial type, Mullin and Rota [17] provided a combinatorial interpretation through counting binomial type structures such as *reluctant functions*. The ideas of binomial enumeration and reluctant functions also provide a combinatorial setting for generalized Gončarov polynomials: We show in this section that generalized Gončarov polynomials are the natural polynomial basis for counting the number of binomial type structures subject to a linear constraint on their order statistics.

To start, let S and X be finite disjoint sets, and $f : S \rightarrow S \cup X$ a function. We say that f is a *reluctant function* from S to X if, for every $s \in S$, there is a positive integer $k = k(s)$ such that $f^k(s) \in X$, in which case we refer to $f^k(s)$ as the *final image* of s (under f). It is easy to see that for any given s, the integer $k(s)$, if it exists, is unique.

Accordingly, we take the *final range* of f, here denoted by $\underline{\mathrm{Im}}(f)$, to be the set of all $\xi \in X$ such that ξ is the final image of some $s \in S$. Given $\xi \in \underline{\mathrm{Im}}(f)$, we let the *final inverse image* of ξ, which we write as $f^{(-1)}(\xi)$, be the set of all the elements in S whose final image is ξ.

From a combinatorial point of view, the final inverse image of an element $\xi \in X$ can be regarded in a canonical way as a rooted forest: The nodes are just the elements of S, and the roots are the elements of the inverse image

$f^{-1}(\xi) = \{s \in S : f(s) = \xi\}$. In a rooted tree we say that a vertex is of depth k if the unique path from u to the root contains k edges. The root itself is of depth 0. Then for each $s_0 \in f^{-1}(\xi)$ and $k \in \mathbb{N}$ the vertices of depth k in a tree of $f^{(-1)}(\xi)$ rooted at s_0 are those elements $s \in S$ such that $f^k(s) = s_0$ and hence $f^{k+1}(s) = \xi$.

The *final coimage* of f is the partition $\{f^{(-1)}(\xi) : \xi \in \underline{\mathrm{Im}}(f)\}$ of S. Based on the foregoing discussion the final coimage carries over a natural structure, T_f, of a rooted forest defined on each block of the partition. Furthermore, each block of the final coimage can be partitioned further into connected components (relative to T_f); the resulting partition is a refinement of the final coimage and has the additional property that each block has the structure of a rooted tree. This finer partition together with the rooted tree structure is called the *final preimage* of the reluctant function f.

Remark 1. What we call "final range," "final coimage," and "final preimage" of a reluctant function were called "range," "coimage," and "preimage" by Mullin and Rota in [17]. However, these latter terms are already used in the everyday practice of mathematics with a different meaning. We add the word "final" to avoid potential misunderstanding.

A *binomial class* $\mathcal{B}$ of reluctant functions is defined as follows. To every pair of finite sets S and X we assign a set $F(S, X)$ of reluctant functions from S to X, where $F(S, X)$ is isomorphic to $F(S', X')$ whenever S is isomorphic to S and X is isomorphic to X'. Consequently, the size of $F(S, X)$ depends only on the sizes of S and X, but not the content of these sets. Let $X \oplus Y$ stand for the disjoint union of X and Y. For every reluctant function f from S to $X \oplus Y$, let $A = \{s \in S : \text{the final image of } s \text{ is in } X\}$ and f_A is the restriction of f to A. Similarly $f_{S \backslash A}$ is the restriction of f to the set $S \backslash A$. The class $\mathcal{B}$ is a *binomial class* if $f_A \in F(A, X)$, $f_{S \backslash A} \in F(S \backslash A, Y)$ and the foregoing decomposition leads to a natural isomorphism

$$\mu : F(S, X \oplus Y) \to \bigcup_{A \subseteq S} (F(A, X) \otimes F(S \backslash A, Y)) \qquad (4.28)$$

by letting $\mu(f) := (f_A, f_{S \backslash A})$, where $\otimes$ denotes the operation of piecing functions f_A and $f_{S \backslash A}$ together, and $\cup$ a disjoint union.

Let $p_n(x)$ denote the size of the set $F(S, X)$ when $|S| = n$ and $|X| = x$. Then for a binomial class, $p_n(x)$ is a well-defined polynomial in the variable x of degree n, and the sequence $(p_n(x))_{n \geq 0}$ is of binomial type:

$$p_n(x + y) = \sum_{k \geq 0} \binom{n}{k} p_k(x) p_{n-k}(y).$$

The foregoing construction gives a family of polynomial sequences of binomial types with combinatorial significance – they count the number of reluctant functions in a binomial class. For example, let B contain all the reluctant functions for which each block of the final preimage is a singleton. Then $p_n(x) = x^n$. Another example is that B contains all possible reluctant functions. In this case $p_n(x) = x(x+n)^{n-1}$, the Abel polynomial $x(x - an)^{n-1}$ with $a = -1$. This result can be proved by using Prüfer codes, see, e.g., [23, Prop. 5.3.2]. More examples of binomial classes are discussed in the next section.

For a sequence of numbers $(a_0, \ldots, a_{n-1})$, let $a_{(0)} \leq \cdots \leq a_{(n)}$ be the non-decreasing rearrangements of the terms a_i. The value of $a_{(i)}$ is called the ith order statistic of the sequence. We will combine the notations of reluctant functions and order statistics. From now on we assume $S = \{s_0, \ldots, s_{n-1}\}$, x is a positive integer, and $X = \{1, \ldots, x\}$. Associated with any reluctant function f from S to X a sequence $\vec{x} = (x_0, \ldots, x_{n-1}) \in X^n$ where x_i is the final image of s_i.

Assume that $z_0 \leq \cdots \leq z_{n-1}$ are integers in X. For a binomial class B of reluctant functions enumerated by $p_n(x)$, define $Ord(z_0, \ldots, z_{n-1})$, *the set of reluctant functions of length n whose order statistics are bounded by Z*, by letting

$$Ord(z_0, \ldots, z_{n-1}) = \{f \in F(S, X) : x_{(0)} \leq z_0, \ldots, x_{(n-1)} \leq z_{n-1}\}.$$

Let $ord(z_0, \ldots, z_{i-1})$ be the size of the set $Ord(z_0, \ldots, z_{n-1})$, i.e.,

$$ord(z_0, \ldots, z_{i-1}) = |Ord(z_0, \ldots, z_{n-1})|.$$

Then we have the following equation.

Theorem 4.4. *With the notation as in the preceding, it holds that, for every* $n \in \mathbb{N}$,

$$p_n(x) = \sum_{i=0}^{n} \binom{n}{i} p_{n-i}(x - z_i) \cdot ord(z_0, \ldots, z_{i-1}). \tag{4.29}$$

Proof. For any reluctant function $f \in F(S, X)$, let $\kappa(f)$ be the maximal index i such that

$$x_{(0)} \leq z_0, \ldots, x_{(i-1)} \leq z_{i-1}. \tag{4.30}$$

The maximality of i implies that

$$z_i < x_{(i)} \leq x_{(i+1)} \leq \cdots \leq x_{(n-1)}.$$

Let X_1 be the subset of X consisting of $\{1, \ldots, z_i\}$, and $X_2 = X \setminus X_1 = \{z_i + 1, \ldots, x\}$. Assume $(x_{r_0}, \ldots, x_{r_{i-1}})$ is the subsequence of $\vec{x}$ from which the sequence $(x_{(0)}, \ldots, x_{(i-1)})$ was obtained by rearrangement, and let $A' = \{s_{r_0}, \ldots, s_{r_{i-1}}\} \subseteq S$. Then the reluctant function f is obtained by piecing two functions, f_A and $f_{S \setminus A}$, the restrictions of f on A and $S \setminus A$, together. Since $\mathcal{B}$ is a binomial class, we have that $f_A \in F(A, X_1)$, $f_{S \setminus A} \in F(S \setminus A, X_2)$. Furthermore, the function f_A has the property that its order statistics are bounded by $\mathcal{Z}$, and hence belongs to $\mathcal{O}rd(z_0, \ldots, z_{i-1})$. The function $f_{S \setminus A}$ can be any reluctant function from $S \setminus A$ to X_2.

Conversely, any pair of subsequences as described in the preceding can be reassembled into a reluctant function from S to X. In other word, the decomposition $f \to (f_A, f_{S \setminus A})$ defines a bijection from the set $F(S, X)$ to

$$\bigcup_{i=0}^{n} \bigcup_{A=\{r_0,\ldots,r_{i-1}\}} F_{ord}(A, \{1, \ldots, z_i\}) \otimes F(S \setminus A, \{z_i + 1, \ldots, x\}), \quad (4.31)$$

where $F_{ord}(A, \{1, \ldots, z_i\})$ is the set of reluctant functions from the set $\{s_i : i \in A\}$ to $\{1, \ldots, z_i\}$, whose order statistics are bounded by $\mathcal{Z}$.

Now counting the number of elements in the disjoint union of (4.31), we get (4.29). □

Comparing (4.29) with the linear recurrence of Proposition 4.6, we obtain a combinatorial interpretation of the generalized Gončarov basis.

Theorem 4.5. *Let $p_n(x)$ count the number of reluctant functions in a binomial class $\mathcal{B}$. Assume $p_n(x)$ is the sequence of basic polynomials of the delta operator $\mathfrak{d}$, and $t_n(x; \mathfrak{d}, \mathcal{Z})$ is the nth generalized Gončarov polynomial associated to the pair $(\mathfrak{d}, \mathcal{Z})$ with $\mathcal{Z} = (z_0, z_1, \ldots)$. Then*

$$ord(z_0, \ldots, z_{n-1}) = t_n(x; \mathfrak{d}, (x - z_i)_{i \geq 0}) = t_n(0; \mathfrak{d}, -\mathcal{Z}). \quad (4.32)$$

That is, $t_n(0; \mathfrak{d}, -\mathcal{Z})$ counts the number of reluctant functions of the binomial class $\mathcal{B}$ whose order statistics are bounded by $\mathcal{Z}$.

4.6 Examples

In this section we give some examples of sequences of polynomials of binomial type that enumerate binomial classes. In each case, we describe the delta operator, the associated generalized Gončarov polynomials, and their combinatorial significance. We also compute $\mathfrak{d}$-Abel polynomials.

In [17] Mullin and Rota introduced two important families of binomial classes of reluctant functions. The first one is class $\mathcal{B}(T)$, where T is a family of rooted trees. The class $\mathcal{B}(T)$ consists of all reluctant functions whose final preimages are labeled rooted forests on S, each of whose components is isomorphic to a rooted tree in the family T. Examples 4.1–4.5 belong to this family. The second family of binomial classes is formed by taking a subclass of $\mathcal{B}(T)$: one allows only those reluctant functions in $\mathcal{B}(T)$ having the property that their final coimage coincides with their final preimage. In other words, each rooted tree in the final preimage is mapped to a distinct element in X. Such a subclass, denoted by $\mathcal{B}_m(T)$, is called the *monomorphic class associated to* $\mathcal{B}(T)$ and is ultimately a generalization of the notion of injective function. Examples 4.6 and 4.7 belong to the monomorphic family.

The combinatorial interpretation of generalized Gončarov polynomials is closely related to vector-parking functions. Hence we recall the necessary notations on parking functions. More results and theories of parking functions can be found in the survey paper [28]. Let $\mathbf{u} = (u_i)_{i \geq 1}$ be a sequence of nondecreasing positive integers. A $\mathbf{u}$-parking function of length n is a sequence $(x_1, \ldots, x_n)$ of positive integers whose order statistics satisfy $x_{(i)} \leq u_i$. When $u_i = i$, we get the classical parking functions, which was originally introduced by Konheim and Weiss [10]. Classical parking functions have a "parking description" as follows.

> There are n cars $C_1, \ldots, C_n$ that want to park on a one-way street with ordered parking spaces $1, \ldots, n$. Each car C_i has a preferred space a_i. The cars enter the street one at a time in the order $C_1, \ldots, C_n$. A car tries to park in its preferred space. If that space is occupied, then it parks in the next available space. If there is no space then the car leaves the street (without parking). The sequence $(a_1, \ldots, a_n)$ is called a *parking function of length* n if all the cars can park; i.e., no car leaves the street.

An equivalent definition for classical parking functions is that at least i cars prefer the parking spaces of labels i or less. Similarly a $\mathbf{u}$-parking function of length n, where $\mathbf{u} = (u_1, \ldots, u_n)$ is a vector of positive integers, can be viewed as a parking preference sequence in which at least i cars prefer the parking spaces of labels $\leq u_i$ (out of a total of $x \geq u_n$ parking spaces).

As explained in the following examples, the classical Gončarov polynomial associated to the differential operator D enumerate $\mathbf{u}$-parking functions, while generalized Gončarov polynomials associated to other delta operators for binomial class $\mathcal{B}(T)$ enumerate variant forms of the parking scheme, in which the cars arrive in groups with certain special structures, cars in the same group have the same preferred space, and there are at least i cars preferring spaces of label u_i or less. Similarly, the generalized Gončarov polynomials associated to other delta operators for the monomorphic class $\mathcal{B}_m(T)$ enumerate those with the additional property that different groups have different preferences.

Example 4.1 (The Standard Power Polynomials). The sequence $(x^n)_{n \geq 0}$ is the basic polynomials of the differential operator D. It enumerates the binomial class $\mathcal{B}(T_0)$, where T_0 consists of a single tree with only one vertex (viz., an isolated vertex which is also the root of the tree).

A reluctant function in this class is just a usual function from S to X, which can be represented by the sequence $\vec{x} = (f(s_1), \ldots, f(s_n))$. Gončarov polynomials $g_n(x; \mathcal{Z})$ associated to $(D, \mathcal{Z})$ are the classical ones studied in [12], for which $(-1)^n g_n(0; \mathcal{Z})$ enumerates the number of $\mathbf{z}$-parking functions of length n, [12, Theorem 5.4].

If $\mathcal{Z}$ is the arithmetic progression $z_i = a + bi$, then $g_n(x; \mathcal{Z})$ is a shift of the classical Abel polynomials, or more explicitly,

$$g_n(x; (a + bi)_{i \geq 0}) = (x - a)(x - a - nb)^{n-1}.$$

It follows that $P_n(a, a + b, \ldots, a + (n-1)b) = a(a + nb)^{n-1}$, where $P_n(z_0, \ldots, z_{n-1})$ is the number of $\mathcal{Z}$-parking functions, i.e., positive integer sequences whose order statistics are weakly bounded by $\mathcal{Z}$. In particular, for $a = b = 1$ we recover the formula for ordinary parking functions $P_n(1, \ldots, n) = (n + 1)^{n-1}$.

Example 4.2 (Abel Polynomials). The Abel polynomial with parameter a, namely $A_n(x; a) = x(x - na)^{n-1}$, is the n-th basic polynomial of the delta operator $\mathfrak{d} = E_a D = D E_a$. When $a = -1$, $A_n(x; -1) = x(x + n)^{n-1}$. This polynomial counts the reluctant functions in the binomial class $\mathcal{B}(T)$ where T contains all possible rooted trees. In fact, such reluctant functions from S to X are represented as rooted forests on the vertex set $S \cup X$, with X being the root set, whose number can be computed by using Prüfer codes. We remark that the rooted forest corresponding to a reluctant function in $\mathcal{B}(T)$ is different from the rooted forest T_f in the final coimage of f. The latter is a rooted tree on S only. The rooted forest corresponding to f can be obtained from T_f by replacing every root s_0 of T_f with an edge from s_0 to $f(s_0) \in X$, and letting every vertex of X be a root.

Similarly, if $a = -k$ for a positive integer k, then $A_n(x; -k) = x(x + nk)^{n-1}$ enumerates the reluctant functions in the binomial class $\mathcal{B}(T_k)$, where T_k contains all the rooted k-trees, which are rooted trees each of whose edge is colored by one of the colors $0, 1, \ldots, k - 1$. Such trees were studied in [22, 27]. In particular the reluctant functions in $\mathcal{B}(T_k)$ can be represented as sequences of rooted k-forests of length x, which are defined in [27] and proved to be in bijection with the (a, b)-parking functions with $a = x$ and $b = k$.

By Proposition 4.5, $t_n(x; E_{-k}D, \mathcal{Z})$, the generalized Gončarov polynomial associated to $(E_{-k}D, \mathcal{Z})$ is the same as $g_n(x; (z_i - ki)_{i \geq 0})$, where $g_n(x; \mathcal{Z})$ is the classical Gončarov polynomial.

The polynomial $t_n(x; E_{-k}D, -\mathcal{Z}) = g_n(x; (-z_i - ki)_{i \geq 0})$, when evaluated at 0, has two combinatorial interpretations. On one hand, it gives the number of

u-parking functions with $\mathbf{u} = (u_i = ki + z_i)_{i \geq 0}$. On the other hand, by Theorem 4.5 $t_n(0; E_{-k}D, -\mathcal{Z})$ also counts the number of ways that n cars form disjoint groups, each group is equipped with a structure of rooted k-tree, cars in the same group have the same preference, and the order statistics of the parking sequence are bounded by $\mathcal{Z}$.

The ∂-Abel polynomial is of the form $t_n(x; E_{-k}D, (a + bi)_{i \geq 0}) = (x - a)(x - a - nb + nk)^{n-1}$.

Example 4.3 (Laguerre Polynomials). The nth Laguerre polynomial $L_n(x)$ is given by the formula

$$L_n(x) = \sum_{k \geq 0} \frac{n!}{k!} \binom{n-1}{k-1}(-x)^k. \tag{4.33}$$

This is the nth basic polynomial of the Laguerre delta operator $K := D(D - I)^{-1} = -\sum_{i \geq 0} D^i$.

The coefficients $\frac{n!}{k!}\binom{n-1}{k-1}$ are called the (unsigned) Lah numbers, which are also the coefficients expressing rising factorials in terms of falling factorials. See Example 4.7.

The polynomial $L_n(-x)$ enumerates the binomial class $\mathcal{B}(T_P)$, where T_P is the set of all rooted trees which is a path rooted at one of its leaves. To see this, consider all such reluctant functions whose final preimage contains exactly k rooted paths. To get such a set of k rooted paths, we can linearly order all the elements of S in a row and then cut it into k nonempty segments; for each segment let the first element be the root. There are $n!\binom{n-1}{k-1}$ ways. Since the paths are unordered, we divide $k!$ to get the number of sets of k rooted paths. We then multiply x^k to get all functions from the set of k paths to X.

Let $t_n(x; K, \mathcal{Z})$ be the generalized Gončarov polynomial associated to $(K, \mathcal{Z})$. By Theorem 4.4 and its proof, we get that the number of reluctant functions in $\mathcal{B}(T_P)$ whose order statistics are bounded by $\mathcal{Z}$ is given by $t_n(-x; K, -x + \mathcal{Z}) = t_n(0; K, \mathcal{Z})$. Equivalently, $t_n(0; K, \mathcal{Z})$ counts the number of parking schemes in which n cars want to park in a parking lot of x spaces such that (1) the cars arrive in disjoint groups, (2) each group forms a queue, (3) all cars in the same queue prefer the same space, and (4) the order statistics of the preference sequence is bounded by $\mathcal{Z}$.

The nth ∂-Abel polynomial associated with the operator $∂ = K$ and the arithmetic grid $\mathcal{Z} = (a + bi)_{i \geq 0}$ is given by

$$t_n(x; K, (a + bi)_{i \geq 0}) = (a - x) \sum_{k=1}^{n} \frac{n!}{k!}\binom{n-1}{k-1}(a + nb - x)^{k-1}.$$

In particular, $t_n(0; K, (a+bi)_{i \geq 0}) = a \sum_{k=1}^{n} \frac{n!}{k!} \binom{n-1}{k-1}(a+nb)^{k-1}$, which, for $a = b = 1$, supplies the sequence $1, 5, 46, 629, 11496, \ldots$, namely A052873 in the On-Line Encyclopedia of Integer Sequences (OEIS) [19], where one can find the exponential generating function and an asymptotic formula.

Example 4.4 (Inverse of the Abel Polynomial $A_n(x; -1)$). Let $p_n(x) := \sum_{k \geq 0} \binom{n}{k} k^{n-k} x^k$. Then $(p_n(x))_{n \geq 0}$ is a sequence of binomial type: This is actually the basic sequence of the delta operator $\eth$ whose D-indicator is the compositional inverse of $f(t) = te^t$ (often referred to as the Lambert w-function), and it is also the inverse sequence of the Abel polynomials $(A_n(x; -1))_{n \geq 0}$ under the umbral composition.

In fact, the sequence $(p_n(x))_{n \geq 0}$ enumerates the binomial class $\mathcal{B}(T_1)$, where T_1 contains all the rooted trees with depth at most 1, i.e., stars $\{S_k : k \geq 0\}$ where S_k is the tree on vertices $\{v_0, \ldots, v_k\}$ with root v_0 and edges $\{v_0, v_i\}$ for $i = 1, \ldots, k$, [17, Section 7].

By Theorem 4.5, the generalized Gončarov polynomial associated to $(\eth, \mathcal{Z})$ gives a formula for the number of parking schemes such that the cars arrive in disjoint groups, each group has a leader, all cars in the same group prefer the same space, and the order statistics of the preference sequence is bounded by $\mathcal{Z}$.

In addition, it follows from the foregoing and Theorem 4.3 that the nth $\eth$-Abel polynomial associated with the arithmetic grid $\mathcal{Z} = (a+bi)_{i \geq 0}$ is given by

$$t_n(x; \eth, (a+bi)_{i \geq 0}) = (x-a) \sum_{k=1}^{n} \binom{n}{k} k^{n-k}(x-a-nb)^{k-1}.$$

In particular, $t_n(0; \eth, (-a-bi)_{i \geq 0}) = a \sum_{k=1}^{n} \binom{n}{k} k^{n-k}(a+nb)^{k-1}$; for $a = b = 1$, this yields the sequence $1, 5, 43, 549, 9341, \ldots$, which is A162695 in OEIS, where one can find the exponential generating function and an asymptotic formula.

Example 4.5 (Exponential Polynomials). The nth exponential polynomial, also called the *Touchard polynomial* or the *Bell polynomial*, is given by $b_n(x) = \sum_{k=1}^{n} S(n, k) x^k$, where the coefficients $S(n, k)$ are the familiar Stirling numbers of the second kind; see [21, pp. 747–750].

Introduced by J. F. Steffensen in his 1927 treatise on interpolation [24] and later reconsidered, in particular, by J. Touchard [25] for their combinatorial and arithmetic properties, the exponential polynomials are the basic polynomials of the delta operator

$$\mathfrak{b} := \log(I + D) := \sum_{i \geq 1} (-1)^{i+1} \frac{1}{i} D^i. \tag{4.34}$$

The exponential polynomials also enumerate a binomial class $\mathcal{B}(T)$ of reluctant functions from S to X, where we require that elements in S are totally ordered, i.e., there is an order such that $s_1 < \cdots < s_n$. Now let T be the family of rooted paths labeled by S such that the labels are monotone along the path, with the root having the largest label. Correspondingly, the generalized Gončarov polynomial $t_n(x; \mathfrak{b}, \mathcal{Z})$ gives the enumeration of parking schemes in which n cars arrive in disjoint groups, all cars in the same group prefer the same space, and the order statistics of the preference sequence are bounded by $\mathcal{Z}$.

Thus, we get from the foregoing and Theorem 4.3 that the nth ∂-Abel polynomial associated with the arithmetic grid $\mathcal{Z} = (a + bi)_{i \geq 0}$ is

$$t_n(x; \mathfrak{b}, (a + bi)_{i \geq 0}) = (x - a) \sum_{k=1}^{n} S(n, k)(x - a - nb)^{k-1}.$$

In particular, $t_n(0; \mathfrak{b}, (-a - bi)_{i \geq 0}) = a \sum_{k=1}^{n} S(n, k)(a + nb)^{k-1}$; for $a = b = 1$, this gives the sequence $1, 4, 29, 311, 4447 \ldots$, which, after a shift of index, is A030019 in OEIS. The sequence also has other combinatorial interpretations, for example, as hypertrees on n labeled vertices [19]. It would be interesting to find bijections between the parking sequences and the hypertrees.

The next two examples correspond to monomorphic classes associated to some binomial class $\mathcal{B}(T)$. As pointed out in [17], if the reluctant functions in $\mathcal{B}(T)$ are counted by $p_n(x)$ where

$$p_n(x) = \sum_{k=0}^{n} a_k x^k,$$

then the basic sequence counting $\mathcal{B}_m(T)$ is

$$q_n(x) = \sum_{k=0}^{n} a_k x_{(k)},$$

where $x_{(k)} = x(x - 1) \cdots (x - k + 1)$ is the lower factorial polynomial.

Example 4.6 (Lower Factorial Polynomials). The lower factorial $x_{(n)} = \prod_{i=0}^{n-1}(x - i)$ is the basic polynomial for the monomorphic class $\mathcal{B}_m(T_0)$, where T_0 is as described in Example 4.1. The corresponding delta operator is the forward difference operator $\Delta_{1,0} = E_1 - E_0 = E_1 - I$. It counts the number of one-to-one functions from S to X.

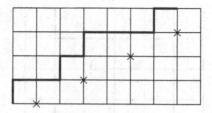

Fig. 4.1. Lattice path corresponding to the sequence $(1, 3, 4, 7)$ with $x = 8$. The stars indicate the right boundary $(1, 3, 5, 7)$.

By Theorem 4.4, the generalized Gončarov polynomials $t_n(x)$ associated to $(\Delta_{1,0}, \mathcal{Z})$ give the number of one-to-one functions from S to X whose order statistics are bounded by $\mathcal{Z}$ via (4.32). Assume $x \geq n$. Note that any sequence $1 \leq x_1 < \cdots < x_n \leq x$ can be represented geometrically as a strictly increasing lattice path in the plane from $(0, 0)$ to $(x - 1, n)$ using only steps $E = (1, 0)$ and $N = (0, 1)$: one simply takes the N-steps from $(x_i - 1, i - 1)$ to $(x_i - 1, i)$, and connects the N-steps with E-steps (see Fig. 4.1).

Clearly there are $n!$ one-to-one functions from S to X whose images are $\{x_1, \ldots, x_n\}$. Thus Theorem 4.5 implies that $\frac{1}{n!}t_n(0; \Delta_{1,0}, -\mathcal{Z})$ counts the number of strictly increasing lattice paths from $(0, 0)$ to $(x - 1, n)$ with strict right boundary $(z_0, \ldots, z_{n-1})$, viz. with right boundary $(z_0, \ldots, z_{n-1})$ that never touch the points $\{(z_i, i) : 0 \leq i < n\}$.

In particular, we have $t_n(x; \Delta_{1,0}, (a + bi)_{i \geq 0}) = (x - a)(x - a - nb - 1)_{(n-1)}$, so the number of strictly increasing lattice paths from $(0, 0)$ to $(x - 1, n)$ with the affine right boundary $(a, a + b, a + 2b, \ldots)$, where $a > 0$ and $b \geq 0$ are integers, is given by

$$\frac{1}{n!}a(a + nb - 1)_{(n-1)} = \frac{a}{a + nb}\binom{a + nb}{n}.$$

When $a = b = 1$, the aforementioned number is $\frac{1}{n+1}\binom{n+1}{n} = 1$, since there is only one strictly increasing lattice path from $(0, 0)$ to $(x - 1, n)$ that is bounded by $(1, \ldots, n)$.

Example 4.7 (Upper Factorial Polynomials). The upper factorial $x^{(n)} = (x + n - 1)_{(n)} = \prod_{i=0}^{n-1}(x + i)$ is related to the Laguerre polynomials by the equation

$$x^{(n)} = \sum_{k \geq 0} \frac{n!}{k!}\binom{n - 1}{k - 1}x_{(k)}.$$

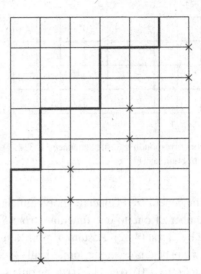

Fig. 4.2. Lattice path corresponding to the images $(1, 1, 1, 2, 2, 4, 4, 6)$ with $x = 6$. The stars indicate the right boundary $(1, 1, 2, 2, 4, 4, 6, 6)$.

Hence $x^{(n)}$ is the basic polynomial for the monomorphic class $\mathcal{B}_m(T_p)$, where T_p is as described in Example 4.3. The corresponding delta operator for $(x^{(n)})_{n \geq 0}$ is the backward difference operator $\Delta_{0,-1} = E_0 - E_{-1} = I - E_{-1}$.

Monomorphic reluctant functions in $\mathcal{B}_m(T_p)$ can also be described by lattice paths. Assume the final preimages of a reluctant function in $\mathcal{B}_m(T_p)$ consists of k paths of lengths $p_1, \ldots, p_k$ whose images are $x_1 < \cdots < x_k$. It corresponds to a lattice path from $(0, 0)$ to $(x - 1, n)$ whose consecutive vertical runs are given by p_1 N-steps at $y = x_1 - 1$, followed by p_2 N-steps at $y = x_2 - 1$, and so on. See Fig. 4.2 for an example.

The labels on each path can be recorded by labeling the N-steps of the lattice path. Again there are $n!$ many labels possible for each lattice path. Hence by Theorem 4.4 we know that if $t_n(x; \Delta_{0,-1}, \mathcal{Z})$ is the generalized Gončarov polynomial associated to the pair $(\Delta_{0,-1}, \mathcal{Z})$, then $\frac{1}{n!} t_n(0; \Delta_{0,-1}, -\mathcal{Z})$ is the number of lattice paths with the right boundary $\mathcal{Z}$, a result first established in [11].

In particular, $t_n(x; \Delta_{0,-1}, (a + bi)_{i \geq 0}) = (x - a)(x - a - nb + 1)^{(n-1)}$; hence the number of lattice paths from $(0, 0)$ to $(x - 1, n)$ with strict affine right boundary $(a, a + b, a + 2b, \ldots)$ for some integers $a > 0$ and $b \geq 0$ is

$$\frac{1}{n!} a(a + nb + 1)^{(n-1)} = \frac{a}{a + n(b+1)} \binom{a + n(b+1)}{n}, \qquad (4.35)$$

a well-known result; see, e.g., [16, p. 9]. In particular, for $a = 1$ and $b = k$ for some positive integer k, it counts the number of lattice paths from the origin to (kn, n) that never pass below the line $x = yk$. In this case, (4.35) gives $\frac{1}{1+(k+1)n}\binom{1+(k+1)n}{n} = \frac{1}{1+kn}\binom{(k+1)n}{n}$, which is the nth k-Fuss–Catalan number.

4.7 Further Remarks

In the theory of binomial enumeration, it is not really necessary to restrict one-self to reluctant functions, and we can in fact consider a more general setting, as outlined in [20, Section 2]. Following Joyal [8], a *species B* is a covariant endofunctor on the category of finite sets and bijections. Given a finite set E, an element $s \in B(E)$ is called a *B-structure on E*. A *k-assembly of B-structures on E* is then a partition π of the set E into k blocks such that each block of π is endowed with a *B*-structure. Let $B_k(E)$ denote the set of all such k-assemblies. For example, when B is a set of rooted trees, a k-assembly of *B*-structures on E is a k-forest of rooted trees with vertex set E. But we can also take B to be other structures, such as permutations, graphs, posets, etc.

To enumerate the number of assemblies of *B*-structures, we define sequences of nonnegative integers by

$$b_{n,k} = \begin{cases} |B_k([n])|, & k \leq n, \\ 0, & k > n. \end{cases}$$

Mullin and Rota's work [17] establishes that if $b_n(x) = \sum b_{n,k}x^k$ is the enumerator for assemblies of *B*-structures on $[n]$, then $(b_n(x))_{n \geq 0}$ is a polynomial sequence of $\mathbb{K}[x]$ of binomial type. Now we can interpret the factor x^k in $b_n(x)$ by considering all functions (or monomorphic functions if one replaces x^k with $x_{(k)}$) from the blocks of a k-assembly to a set X of size x. When X is totally ordered, i.e., X is isomorphic to the poset $[s]$ with numerical order, where $s = |X|$, we can consider all such functions whose order statistics are bounded by a given sequence. The enumeration for such k-assemblies with an order-statistics constraint is captured by the associated generalized Gončarov polynomials.

In principle, the foregoing combinatorial description applies only to polynomial sequences $(p_n(x))_{n \geq 0}$ of binomial type with nonnegative integer coefficients. Mullin and Rota hinted at a generalization of their theory to include polynomials with negative coefficients, and Ray [20] developed a concept of weight functions on the partition category which allows one to realize any binomial sequence, over any commutative ring with identity, as the enumerator of weights. It would be interesting to investigate the role of generalized Gončarov polynomials in such weighted counting, as well as in other dissecting schemes as described in Henle [7], and to find connections to rook polynomials, order invariants, Tutte invariants of combinatorial geometries, and symmetric functions.

Acknowledgments

This publication was made possible by NPRP grant No. [5-101-1-025] from the Qatar National Research Fund (a member of Qatar Foundation). The statements made herein are solely the responsibility of the authors.

We thank Professor Graham for his continual support and encouragement throughout our research.

References

1. R. P. Boas and R. C. Buck. *Polynomial Expansion of Analytic Functions*. Springer, Heiderberg, 1958.
2. P. J. Davis. *Interpolation and Approximation*. Dover, New York, 1975 (reprint edition).
3. J. L. Frank and J. K. Shaw. Abel-Gončarov polynomial expansions. *J. Approx. Theory* **10**, no. 1 (1974), 6–22.
4. V. Gončarov. Recherches sur les dérivées successives des fonctions analytiques. Etude des valeurs que les dérivées prennent dans une suite de domaines. *Bull. Acad. Sc. Leningrad* (1930), 73–104 (in French).
5. V. L. Gončarov. *The Theory of Interpolation and Approximation of Functions*. Gostekhizdat, Moscow, 1954 (2nd edition, in Russian).
6. F. Haslinger. Abel-Gončarov polynomial expansions in spaces of holomorphic functions. *J. Lond. Math. Soc.* (2) **21**, no. 3 (1980), 487–495.
7. M. Henle. Binomial enumeration on dissects. *Trans. Amer. Math. Soc.* **202** (1975), 1–39.
8. A. Joyal. Une théorie combinatoire des séries formelles. *Adv. Math.* **42** (1981), 1–82.
9. N. Khare, R. Lorentz, and C. Yan. Bivariate Gončarov polynomials and integer sequences. *Sci. China Math.* **57** (2014), No. 8, 1561–1578.
10. A. G. Konheim and B. Weiss. An occupancy discipline and applications. *SIAM J. Appl. Math.* **14**, no. 6 (1966), 1266–1274.
11. J. P. S. Kung, X. Sun, and C. Yan. *Gončarov-Type Polynomials and Applications in Combinatorics*, unpublished manuscript (2006).
12. J. P. S. Kung and C. Yan. Gončarov polynomials and parking functions. *J. Combin. Theory Ser. A* **102**, no. 1 (2003), 16–37.
13. N. Levinson. The Gontcharoff polynomials. *Duke Math. J.* **11**, no. 4 (1944), 729–733.
14. R. Lorentz, S. Tringali, and C. Yan. Multivariate Delta Gončarov and Abel polynomials. *J. Math. Anal. Appl.* 446, no. 1 (2017), 663–680.
15. R. Lorentz and C. Yan. Bivariate affine Gončarov polynomials. *Discrete Math.* 339, no. 9 (2016), 2371–2383.
16. S. G. Mohanty. *Lattice Path Counting and Applications*. Academic Press, New York, 1979.
17. R. Mullin and G.-C. Rota. "On the Foundations of Combinatorial Theory. III. Theory of Binomial Enumeration", pp. 167–213. In B. Harris (ed.), *Graph Theory and Its Applications*. Academic Press, New York, 1970.

18. H. Niederhausen. Rota's umbral calculus and recursions. *Algebra Universalis* **49**, no. 4 (2003), 435–457.
19. The On-Line Encyclopedia of Integer Sequences, https://oeis.org.
20. N. Ray. Umbral calculus, binomial enumeration and chromatic polynomials. *Trans. Amer. Math. Soc.* **309**, no. 1 (1988), 191–213.
21. G.-C. Rota, D. Kahaner, and A. Odlyzko. On the foundations of combinatorial theory. VII. Finite operator calculus. *J. Math. Anal. Appl.* **42**, no. 3 (1973), 684–760.
22. R. P. Stanley. Hyperplane arrangements, parking functions and tree inversions, pp. 359–375. In B. Sagan and R. Stanley (eds.), *Mathematical Essays in Honor of Gian-Carlo Rota*. Birkhäuser, Boston, 1998.
23. R. P. Stanley. *Enumerative Combinatorics:* Vol. 2, Cambridge Stud. Adv. Math. **62**, Cambridge University Press, Cambridge, 1999.
24. J. F. Steffensen. *Interpolation*, 2nd edn. Dover, New York, 2006.
25. J. Touchard. Nombres exponentiels et nombres de Bernoulli. *Canad. J. Math.* **8** (1956), 305–320.
26. J. M. Whittaker. *Interpolatory function theory*. Cambridge Tracts in Math. and Math. Physics **33**, Cambridge University Press, Cambridge, 1935.
27. C. Yan. Generalized Tree Inversions and *k*-Parking Functions. *J. Combin. Theory Ser. A* **79**, no. 2 (1997), 268–280.
28. C. Yan. "Parking Functions", pp. 835–893. In M. Bóna (ed.), *Handbook of Enumerative Combinatorics*, Discrete Math. Appl., Chapman and Hall/CRC, Boca Raton, FL, 2015.

5
The Digraph Drop Polynomial

Fan Chung and Ron Graham

Abstract

For a weighted directed graph (or digraph, for short), denoted by $D = (V, E, w)$, we define a two-variable polynomial $B_D(x, y)$, called the *drop polynomial* of D, which depends intimately on the cycle structure of the permutations on vertices of D. This polynomial generalizes several other digraph polynomials that have been studied in the literature previously, such as the binomial drop polynomial for posets [2], the binomial drop polynomial for digraphs [7], the path/cycle cover polynomial for digraphs [5], and the matrix cover polynomial [6]. We show that $B_D(x, y)$ satisfies a reduction/contraction recurrence as well as the (somewhat mysterious) reciprocity relation $B_{\bar{D}}(x, y) = (-1)^n B_D(-x - y, y)$, where D has n vertices and $\bar{D} = (V, E, \bar{w})$ is the digraph formed from D by defining $\bar{w}(e) = 1 - w(e)$ for each edge $e \in E$.

5.1 Introduction

Suppose $P = (V, \prec)$ is a partially ordered set on a set V of cardinality n. In [2], a polynomial $B_P(x)$ called the *binomial drop polynomial* for P was introduced that was defined as follows. If $\pi : V \to V$ is a permutation on V, we say that π has a *drop* at u if $\pi(u) = v$ and $v \prec u$. Let $drop(\pi)$ denote the number of drops that π has. Then define

$$B_P(x) = \sum_{\pi} \binom{x + drop(\pi)}{n},$$

where π runs over all $n!$ permutations on V. The polynomial $B_P(x)$ turns out to have many connections to a variety of subjects in combinatorics. For example, suppose we define the *incomparability graph* $Inc(P)$ of P to be the graph on V where $\{u, v\}$ is an edge of $Inc(p)$ if and only if u and v are incomparable in P.

Let $\chi_{Inc(P)}(x)$ denote the chromatic polynomial of $Inc(P)$ (e.g., see [12]). Then for any positive integer b, $B_P(b) = \chi_{Inc(P)}(b)$. Further, $B_P(-1)$ is equal to the number of permutations π on V that have no drops. This is also equal to the number of acyclic orientations of $Inc(P)$ by a classic theorem of Stanley [12]. In the case that P is a chain, i.e., a linearly ordered set, then $Inc(P)$ has no edges and we have

$$B_P(x) = x^n = \sum_k \left\langle {n \atop k} \right\rangle \binom{x+k}{n} \tag{5.1}$$

where $\left\langle {n \atop k} \right\rangle$ denotes the usual Eulerian number. This is usually known as Worpitzky's identity (see [9]).

There is a natural digraph $D = (V, E)$ associated to a poset $P = (V, \prec)$. Namely, the vertex set for D is V. For $u, v \in V$ we have a directed edge $(u, v) \in E$ if and only if $v \prec u$ in P. Thus, one might ask if (5.1) could be extended to more general digraphs. In fact, this can be done as follows for any given digraph D. Suppose $\pi : V \to V$ is a permutation on V. Let us say that π has a *drop* at u if $(u, \pi(u)) \in E$. Denote by $drop(\pi)$ the *number of drops* that π has. The following polynomial $B_D(x)$, called the *binomial drop polynomial* for D, which generalizes (5.1), was studied in [7]:

$$B_D(x) = \sum_\pi \binom{x + drop(\pi)}{n} \tag{5.2}$$

where $|V| = n$ and π runs over all $n!$ permutations on V. This polynomial can be applied to a more general class of digraphs, called *weighted* digraphs. A weighted digraph $D = (V, E, w)$ is a digraph on the vertex set V with an edge set $E \subseteq V \times V$ in which each edge $e \in E$ is assigned a real-valued *weight* $w(e)$ (possibly 0). In this case, if π is a permutation on V, we generalize the definition of a $drop(\pi)$ by defining

$$drop(\pi) = \sum_{\substack{u \in V \\ e=(u,\pi(u))}} w(e). \tag{5.3}$$

We can then extend the definition of $B_D(x)$ to weighted digraphs as follows:

$$B_D(x) = \sum_\pi \binom{x + drop(\pi)}{n}, \tag{5.4}$$

where, as before, the vertex set V of D has size n. Here, we interpret the binomial coefficient in the sum as

$$\binom{x+\alpha}{n} = n!(x+\alpha)^{\underline{n}} = n!(x+\alpha)(x+\alpha-1)\dots(x+\alpha+n-1),$$

where we use the *falling factorial* notation $z^{\underline{n}} = z(z-1)(z-2)\ldots(z-n+1)$. It was shown in [7] that $B_D(x)$ has a number of interesting properties, such as a deletion/contraction recurrence similar in form to the well-known deletion/contraction recurrence for the Tutte polynomial (see [14]).

The polynomial we will consider in this chapter is a generalization of $B_D(x)$ to a two-variable polynomial $B_D(x, y)$, called the *drop polynomial* for D.

However, before defining $B_D(x, y)$, we first mention a related digraph polynomial. This is the so-called (path/cycle) *cover* polynomial $C_D(x, y)$ for D (cf. [5]). For a *simple* digraph $D = (V, E)$, that is, one in which all edge weights are 1, we normally think of an edge $e = (u, v) \in E$ as represented by a directed arc going from u to v. A *path/cycle cover* $\mathbf{C}$ of E is a collection of vertex-disjoint paths and cycles that cover all the vertices in V, where we consider a single vertex to be a path of length 0, and a loop (u, u) to be a cycle of length 1. Let $c_D(i, j)$ denote the number of path/cycle covers of E consisting of i paths and j cycles. The (path/cycle) *cover polynomial* for D is defined by

$$C_D(x, y) = \sum_{i,j} c_D(i, j)x^i y^j. \tag{5.5}$$

The cover polynomial $C_D(x, y)$ was generalized in [6] to weighted digraphs $D = (V, E, w)$ as follows. For a path/cycle cover $\mathbf{C}$ of D, let $p(\mathbf{C})$ denote the *number of paths* in $\mathbf{C}$, and let $cyc(\mathbf{C})$ denote the *number of cycles* in $\mathbf{C}$. By the *weight* $w(\mathbf{C})$ of $\mathbf{C}$ we mean the *product* of all the weights of the edges in $\mathbf{C}$. Then the generalization of $C_D(x, y)$ to weighted digraphs is

$$C_D(x, y) = \sum_{\substack{\mathbf{C} \\ \text{path/cycle cover}}} x^{\underline{p(\mathbf{C})}} y^{cyc(\mathbf{C})} w(\mathbf{C}). \tag{5.6}$$

It was shown in [6] that in addition to a deletion/contraction recurrence, $C_D(x, y)$ satisfies a number of other interesting properties such as:

(1) If $D_1 = (V_1, E_1, w_1)$ and $D_2 = (V_2, E_2, w_2)$ are vertex disjoint digraphs and the digraph D is formed from D_1 and D_2 by joining every $v_1 \in V_1$ to every $v_2 \in V_2$ by an edge (v_1, v_2) of weight 1, then $C_D(x, y) = C_{D_1}(x, y)C_{D_2}(x, y)$.
(2) If $\bar{D} = (V, E, \bar{w})$ denotes the weighted digraph formed from $D = (V, E, w)$ by subtracting the weight of each edge $e \in E$ from 1, i.e., $\bar{w}(e) = 1 - w(e)$, and $|V| = n$, then we have the surprising reciprocity formula:

$$C_{\bar{D}}(x, y) = (-1)^n C_D(-x - y, y). \tag{5.7}$$

This chapter is organized as follows: In Section 5.2, we will give the definition of drop polynomial $B_D(x, y)$ for a weighted digraph $D = (V, E, w)$

in which each edge $e \in E$ is assigned some *weight* $w(e)$. We will ordinarily assume that $w(e)$ is a real number (possibly 0) although all of our results are valid if $w(e)$ just lies in some commutative ring with identity. Basically, $B_D(x, y)$ is a sum over all permutations π on V, where each term of the sum depends on $drop(\pi)$ and the cycle structure of π is restricted to D. In Section 5.3, we discuss a number of useful facts about $B_D(x, y)$. An alternative definition of $B_D(x, y)$ is given in Section 5.4. The drop polynomials $B_D(x, y)$ are similar to but different from the path-cycle polynomials $C_D(x, y)$. We will show in Section 5.6 that they coincide if D is a simple digraph. However, for general digraphs, they satisfy different deletion/contraction rules as seen in Section 5.5. Nevertheless, for both polynomials the same reciprocity theorems hold as shown in Section 5.7. The drop polynomials have direct connections with permutations on the vertex set and in turn lead to intriguing questions concerning enumeration problems for permutations and partial permutations, some of which are mentioned in Section 5.8.

5.2 The Drop Polynomial $B_D(x, y)$

As usual, we start with a weighted digraph $D = (V, E, w)$ on a set V of size n (possibly having some edges of weight 0.) For a permutation $\pi : V \to V$ we define $E(\pi)$ to be the set of edges $e = (u, v)$ such that $\pi(u) = v$. Thus, $E(\pi)$ consists of paths and cycles. In this case, *loops*, i.e., edges of the form (u, u), are still considered to be cycles (of length 1). Let $P(\pi)$ denoted the set of edges in paths in $E(\pi)$ and let $Cyc(\pi)$ denote the set of cycles in $E(\pi)$. Also, let $cyc(\pi)$ denote $|Cyc(\pi)|$. More generally, for any subset $S \subseteq E$, we can define $cyc(S)$ in the obvious way, i.e., as the number of cycles formed by the edges in S. Further, we define the *weight* $w(S)$ by

$$w(S) = \sum_{e \in S} w(e). \tag{5.8}$$

We finally come to the definition of the drop polynomial $B_D(x, y)$:

$$B_D(x, y) = \sum_{\pi} \sum_{P(\pi) \subseteq S \subseteq E(\pi)} (-1)^{|E(\pi) \backslash S|} \binom{x + w(S)}{n} (y - 1)^{cyc(\pi) - cyc(S)} y^{cyc(S)},$$

$$\tag{5.9}$$

where π ranges over all permutation of the vertex set V of D and $|V| = n$.

For example, suppose that $D = I(n)$, the digraph on n vertices with *no* edges. Then it is easy to see that in this case, $B_{I(n)}(x, y) = x^{\underline{n}}$, where we recall the falling factorial notation $z^{\underline{n}} = z(z - 1)(z - 2) \ldots (z - n + 1)$.

Note that if we substitute $y = 1$ in (5.9), then $B_D(x, y)$ reduces to the *binomial drop* polynomial $B_D(x)$ in (5.4). For in this case, the only term in the inner sum in (5.9) that doesn't vanish is for the choice $S = E(\pi)$, and in this case,

$$B_D(x, 1) = \sum_{\pi} \binom{x + w(S)}{n}$$

$$= \sum_{\pi} \binom{x + \text{drop}(\pi)}{n}$$

$$= B_D(x),$$

which is (5.4), as desired.

5.3 Adding Weight 0 Edges

For the general weighted digraphs $D = (V, E, w)$ we have been considering up to now, E is some subset of $V \times V$ with various weights assigned to the edges in E. However, there will be some advantages in being able to assume that E is *all* of $V \times V$, where pairs (u, v) that were not originally in E are replaced by edges $e = (u, v)$ with weight $w(e) = 0$. Of course, the downside of doing this is that now the sums involved in evaluating (5.9) may have many more terms than before. One upside is that equivalent forms of (5.9) become simpler (cf. Section 5.4).

We next show that this modification of D (replacing "non-edges" by edges of weight 0) does not change the value of $B_D(x, y)$. We first show that $B_D(x, y)$ is unchanged if a single pair $(u, v) \notin E$ is replaced by an edge $e = (u, v)$ with $w(e) = 0$. Equivalently, we show that removing an edge of weight 0 does not change $B_D(x, y)$.

Theorem 5.1. *Suppose $D = (V, E, w)$ is a weighted digraph, and $e_0 \in E$ with $w(e_0) = 0$. Form the digraph $D' = (V, E', w)$ by deleting the edge e_0, i.e., $E' = E \setminus e_0$. Then*

$$B_{D'}(x, y) = B_D(x, y).$$

Proof. When referring to D', we will use $E'(\pi)$, $cyc'(\pi)$, etc., to denote the corresponding parameters for D'. Thus, $|E'| = |E| - 1$. From the definition in (5.9), we have

$$B_{D'}(x, y) = \sum_{\pi} \sum_{P'(\pi) \subseteq S' \subseteq E'(\pi)} (-1)^{|E'(\pi) \setminus S'|} \binom{x + w(S')}{n} (y - 1)^{cyc'(\pi) - cyc'(S')} y^{cyc'(S')}.$$

The overall plan is to show that $B_{D'}(x, y)$ reduces to $B_D(x, y)$ when $w(e_0) = 0$. To do this, we are going to expand the terms of $B_{D'}(x, y)$ for each π and show how in each case they either cancel each other out, or correspond to unique terms in $B_D(x, y)$. So consider some fixed permutation π on V.

Case (1). $e_0 \notin E(\pi)$. In this case, for each choice of S' in the sum for $B_{D'}$ we can choose $S = S'$ in the corresponding term in the sum for B_D. Since in this case, $|E'(\pi)| = |E(\pi)|$, $c'(\pi) = c(\pi)$, $c'(S') = c(S)$, etc., the corresponding terms in $B_{D'}$ and B_D are equal.

Case (2). $e_0 \in P(\pi)$. In this case, we will always have $e_0 \in S$, since $P(\pi) \subseteq S$. Hence, we can write $S = S' \cup e_0$ for some S' with $P'(\pi) \subseteq S' \subseteq E'(\pi)$. Now we have $|E'(\pi)| = |E(\pi)| - 1$, $|S'| = |S| - 1$, $|E'(\pi) \setminus S'| = |E(\pi) \setminus S|$, $cyc'(\pi) = cyc(\pi)$ and $cyc'(S') = cyc(S)$. Since $w(S) = w(S') + w(e_0)$, then when $w(e_0)$ is 0, the corresponding terms in $B_{D'}$ and B_D are equal in this case as well.

Case (3). $e_0 \in C(\pi)$. Thus, e_0 belongs to some *cycle* $C = \{e_0, e_1, e_2, \ldots, e_r\}$ in D (where $r = 0$ is allowed). Let C^- denote the set $C \setminus \{e_0\}$. There are now two possibilities for S:

(a) $C \nsubseteq S$. In this case, we will find a unique "mate" S'' for S as follows. If $e_0 \notin S$ then set $S'' = S \cup e_0$; if $e_0 \in S$ then set $S'' = S \setminus e_0$. We will examine the sum of the two terms

$$\sum_{P(\pi) \subseteq S \subseteq E(\pi)} (-1)^{|E(\pi) \setminus S|} \binom{x + w(S)}{n} (y - 1)^{cyc(\pi) - cyc(S)} y^{cyc(S)} \qquad (5.10)$$

and

$$\sum_{E(\pi) \subseteq S'' \subseteq E(\pi)} (-1)^{|E(\pi) \setminus S''|} \binom{x + w(S'')}{n} (y - 1)^{cyc(\pi) - cyc(S'')} y^{cyc(S'')} \qquad (5.11)$$

in B_D. Suppose $S'' = S \setminus e_0$ (the other case is symmetric). Then $|S| = |S''| + 1$, $cyc(S'') = cyc(S)$ and $w(S) = w(S'') + w(e_0) = w(S'')$. Thus, $(-1)^{|E(\pi) \setminus S''|} = -(-1)^{|E(\pi) \setminus S|}$ so that the sum of the two terms (5.10) and (5.11) is 0.

(b) $C \subseteq S$. In this case we can write $S = C \cup T$ where $T \subseteq E(\pi) \setminus C$. We have $P'(\pi) = P(\pi) \cup C^+$ and we set $S'' = C^- \cup T$. Now we have $cyc(S) = cyc(S'') + 1$. Thus, since $|S| = |S''| + 1$ then adding (5.10) and (5.11), we get

$$\sum_{P(\pi)\subseteq S\subseteq E(\pi))} (-1)^{|E(\pi)\setminus S|} \binom{x+w(S)}{n}(y-1)^{cyc(\pi)-cyc(S)}y^{cyc(S)}$$

$$+ \sum_{P(\pi)\subseteq S''\subseteq E(\pi)} (-1)^{|E(\pi)\setminus S''|} \binom{x+w(S'')}{n}(y-1)^{cyc(\pi)-cyc(S'')}y^{cyc(S'')}$$

$$= \sum_{P(\pi)\subseteq S''\subseteq E(\pi)} -(-1)^{|E(\pi)\setminus S''|} \binom{x+w(S'')+w(e_0)}{n}$$

$$\times (y-1)^{cyc(\pi)-cyc(S'')-1}y^{cyc(S'')+1}$$

$$+ \sum_{P(\pi)\subseteq S''\subseteq E(\pi)} (-1)^{|E(\pi)\setminus S''|} \binom{x+w(S'')}{n}(y-1)^{cyc(\pi)-cyc(S'')}y^{cyc(S'')}$$

$$= \sum_{P(\pi)\subseteq S''\subseteq E(\pi)} (-1)^{|E(\pi)\setminus S''|} \binom{x+w(S'')}{n}$$

$$\times (y-1)^{cyc(\pi)-cyc(S'')-1}y^{cyc(S'')}((y-1)-y)$$

$$= \sum_{P(\pi)\subseteq S''\subseteq E(\pi)} -(-1)^{|E(\pi)\setminus S''|} \binom{x+w(S'')}{n}(y-1)^{cyc(\pi)-cyc(S'')-1}y^{cyc(S'')}.$$

Now, taking $S' = S''$ in $B_{D'}$ and noting that $cyc'(\pi) = cyc(\pi)+1$ and $|E(\pi)| = |E'(\pi)|+1$, we see that the term in this last sum becomes

$$(-1)^{|E'(\pi)\setminus S'|} \binom{x+w(S')}{n}(y-1)^{cyc'(\pi)-c'(S)}y^{cyc'(S)} \qquad (5.12)$$

This is exactly the term corresponding to the choice of S', with $P'(\pi) \subseteq S' \subseteq E'(\pi)$, in $B_{D'}$.

The preceding arguments show there is a bijection between the terms of B_D and the (noncanceling) terms of $B_{D'}$. Thus, we have proved $B_D(x, y) = B_{D'}(x, y)$. $\qquad\square$

Corollary 5.1. *Suppose $D = (V, E, w)$ is a weighted digraph, and $D' = (V, E', w)$ is formed by adding an edge $e = (u, v)$ to E with $w(e) = 0$ for every pair $(u, v) \notin E$. Then*

$$B_{D'}(x, y) = B_D(x, y).$$

Proof. Just apply Theorem 5.1 recursively. $\qquad\square$

5.4 An Alternative Form for $B_D(x, y)$

Suppose $D = (V, E, w)$ is a weighted digraph where $E = V \times V$ (so that there may be many weight 0 edges). A *partial permutation* σ on V is an injective mapping $\sigma : U \to V$ for some subset $U \subseteq V$. There are several ways to represent σ. One is by the familiar two-line notation:

$$
\sigma = \begin{pmatrix} u_1 & u_2 & \cdots & u_k \\ \sigma(u_1) & \sigma(u_2) & \cdots & \sigma(u_k) \end{pmatrix}
$$

where u_i's denote distinct vertices of D. Here, we assume that $|U| = k$ and we will say that σ has size $|\sigma| = k$.

Another way to represent σ is to identify it with the corresponding entries in an associated matrix $M = M(D)$ indexed by elements of V. Here, the entry $M(u_i, \sigma(u_i))$ of M is just the weight $w(e_i)$ of the edge $e_i = (u_i, \sigma(u_i))$. If $S(\sigma)$ denotes the set of edges formed by σ, then it is clear that $S(\sigma)$ consists of vertex disjoint paths and cycles. Since we assume $E = V \times V$, we can abuse notation slightly by just saying that $S = S(\sigma)$ is a partial permutation, since specifying the entries in the matrix M automatically determines the actual partial permutation σ. Denote by $w(S)$ denote the sum of the weights of the edges in S and let $cyc(S)$ the number of cycles in D formed by the edges in S. Also, let $PP(V)$ denote the set of all partial permutations on V. We then have the following alternative expression for $B_D(x, y)$.

Theorem 5.2. *For any weighted digraph D, we have*

$$
B_D(x, y) = \sum_{S \in PP(V)} \binom{x + w(S)}{n} (1 - y)^{n - |S|} y^{cyc(S)}. \tag{5.13}
$$

Proof. From Theorem 5.1, we may assume $E = V \times V$ by adding edges of weight 0 while the drop polynomial $B_D(x, y)$ remains unchanged. Consequently, for any permutation $\pi : V \to V$, we have $P(\pi) = \emptyset$. Hence, we can interchange the order of summation in (5.9) to obtain

$$
B_D(x, y) = \sum_{\pi} \sum_{S \subseteq E(\pi)} (-1)^{|E(\pi) \setminus S|} \binom{x + w(S)}{n} (y - 1)^{cyc(\pi) - cyc(S)} y^{cyc(S)}
$$

$$
= \sum_{S \in PP(V)} \sum_{\pi} (-1)^{|E(\pi) \setminus S|} \binom{x + w(S)}{n} (y - 1)^{cyc(\pi) - cyc(S)} y^{cyc(S)},
$$

where π ranges over all permutations of V for which $S \subseteq E(\pi)$ so that S ranges over all partial permutations. The number of π with k cycles is given by $\begin{bmatrix} n \\ k \end{bmatrix}$, a Stirling number of the first kind (see [9]). More generally, if $|S| = s$, then π has $n - s$ free blocks from which to form $k - cyc(S)$ cycles. Thus, there are just

$\left[\begin{smallmatrix} n-|S| \\ k-cyc(S) \end{smallmatrix}\right]$ such π with $S \subseteq E(\pi)$. A basic identity for Stirling numbers of the first kind is the following (see [9]):

$$\sum_k \begin{bmatrix} m \\ k \end{bmatrix} z^k = z^{\overline{m}}, \qquad (5.14)$$

where $z^{\overline{m}}$ denotes the rising factorial $z^{\overline{m}} = z(z+1)\ldots(z+m-1)$. Hence our expression for $B_D(x, y)$ becomes

$$
\begin{aligned}
B_D(x, y) &= \sum_{S \in PP(V)} (-1)^{n-|S|} \binom{x+w(S)}{n} \sum_k \begin{bmatrix} n-|S| \\ k-cyc(S) \end{bmatrix} (y-1)^{k-cyc(S)} y^{cyc(S)} \\
&= \sum_{S \in PP(V)} (-1)^{n-|S|} \binom{x+w(S)}{n} (y-1)^{\overline{n-|S|}} y^{cyc(S)} \\
&= \sum_{S \in PP(V)} \binom{x+w(S)}{n} (1-y)^{\underline{n-|S|}} y^{cyc(S)},
\end{aligned}
$$

which is (5.13). This completes the proof of Theorem 5.2. $\qquad\qquad\square$

As an example, consider the digraph $D = (V, E, w)$ where $V = [n]$, $E = [n] \times [n]$ and $w(e) = 0$ for all $e \in E$. Since in this case the "reduced" digraph is just $I(n)$, a digraph with no edges, and we have seen that $B_{I(n)}(x, y) = x^{\underline{n}}$. Thus, by (5.13), we get the interesting identity:

Corollary 5.2.

$$\sum_{S \in PP(V)} (1-y)^{\underline{n-|S|}} y^{cyc(S)} = n!.$$

It doesn't appear obvious (to us) how to prove this directly.

5.5 A Reduction/Contraction Rule for $B_D(x, y)$

In order to manipulate $B_D(x, y)$, we will first need to define the operations of *reduction* and *contraction* in a digraph $D = (V, E, w)$ (see Figs. 5.1 and 5.2).

First, suppose $e = (u, v)$ is a non-loop edge of D (so $u \neq v$). The *e-reduced* digraph $D' = (V, E, w')$ has the same set of vertices and edges as D, with the only change being that the weight of e is reduced by 1, i.e., $w'(e) = w(e) - 1$. The *e-contracted* digraph D'' is slightly more complicated. For the vertex set of D'' we replace the two vertices u and v by a single vertex uv. Any edge in D of the form (x, u) becomes an edge (x, uv) in D''. Also, any edge in D of the form (v, y) becomes an edge (uv, y) in D''. All other edges incident to either u or v (including loops) are deleted. If (v, u) happens to be an edge in D, it becomes a

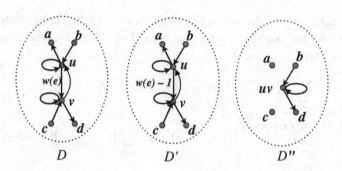

Fig. 5.1. Reduction and contraction of a non-loop edge e in a weighted digraph

loop at uv in D''. All other edges (not incident to u or v) in D are retained in D''. No edge weights of surviving edges are changed by the contraction operation.

Next, suppose $e = (u, u)$ is a loop at u in D. Then we do the following. As before, in the e-reduced digraph D', the weight of e is reduced by 1. To form the e-contracted digraph D'', we simply delete the vertex u and *all* edges incident to u.

Using the representation of $B_D(x, y)$ in (5.13), we will now show that drop polynomials obey a reduction/contraction rule.

Theorem 5.3. *Let D be a weighted digraph and let e be an edge of D. Denote the e-reduced and e-contracted digraphs by D' and D'', respectively.*

(1) If e is a non-loop edge then

$$B_D(x, y) = B_{D'}(x, y) + B_{D''}(x + w(e) - 1, y); \qquad (5.15)$$

(2) If e is a loop then

$$B_D(x, y) = B_{D'}(x, y) + y B_{D''}(x + w(e) - 1, y). \qquad (5.16)$$

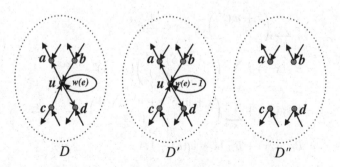

Fig. 5.2. Reduction and contraction of a loop edge e in a weighted digraph

Proof. Let w, w', w'' denote the edge weights in D, D', D'', respectively. We first consider the case that $e = (u, v)$ is a non-loop edge. From the definition (5.13), we have

$$
\begin{aligned}
B_D(x, y) &= \sum_{S \in PP(V)} \binom{x + w(S)}{n}(1 - y)^{n-|S|} y^{c(S)} \\
&= \sum_{e \notin S} \binom{x + w(S)}{n}(1 - y)^{n-|S|} y^{c(S)} + \sum_{e \in S} \binom{x + w(S)}{n}(1 - y)^{n-|S|} y^{c(S)} \\
&= \left(\sum_{e \notin S} \binom{x + w(S)}{n}(1 - y)^{n-|S|} y^{c(S)} \right. \\
&\quad + \left. \sum_{e \in S} \binom{x + w'(S)}{n}(1 - y)^{n-|S|} y^{c(S)} \right) \\
&\quad - \sum_{e \in S} \binom{x + w'(S)}{n}(1 - y)^{n-|S|} y^{c(S)} \\
&\quad + \sum_{e \in S} \binom{x + w(S)}{n}(1 - y)^{n-|S|} y^{c(S)} \\
&= B_{D'}(x, y) - \sum_{e \in S} \binom{x + w'(S)}{n}(1 - y)^{n-|S|} y^{c(S)} \\
&\quad + \sum_{e \in S} \binom{x + w(S)}{n}(1 - y)^{n-|S|} y^{c(S)},
\end{aligned}
$$

since for $e \notin S$, we have $w(S) = w'(S)$ for any partial permutation S. We now use the fact that if $e \in S$, we have $w'(S) = w(S) - 1$. Hence

$$
\begin{aligned}
B_D(x, y) &= B_{D'}(x, y) - \sum_{e \in S} \binom{x + w(S) - 1}{n}(1 - y)^{n-|S|} y^{c(S)} \\
&\quad + \sum_{e \in S} \binom{x + w(S)}{n}(1 - y)^{n-|S|} y^{c(S)} \\
&= B_{D'}(x, y) + \sum_{e \in S} \binom{x + w(S) - 1}{n - 1}(1 - y)^{n-|S|} y^{c(S)} \\
&= B_{D'}(x, y) + \sum_{S''} \binom{x + w(S'') + w(e) - 1}{n - 1}(1 - y)^{n-1-|S''|} y^{c(S'')} \\
&= B_{D'}(x, y) + B_{D''}(x + w(e) - 1, y),
\end{aligned}
$$

as desired.

Next, we consider the case that $e = (u, u)$ is a loop. The first part of the derivation is very similar to the preceding case and we won't repeat it. For the second part, we have

$$B_D(x, y) = \sum_{e \notin S} \binom{x + w(S)}{n}(1 - y)^{n - |S|} y^{c(S)} + \sum_{e \in S} \binom{x + w(S)}{n}(1 - y)^{n - |S|} y^{c(S)}$$

$$= B_{D'}(x, y) - \sum_{e \in S} \binom{x + w'(S)}{n}(1 - y)^{n - |S|} y^{c(S)}$$

$$+ \sum_{S} \binom{x + w(S)}{n}(1 - y)^{n - |S|} y^{c(S)}$$

$$= B_{D'}(x, y) - \sum_{\substack{S'' \\ S = e \cup S''}} \binom{x + w''(S) + w(e) - 1}{n}(1 - y)^{n - 1 - |S''|} y^{c(S'') + 1}$$

$$+ \sum_{\substack{S'' \\ S = e \cup S''}} \binom{x + w(S'') + w(e)}{n}(1 - y)^{n - 1 - |S''|} y^{c(S'') + 1}$$

$$= B_{D'}(x, y) + y \sum_{S''} \binom{x + w(S'') + w(e) - 1}{n - 1}(1 - y)^{n - 1 - |S''|} y^{c(S'')}$$

$$= B_{D'}(x, y) + y B_{D''}(x + w(e) - 1, y).$$

This completes the proof of Theorem 5.3. $\square$

5.6 Relating the Drop Polynomial to the Cover Polynomial

As mentioned in the introduction, the cover polynomial $C_D(x, y)$ for a weighted digraph $D = (V, E, w)$ is defined by (5.6). In the case that D is *simple*, that is, $w(e) = 1$ for every $e \in E$, then (5.6) reduces to (5.5)

$$C_D(x, y) = \sum_{i,j} c_D(i, j) x^{\underline{i}} y^j,$$

where we recall that $c_D(i, j)$ counts the number of path/cycle covers of D consisting of i paths and j cycles.

It was shown in [5] that for $e \in E$, if $D \setminus e$ denotes the digraph obtained from the simple digraph D by *deleting* the edge e, and D/e denotes the corresponding e-contracted digraph then we have the deletion/contraction recurrences:

(1) If e is a non-loop edge then

$$C_D(x, y) = C_{D \setminus e}(x, y) + C_{D/e}(x, y).$$

(2) If e is a loop then

$$C_D(x, y) = C_{D \setminus e}(x, y) + y C_{D/e}(x, y).$$

Also, it is easy to see that $C_{I(n)} = x^{\underline{n}}$, where $I(n)$ denotes the n-vertex digraph having no edges.

However, these are the same recurrences that $B_D(x, y)$ satisfies for simple digraphs D, since every edge $e \in E$ has $w(e) = 1$, so that when we reduce its weight by 1, it becomes a weight 0 edge, which we know by Theorem 5.1 we can remove. Thus, we have

Theorem 5.4. *For simple digraphs $D = (V, E)$, we have*

$$B_D(x, y) = C_D(x, y).$$

Proof. The proof precedes by induction on the number of edges using the reduction/contraction rule in Theorem 5.3. We have seen that for the base case $D = I(n)$ with no edges:

$$B_{I(n)}(x, y) = x^{\underline{n}} = C_{I(n)}(x, y).$$

Since $B_D(x, y)$ and $C_D(x, y)$ satisfy the same recurrences for reduction and contraction, then they are equal for all simple digraphs D.   □

For digraphs with arbitrary edge weights, the deletion/contraction rules for C_D are the following, where $D \setminus e$ and D/e are defined in the preceding text (see [6]):

(1) If e is a non-loop edge then

$$C_D(x, y) = C_{D \setminus e}(x, y) + w(e) C_{D/e}(x, y).$$

(2) If e is a loop then

$$C_D(x, y) = C_{D \setminus e}(x, y) + y\, w(e) C_{D/e}(x, y).$$

Thus, the deletion/contraction recurrences for $C_D(x, y)$ and the reduction/contraction recurrences for $B_D(x, y)$ are quite different for general weighted digraphs. However we have seen in Theorem 5.4 that $B_D(x, y)$ and $C_D(x, y)$ are equal when D is a *simple* digraph. However, this is not the only time they can be equal. We show an example in Fig. 5.3 for a general weighted digraph D having two vertices. A little computation shows

$$B_D(x, y) = x^2 + ((\alpha + \beta)y + \gamma + \delta - 1)x + \alpha\beta y^2$$
$$+ ((1/2)\alpha^2 + (1/2)\beta^2 + \gamma\delta - (1/2)\alpha - (1/2)\beta)y$$
$$+ (1/2)\gamma^2 + (1/2)\delta^2 - (1/2)\gamma - (1/2)\delta,$$
$$C_D(x, y) = x^2 + ((\alpha + \beta)y + \gamma + \delta - 1)x + \alpha\beta y^2 + \gamma\delta y.$$

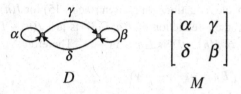

Fig. 5.3. A small general weighted digraph D and its associated matrix M.

Taking the difference we find

$$B_D(x, y) - C_D(x, y) = (1/2)((\alpha(\alpha - 1) + \beta(\beta - 1))y$$
$$+ \gamma(\gamma - 1) + \delta(\delta - 1)).$$

Certainly this difference is 0 if each of α, β, γ and δ is either 0 or 1. However, it is also 0 when $\alpha = 2/5, \beta = 6/5, \gamma = -1/5, \delta = 3/5$, for example.

5.7 A Reciprocity Theorem

The goal here is to prove a reciprocity theorem for weighted digraphs. Let $J(n)$ denote the n-by-n matrix of all 1's.

Theorem 5.5. *For a weighted digraph D on n vertices, we have*

$$B_{J(n)-D}(x, y) = (-1)^n B_D(-x - y, y). \tag{5.17}$$

Proof. It follows from the alternative definition of $B_D(x, y)$ in Theorem 5.2 that it is a polynomial in x, y and the weights $w(e), e \in E$. Thus, to prove the reciprocity theorem in (5.17) for general digraphs, it suffices to prove it for digraphs having nonnegative integer weights. This will then imply that (5.17) holds for arbitrary weights. We will proceed by induction on the number of edges in D. The base case for $I(n)$ holds since the reciprocity theorem holds for the cover polynomial $C_D(x, y)$ for simple digraphs (see [6] for a detailed proof) and $C_D(x, y) = B_D(x, y)$ for simple digraphs D as shown in Theorem 5.4.

Suppose $e = (u, v)$ is an edge in D. We first consider the case that e is a nonloop edge. We apply the reduction/contraction rule with D' and D'' as given in Theorem 5.3.

$$B_D(x, y) = B_{D'}(x, y) + B_{D''}(x + w(e) - 1, y)$$
$$= (-1)^n B_{J(n)-D'}(-x - y, y)$$
$$+ (-1)^{n-1} B_{J(n-1)-D''}(-x - y + w(e) - 1, y).$$

We now apply the reduction/contraction rule (5.15) for $J(n) - D'$ using the edge e. Note that the e-reduction of $J(n) - D'$ is just $J(n) - D$. Furthermore, the e-contraction of $J(n) - D'$ is $J(n-1) - D''$. Therefore,

$$
\begin{aligned}
B_D(x, y) &= (-1)^n B_{J(n)-D'}(-x - y, y) \\
&\quad + (-1)^{n-1} B_{J(n-1)-D''}(-x - y + w(e) - 1, y) \\
&= (-1)^n \Big(B_{J(n)-D}(-x - y, y) + B_{J(n-1)-D''}(-x - y + w(e) - 1, y) \Big) \\
&\quad + (-1)^{n-1} B_{J(n-1)-D''}(-x - y + w(e) - 1, y) \\
&= (-1)^n B_{J(n)-D}(-x - y, y),
\end{aligned}
$$

as desired.

For the case that e is a loop, the proof is quite similar and is omitted. □

An immediate consequence of (5.17) is the following interesting identity:

Corollary 5.3.

$$
B_{J(n)}(x, y) = \sum_{S \in PP(v)} \binom{x + |S|}{n} (1 - y)^{n - |S|} y^{cyc(S)} = (x + y)^{\overline{n}},
$$

5.8 Concluding Remarks

There are many questions that are suggested by the preceding results. For example, although $B_D(x, y)$ and $C_D(x, y)$ are quite different when evaluated on general weighted digraphs, they agree when D is simple. Can those D for which $B_D(x, y) = C_D(x, y)$ be characterized? The fact that $B_D(x, y)$ and $C_D(x, y)$ both satisfy the same reciprocity formula for general D (see (5.7)) suggests that something more fundamental is responsible for this behavior.

It would be of interest to understand the computational complexity of evaluating $B_D(x, y)$. For example, for simple D, the coefficient of $x^0 y^1$ term is just the number of Hamiltonian cycles in D, something that is known to be #P-hard to compute. (This follows from the fact that $B_D(x, y) = C_D(x, y)$ for simple digraphs D.) How hard is it to evaluate $B_D(x, y)$ at specific points in the (x, y) plane? For example, in the case that D is simple, $B_D(0, 0) = 0$. However, for general weighted D, even the value of $B_D(0, 0)$ does not seem to be easy to compute.

For the case of the Tutte polynomial, we mention the following result of Jaeger, Vertigan, and Welsh [10]:

Theorem 5.6. *The problem of evaluating the Tutte polynomial $T_G(x, y)$ of a general graph G at a point (a, b) is #P-hard except when (a, b) is on the special hyperbola $(x - 1)(y - 1) = 1$ or when (a, b) is one of the eight special points $(1, 1), (-1, -1), (0, -1), (-1, 0), (i, -i), (-i, i), (j, j^2)$, and (j^2, j), where $j = e^{\frac{2\pi i}{3}}$. In each of these exceptional cases, the evaluation can be done in polynomial time.*

In the case of the cover polynomial $C_D(x, y)$ (see [5]), the situation is even worse! In this case, Bläser and H. Dell [1] have shown that it is #P-hard to evaluate $C_D(x, y)$ for general digraphs (with multiple edges and loops) at a point (a, b) unless (a, b) is one of the *three* special points $(0, 0), (0, -1)$, and $(1, -1)$ (and for these three points, the computation can be done in polynomial time). Presumably, the same result also may hold when restricted to simple digraphs D.

We make the point that the results for weighted digraphs presented in this chapter can be restated entirely in terms of matrices with entries in some commutative ring **R** with identity. For if $D = (V, E, w)$ is our digraph on n vertices, then as we have mentioned earlier, we can define a matrix $M = M(D)$ that is indexed by the elements of V. For $u, v \in V$, the (u, v) entry of M is $M(u, v) = w(e)$ if $e = (u, v) \in E$, and is 0 if $(u, v) \notin E$. In terms of matrix operations on the matrix $M = M(D)$, we can define the reduction and contraction rules in the same way as those for weighted digraphs. Then the same reciprocity theorems for digraphs (i.e., (5.3) and (5.17)) apply to the corresponding matrix drop polynomial $B_M(x, y)$ as well.

In another direction, it would be quite interesting to extend many of the known results for permutation statistics (e.g., see [13] or [11]) to the case of partial permutations. Some results in this direction have recently appeared in [4], for example. However, even the most basic questions concerning partial permutations remain unanswered at present. For example, suppose we let $N_n(s, c)$ denote the number of partial permutations on $[n]$ that have size s and contain c cycles. That is,

$$N_n(s, c) = |\{S \subseteq [n] : S \text{ is a partial permutation}, |S| = s, cyc(S) = c\}|.$$

When $s = n$ then $N_n(n, c)$ is just equal to the familiar (unsigned) Stirling number $\begin{bmatrix} n \\ c \end{bmatrix}$ (see [9]). For example, $\begin{bmatrix} n \\ 1 \end{bmatrix} = (n - 1)!$ and $\begin{bmatrix} n \\ 2 \end{bmatrix} = (n - 1)!(\frac{1}{1} + \frac{1}{2} + \cdots + \frac{1}{n-1})$. However, for $0 < s < n$, it is not hard to show that $N_n(s, 0) = \binom{n}{s}(n - 1)^{\underline{s}}$ and $N_n(s, 1) = \binom{n}{s}(n - 1)^{\underline{s}}(\frac{1}{n-s} + \frac{1}{n-s+1} + \cdots + \frac{1}{n-1})$. It is possible to derive explicit expressions for $N_n(s, 2)$ and $N_n(s, 3)$ but the formulas seem to be getting increasingly unpleasant! Is there nice way of expressing $N_n(s, k)$ in general?

Of course, since there are just $\binom{n}{s}n^{\underline{s}}$ possible partial permutations on $[n]$ of size s, then

$$\sum_{0 \le c \le n} N_n(s, c) = \binom{n}{s}n^{\underline{s}}.$$

However, the following identity (equivalent to Corollary 2) seems less obvious.

$$\sum_{s,c} N_n(s, c)(1 - y)^{\underline{n-s}}y^c = n!.$$

If the vertex set V of D is totally ordered, say $V = [n]$, we can define $exc(S)$, the number of *exceedences* of a partial permutation S as follows. Namely, define $exc(S)$ to be the number of indices i such that $\sigma(i) > i$ for the associated mapping $\sigma : [n] \to [n]$. From the matrix perspective, this is the number of entries of S that lie above the diagonal of M. How many partial permutations S of size k have $exc(S) = d$? For the usual case of permutations of $[n]$, these numbers are the usual Eulerian numbers (see [9]). Of course, the same questions can be asked when V is just partially ordered, or more generally, when D is an arbitrary simple digraph. There is clearly much more to do.

References

1. M. Bläser and H. Dell. Complexity of the Cover Polynomial. *Automata, Languages and Programming*. Lecture Notes in Computer Science **4596**, (2007), 801–812.
2. J. P. Buhler and R. L. Graham. A note on the drop polynomial of a poset. *Combinat. Theory* (A) **66** (1994), 321–326.
3. J. P. Buhler and R. L. Graham. Juggling patterns, passing and posets. *Mathematical Adventures for Students and Amateurs*, MAA (2004), 99–116.
4. A. Claesson, V. Jelínek, E. Jelínková, and S. Kitaev. Pattern avoidance in partial partitions. *Electron. J. Combinat.* **18**, no. 2, (2011), #P5.
5. F. R. K. Chung and R. L. Graham. On the cover polynomial of a digraph. *Combinat. Theory* (B) **65** (1995), 273–290.
6. F. R. K. Chung and R. L. Graham. The matrix cover polynomial. *J. Combinat.* **7** (2016), 375–412.
7. F. R. K. Chung and R. L. Graham. The binomial drop polynomial of a digraph. (preprint), http://www.math.ucsd.edu/~fan/wp/diw_2.pdf.
8. O. M. D'Antona and E. Munarini. The cycle-path indicator polynomial of a digraph. *Adv. Appl. Math.* **25** (2000), 41–56.
9. R. L. Graham, D. E. Knuth, and O. Patashnik. *Concrete Mathematics, A Foundation for Computer Science*. Addison-Wesley, Boston, 1994.
10. F. Jaeger, D. L. Vertigan, and D. J. A. Welsh. On the computational complexity of the Jones and Tutte polynomials. *Math. Proc. Cambridge Philos. Soc.* **108** (1990), 35–53.

11. A. Mendes and J. Remmel. *Counting with Symmetric Functions*. Springer, Heidelberg, 2015.
12. R. P. Stanley. Acyclic orientations of graphs. *Discrete Math.* **5** (1973), 171–178.
13. R. P. Stanley. Enumerative Combinatorics. Vol 1, 2nd ed., *Cambridge Studies in Advanced Mathematics*, **62** Cambridge University Press, Cambridge, 2012.
14. D. J. A. Welsh. The Tutte polynomial. *Random Struct. Algorithms* **15** (1999), 210–228.

6

Unramified Graph Covers of Finite Degree

Hau-Wen Huang and Wen-Ching Winnie Li

Abstract

Regarding the fundamental group of a finite connected undirected graph X as the absolute Galois group of X, in this chapter we explore graph theoretical counterparts of several important theorems for number fields. We first characterize finite-degree unramified normal covers of X for which the Chebotarëv density theorem holds in natural density. Then we give finite necessary and sufficient conditions to classify finite-degree unramified covers of X up to equivalence. Similar to the reciprocity law for finite Galois extensions of a number field, it is shown that the unramified normal covers of X of degree d, up to isomorphism, are determined by the primes of X of length $\leq (4|X| - 1)d - 1$ which split completely. Finally we obtain a finite criterion for Sunada equivalence, improving a result of Somodi.

6.1 Introduction

Many results in number theory have parallel statements in graph theory. In this chapter we focus on graph covers. A connected undirected graph X can be thought of as a combinatorial counterpart of a number field F. Viewed as a one-dimensional topological space, it has a simply connected universal cover $\widetilde{X}$, unique up to isomorphism, along with the canonical covering map $pr(X) : \widetilde{X} \to X$. The tree $\widetilde{X}$ plays the role of the algebraic closure $\bar{F}$ of F and the fundamental group $\pi_1(X, x)$ plays the role of the absolute Galois group $Gal(\bar{F}/F)$. Since X is connected, different choices of the base point x result in isomorphic fundamental groups. An unramified cover is a surjective local isomorphism between two connected graphs. Given an unramified cover $\alpha : Y \to X$, the universal cover $\widetilde{Y}$ of Y is also a universal cover of X. Then the composition of $pr(Y) : \widetilde{Y} \to Y$ followed by $\alpha : Y \to X$ is isomorphic to $pr(X) : \tilde{X} \to X$. Hence there is an unramified cover $\beta : \tilde{X} \to Y$ such that $\alpha \circ \beta = pr(X)$. Note that α is uniquely determined by β. This is parallel to embedding an algebraic extension of F into $\bar{F}$. We will call β a *subcover* of $pr(X)$ for brevity.

A subcover $\beta : \tilde{X} \to Y$ of $pr(X)$ is said to *contain* another subcover β' : $\tilde{X} \to Z$ if there exists an unramified cover $\sigma : Y \to Z$ such that $\beta' = \sigma \circ \beta$. Two subcovers of $pr(X)$ are said to be *equivalent* if they contain each other. The equivalence classes of subcovers of $pr(X)$ correspond to the extensions of F in $\bar{F}$. An unramified cover $\alpha : Y \to X$ is called *normal* if the automorphism group of α acts transitively on one and hence all fibers $\alpha^{-1}(x)$ for any vertex x of X. In this case the automorphism group of α is called the Galois group of α, denoted G_α. There is a graph theoretical Galois correspondence between equivalence classes of subcovers of $pr(X)$ and subgroups of $\pi_1(X, x) = G_{pr}$, parallel to the correspondence between intermediate fields of $\bar{F}/F$ and closed subgroups of $Gal(\bar{F}/F)$. See [13] for details.

In view of this analogy, several questions concerning unramified graph covers naturally arise. Our work [7] gives two algorithms to find the Galois closure of a finite unramified cover. In fact, the categorical setting employed in [7] provides a unified approach to the problem of finding the Galois closure of a finite cover, which includes as special cases the familiar finite separate field extensions, finite unramified covers of a connected undirected graph, finite covering spaces of a locally connected topological space, and finite étale covers of a smooth projective irreducible algebraic variety.

In this chapter we explore graph theoretical counter parts of several important theorems for number fields. Since our base graph X is finite, we obtain stronger results, when appropriate, in terms of finite conditions. The behavior of (nonzero integral) prime ideals play an essential role in number theory. The "primes" in a graph X are equivalence classes of primitive closed geodesics in X. Alternatively, they are conjugacy classes of primitive elements in the fundamental group $\pi_1(X, x)$. Analogous to the Dedekind zeta function for a number field, we have the Ihara zeta function for a graph X, denoted $Z(X, u)$, defined as an Euler product over the primes of X. The zeta function $Z(X, u)$ counts the number of closed geodesics in X of given length. Our first result, Theorem 6.2 and its restatement Corollary 6.1, shows that $Z(X, u)$ is actually determined by the number of primes with length $\leq 2|X| - 1$. The prime number theorem for graphs, which estimates the number of primes of given length, was proved in [17, Theorem 10.1] and [6, Theorem 4], respectively. The analogous Artin L-functions $L(X, \rho, u)$ attached to finite-dimensional irreducible representations ρ of $\pi_1(X, x)$, which generalize the Ihara zeta function $Z(X, u)$, were studied in [2, 4, 5, 8, 12]. They are the reciprocal of polynomials in u.

Given a finite Galois extension K over F, to each nonzero prime ideal $\mathfrak{p}$ of F unramified in K, we associate a Frobenius conjugacy class Frob($\mathfrak{p}$) of the Galois group $Gal(K/F)$. The celebrated Chebotarëv density theorem says that the Frobenius conjugacy classes are uniformly distributed with respect to the natural density. (See [9] for this together with an estimate of the error term.) More precisely, given a conjugacy class $\mathcal{C}$ of $Gal(K/F)$, the set of prime ideals $\mathfrak{p}$ with Frob($\mathfrak{p}$) = $\mathcal{C}$ has natural (and hence Dirichlet) density $\frac{|\mathcal{C}|}{|Gal(K/F)|}$. For a finite unramified normal cover $\alpha : Y \to X$, the behavior of prime decomposition was studied by Hashimoto [5] and Stark and Terras [13]. We can similarly associate the Frobenius conjugacy class to each prime of X, and the Chebotarëv density theorem in Dirichlet density is shown to hold in the book by Terras [17, Theorem 22.3]. Using the analytic behavior of the Artin L-functions $L(X, \rho, u)$, Hashimoto [6, Theorem 4(i)] explored the Chebotarëv density theorem in natural density. It turns out that the result is in fact conditional, depending on the greatest common divisor of the lengths of the primes in a graph Z, denoted Δ_Z. We prove in Theorem 6.2 that the Chebotarëv density theorem holds in natural density for any conjugacy class of the Galois group of α if and only if $\Delta_Y = \Delta_X$. Stark and Terras [14] showed that either $\Delta_Y = \Delta_X$ or $\Delta_Y = 2\Delta_X$. Hashimoto's argument in [6, Theorem 4(i)] works for the case $\Delta_Y = \Delta_X$. An example of the case $\Delta_Y = 2\Delta_X$ is exhibited to illustrate the failure of the Chebotarëv density theorem in natural density.

One version of the reciprocity law in number theory states that finite Galois extensions K of F are determined by the prime ideals of F that split completely in K. Recall that if K is a degree d extension of F, then the set of prime ideals of F that split completely in K has Dirichlet density $1/d$. In particular this is an infinite set. In Section 6.5 two analogous, but stronger, criteria for graph covers are obtained. Theorem 6.5 says that finite unramified covers $Z \to X$ of degree d are determined, up to equivalence, by the cycles in X starting at x of length $\leq 2|X|d - 1$ that lift to a closed cycle in Z, while Theorem 6.6 asserts that finite unramified normal covers $Z \to X$ of degree d are determined, up to equivalence, by the primes of X of length $\leq (4|X| - 1)d - 1$ that split completely in Z provided that the rank of $\pi_1(X, x)$ is at least 2. In both cases it is a finite check. It is worth pointing out that when $\pi_1(X, x)$ has rank < 2, X has a unique unramified cover of degree d up to homotopy, so it is not an interesting case to consider.

Finally we consider a weaker concept, called Sunada equivalence. This originates from the problem in number theory to construct two nonisomorphic number fields K and K' that have the same Dedekind zeta function. The strategy is to first embed K and K' as subfields of a finite Galois extension M of F. Denote by H and H' the corresponding Galois subgroups of M over K and K', respectively.

Then K and K' have the same Dedekind zeta function if and only if H and H' are locally conjugate in $Gal(M/F)$; that is, every conjugacy class of $Gal(M/F)$ meets H and H' in the same number of elements. Sunada [15] extended this to a criterion for isospectral compact Riemannian manifolds. On the graph side, let $\beta : Y \to Z$ and $\beta' : Y \to Z'$ be two subcovers of a finite unramified normal cover $\alpha : Y \to X$. We call β and β' Sunada equivalent if the Galois groups of β and β' are locally conjugate in the Galois group of α. Somodi [11] showed that β and β' are Sunada equivalent if and only if for every prime $\mathfrak{p}$ of X the number of primes of Z above $\mathfrak{p}$ with the same length as $\mathfrak{p}$ agrees with that in Z'. As a consequence of Sunada equivalence, Z and Z' will have the same Ihara zeta function. Theorem 6.9 improves Somodi's result by requiring only primes $\mathfrak{p}$ of X with length $\ell(\mathfrak{p}) \leq 2|X| \deg \alpha$ if $\pi_1(X, x)$ has rank at least 2.

The graphs in this chapter are connected and undirected. A cover between two graphs considered in this chapter is a surjective graph homomorphism that is a local isomorphism. As such, it is always an unramified cover. The word "unramified" will be omitted from now on.

6.2 The Ihara Zeta Function

Let X be a finite connected undirected graph. A *cycle* (or walk) on X is a closed path with no backtracking; it has a starting point and an orientation. Two cycles are *equivalent* if one is obtained from the other by changing the starting point and retaining the orientation. A cycle is called *geodesic*[1] if all closed paths equivalent to it are backtrackless. It is called *primitive* if it is not obtained from a shorter cycle by traveling along it more than once. The length $\ell(C)$ of a geodesic cycle C is the number of edges in the path. Denote the equivalence class of C by $[C]$. The equivalence classes of primitive geodesic cycles in X are called the "primes" of X.

6.2.1 The Ihara Zeta Function

For $\ell \geq 1$, let $N_\ell = N_\ell(X)$ be the number of geodesic cycles on X of length ℓ. The Ihara zeta function of X counts the number of geodesic cycles on X. More precisely it is defined as

$$Z(X, u) := \prod_{\text{primes } [C]} \frac{1}{1 - u^{\ell(C)}}$$

$$= \exp\left(\sum_{\ell=1}^{\infty} \frac{N_\ell}{\ell} u^\ell\right),$$

---

[1] It is called a tailless geodesic cycle in [12].

which converges for $|u|$ small. Here the second equality follows from taking logarithmic derivatives of both sides. Generalizing the Selberg zeta function from the field of real numbers to fields of p-adic numbers, Ihara [8] first defined the zeta functions using group theoretical language for the graphs arising as finite quotients of the $(q + 1)$-regular tree, which can be identified with the homogeneous space $PGL_2(F)/PGL_2(\mathcal{O}_F)$ for a nonarchimedean local field F with the ring of integers $\mathcal{O}_F$ and q elements in its residue field. Serre [10] observed that the zeta function Ihara defined can be phrased as earlier for a general graph.

Observe that if X is a tree, then its zeta function is 1 since there are no cycles in X. Since the zeta function counts geodesic cycles in X, it remains the same if vertices of degree 1 are removed, so *we shall assume that all vertices of X have degree at least 2 from now on.*

Ihara [8] obtained an expression of the Ihara zeta function as a rational function in u, which holds for regular graphs, in terms of the vertex adjacency matrix of the graph. This was later extended by Bass [2] to include irregular graphs. Stark and Terras [12] provided several different proofs of this fact.

Theorem 6.1 (Ihara [8], Bass [2]). *Let X be a finite connected undirected graph with n vertices and m edges. Then its zeta function $Z(X, u)$ is a rational function in u of the following form:*

$$Z(X, u) = \frac{(1 - u^2)^{\chi(X)}}{\det(I - A_X u + Q_X u^2)},$$

where $\chi(X) = n - m$ is the Euler characteristic of X, A_X is the vertex adjacency matrix of X, and Q_X is the degree matrix of X minus the identity matrix.

Hence the numerator of $Z(X, u)$ gives topological information of X, while the denominator provides spectral information of X. Since the Euler characteristic of X is in fact negative, we see that $Z(X, u)$ is actually the reciprocal of a polynomial.

In [4, Theorem 2] Hashimoto gave a different expression for the Ihara zeta function of X using the edge adjacency matrix T_X defined as follows. Equip two opposite directions on each edge in X. The out-neighbors of a directed edge e from vertex v to vertex v' are the directed edges from the vertex v' to vertices v'' that are not e traveled in the reverse direction. Hence if there is only one edge between v and v', this means simply $v'' \neq v$. Assume that X has m edges so that it has $2m$ directed edges. The matrix T_X has its rows and columns indexed by the $2m$ directed edges such that the ee' entry is 1 if e' is an out-neighbor of

e, and 0 otherwise. Hashimoto proved the relation

$$Z(X, u) = \frac{1}{\det(I - T_X u)}. \qquad (6.1)$$

Denote by $r(X)$ the rank $m - n + 1$ of the fundamental group of X. When $r(X) = 0$, the graph is a tree and hence its zeta function is 1. If $r(X) = 1$, then X is an N-gon and

$$Z(X, u) = \frac{1}{(1 - u^N)^2},$$

with the two primes being the N-gon with two opposite orientations. There is not much to say about its zeta function. The situation is much more interesting when $r(X) \geq 2$, in which case X contains infinitely many primes. However, since the zeta function is the reciprocal of a polynomial of degree $2m$, we expect the zeta function to be determined by finite data. This is confirmed in Theorem 6.2.

Theorem 6.2. *Suppose that X has n vertices and m edges. The Ihara zeta function of X is determined by the numbers N_ℓ of geodesic cycles in X of length ℓ for $1 \leq \ell \leq 2n - 1$. In particular, if X is bipartite, then the Ihara zeta function $Z(X, u)$ is determined by N_ℓ for all even ℓ satisfying $2 \leq \ell \leq 2n - 2$.*

Proof. Consider the identities

$$Z(X, u) = \exp\left(\sum_{\ell=1}^{\infty} \frac{N_\ell}{\ell} u^\ell\right) = \frac{1}{\det(I - T_X u)}.$$

Taking logarithmic derivatives yields

$$N_\ell = \operatorname{tr}((T_X)^\ell) \qquad \text{for all integers } \ell \geq 1.$$

Let $\lambda_1, \lambda_2, \ldots, \lambda_{2m}$ be the eigenvalues of T_X. Then

$$N_\ell = \sum_{i=1}^{2m} \lambda_i^\ell \qquad \text{for all integers } \ell \geq 1.$$

Let $u_1, u_2, \ldots, u_{2m}$ denote $2m$ indeterminates. The Newton polynomials $\sum_{i=1}^{2m} u_i^\ell$ for $1 \leq \ell \leq 2m$ generate the ring of the symmetric polynomials over $\mathbb{C}$ in $2m$ variables. In particular, the numbers N_ℓ for all $\ell \geq 2m + 1$ are uniquely determined by N_ℓ for $1 \leq \ell \leq 2m$; hence they determine the Ihara zeta function of X.

Actually, we can do better by using information on the eigenvalues of T_X. The bound is clear if X is an N-gon. Suppose the rank $r(X) \geq 2$. Hashimoto in [4] has determined the multiplicity of ± 1 as eigenvalues of T_X. More precisely,

1 has multiplicity $r(X)$, while -1 has multiplicity $r(X) - 1$ if X is not bipartite and multiplicity $r(X)$ if X is bipartite. Hence for a nonbipartite X, the number N_ℓ is a symmetric function of degree ℓ on the remaining $2m - 2r(X) + 1 = 2n - 1$ unknown eigenvalues of T_X; the foregoing argument shows that the Ihara zeta function is determined by N_ℓ for $1 \leq \ell \leq 2n - 1$. For a bipartite X, it contains no cycles of odd length and the eigenvalues of T_X are symmetric with respect to 0, so the Ihara zeta function of X is determined by N_ℓ for positive even $\ell \leq 2m - 2r(X) = 2n - 2$. This proves the theorem.   □

Remark. We can further reduce the upper bound on ℓ if more eigenvalues of T_X are known. For instance, if X is a $(k + 1)$-regular graph, then k is also an eigenvalue of T_X of multiplicity 1. If X is also bipartite, then $-k$ is also an eigenvalue of multiplicity 1.

For integers $\ell \geq 1$, let $M_\ell = M_\ell(X)$ be the number of primes of X of length ℓ. Then

$$N_\ell = \text{tr}\left((T_X)^\ell\right) = \sum_{d|\ell} d \cdot M_d.$$

Using the Möbius μ-function, we invert the foregoing expression as

$$M_\ell = \frac{1}{\ell} \sum_{d|\ell} \mu\left(\frac{\ell}{d}\right) N_d.$$

This shows that the numbers N_ℓ with $\ell \leq 2n - 1$ and the numbers M_ℓ with $\ell \leq 2n - 1$ determine each other. So we can restate Theorem 6.2 as

Corollary 6.1. *Suppose that X has n vertices and m edges. The Ihara zeta function of X is determined by the number M_ℓ of primes of X of length ℓ for $1 \leq \ell \leq 2n - 1$. Further, if X is bipartite, its Ihara zeta function is determined by the number M_ℓ for all even ℓ with $2 \leq \ell \leq 2n - 2$.*

6.2.2 Prime Number Theorem for Graphs

Let X be a finite connected undirected graph on n vertices and m edges as before. By [17, Corollary 11.12] the matrix T_X is irreducible, which means that all entries of $(I + T_X)^{2m-1}$ are positive. Let λ_X be $\max\{|\lambda|\}$ as λ runs through all eigenvalues of T_X. For instance, if X is $(k + 1)$-regular, then $\lambda_X = k$. The irreducibility of T_X implies that λ_X is also an eigenvalue of T_X, called the Perron–Frobenius eigenvalue. We conclude from (6.1) that the *radius of convergence around* 0, R_X, of the zeta function $Z(X, u)$ is equal to $R_X = \lambda_X^{-1}$. Denote by Δ_X the greatest common divisor of the lengths $\ell(\mathfrak{p})$ for all primes $\mathfrak{p}$ of X. Then

Δ_X is equal to the greatest common divisor of all positive integers h such that $\text{tr}\,((T_X)^h)$ are positive. By the Perron–Frobenius theorem we have

Lemma 6.1. *The eigenvalues of T_X with absolute value λ_X are exactly $\lambda_X e^{2\pi i h / \Delta_X}$ for all $1 \leq h \leq \Delta_X$, where $i = \sqrt{-1}$. Moreover, each of these eigenvalues occurs with multiplicity 1.*

As observed before,

$$M_{\ell \Delta(X)} = \frac{1}{\ell \Delta_X} \sum_{d \mid \ell} \mu\left(\frac{\ell}{d}\right) \text{tr}\,((T_X)^{d\Delta_X}) \qquad \text{for all integers } \ell \geq 1.$$

Lemma 6.1 implies $\text{tr}\,((T_X)^{\ell \Delta_X}) \sim \Delta_X (\lambda_X)^{\ell \Delta_X}$ as $\ell \to \infty$. This then implies

$$M_{\ell \Delta_X} \sim \frac{(\lambda_X)^{\ell \Delta_X}}{\ell} \qquad \text{as} \quad \ell \to \infty. \tag{6.2}$$

Consequently,

$$\left| \{\text{primes } \mathfrak{p} \text{ in } X \,:\, \ell(\mathfrak{p}) < \ell \Delta_X \} \right| \sim \sum_{d=1}^{\ell-1} \frac{(\lambda_X)^{d\Delta_X}}{d}$$

$$\sim \frac{(\lambda_X)^{\ell \Delta_X}}{\ell((\lambda_X)^{\Delta_X} - 1)} \qquad \text{as} \quad \ell \to \infty. \tag{6.3}$$

The asymptotic formulae (6.2) and (6.3) are regarded as the prime number theorem for graphs, proved in [17, Theorem 10.1] and [6, Theorem 4], respectively.

6.3 The Fundamental Galois Theory for Graph Covers

All graphs in this chapter are connected and undirected. A graph homomorphism $\alpha : Y \to X$ is called an *unramified cover* if it is a surjective local isomorphism. Since only unramified covers are considered in this chapter, the word "unramified" will be omitted. The group of automorphisms of the cover $\alpha : Y \to X$, denoted by $Aut(\alpha)$, consists of graph automorphisms $\gamma : Y \to Y$ such that $\alpha = \alpha \circ \gamma$. Hence $\gamma \in Aut(\alpha)$ permutes vertices on a fiber $\alpha^{-1}(x)$ for all vertices x of X. As X is connected, γ is determined by its action on any one fiber $\alpha^{-1}(x)$. Call α a *normal* cover if $Aut(\alpha)$ acts transitively on a fiber $\alpha^{-1}(x)$. In this case the group of automorphisms $Aut(\alpha)$ is called the *Galois group* of α, written G_α. The cardinality of a fiber $\alpha^{-1}(x)$ is independent of the choice of the vertex x of X, and it is called the *degree* of the cover α. When α has a finite degree, it is normal if and only if $|Aut(\alpha)| = \deg \alpha$.

Let $\alpha : Y \to X$ be a cover. A cover $\beta : Y \to Z$ is called a *subcover* of α if α factors through β, that is, there is a cover $\gamma : Z \to X$ such that $\alpha = \gamma \circ \beta$.

Since the cover γ is uniquely determined by β and α, we denote it by α/β. Two subcovers β and β' of α are said to satisfy $\beta \preceq \beta'$ if β factors through β'. The binary relation $\preceq$ is irreflexive and transitive. We review some properties of covers. The reader is referred to [13] for details.

Proposition 6.1. *Any subcover β of a normal cover α with Galois group G_α is normal, and its Galois group G_β is a subgroup of G_α.*

Proposition 6.2. *Two subcovers β and β' of a normal cover α satisfy $\beta \preceq \beta'$ if and only if $G_{\beta'} \subseteq G_\beta$.*

Two subcovers β and β' of α are said to be *equivalent* if $\beta \preceq \beta'$ and $\beta' \preceq \beta$. Denote by $[\beta]_\alpha$ the class of subcovers of α equivalent to β. Like in number theory, there is a corresponding Galois theory for graph covers.

Theorem 6.3. *Let $\alpha : Y \to X$ be a normal cover with Galois group G_α. Then*

(1) *The map sending $[\beta]_\alpha$ to G_β is a bijection from the set of equivalence classes of subcovers of α to the set of subgroups of G_α.*
(2) *Let β be a subcover of α. Then α/β is a normal cover if and only if G_β is a normal subgroup of G_α. In this case $G_{\alpha/\beta}$ is isomorphic to the quotient G_α/G_β.*

Regard X as a one-dimensional topological space. Denote by $\widetilde{X}$ its universal cover and $pr = pr(X) : \widetilde{X} \to X$ the projection (covering) map. The fundamental group $\pi_1(X, x)$ with base point a vertex x of X acts on $\widetilde{X}$ as deck transformations in the following way. Fix a vertex y of $\widetilde{X}$ lying in the fiber $pr^{-1}(x)$ at x. If C is a closed path in X starting and ending at x, then C has a unique lifting to a path C' in $\widetilde{X}$ starting at y. The ending point y' of C' also projects to x, and it depends only on the homotopy class of C, denoted by $\langle C \rangle$. Define the action of $\langle C \rangle$ on $\widetilde{X}$ by mapping y to y'. For any other vertex z of $\widetilde{X}$, there is a path q from y to z and $\langle C \rangle$ maps z to the end point z' of the lifting of $pr(q)$ starting at y'. Since $\widetilde{X}$ is simply connected, z' is independent of the choice of the path q. This defines the action of $\pi_1(X, x)$ on $\widetilde{X}$ as graph automorphisms. Conversely, given any vertex y' in the fiber $pr^{-1}(x)$, since $\widetilde{X}$ is path connected, there is a path C' in $\widetilde{X}$ from y to y'. Its projection to X is a closed path C starting at x. The homotopy class $\langle C \rangle$ lies in $\pi_1(X, x)$ and, as described earlier, $\langle C \rangle$ then maps y to y'. This shows that the action of $\pi_1(X, x)$ on $pr^{-1}(x)$ is transitive. Therefore the cover pr is normal with Galois group G_{pr} isomorphic to $\pi_1(X, x)$. Note that the isomorphism depends on the choice of the vertex y of $\widetilde{X}$ projected to x.

Let $\beta : \widetilde{X} \to Y$ be a subcover of pr. Since $\widetilde{X}$ is also a universal cover of Y, the Galois group G_β is isomorphic to the fundamental group $\pi_1(Y, y)$ for any vertex y of Y lying in the fiber $(pr/\beta)^{-1}(x)$. Identify G_β with a subgroup of $\pi_1(X, x)$

by Theorem 6.3. The isomorphism from $\pi_1(Y, y)$ to G_β is given by projecting a cycle C' in Y starting at y to a cycle C in X starting at x via the covering map pr/β, which induces the map on homotopy classes sending $\langle C' \rangle$ to $\langle C \rangle$. If β is a normal subcover, then $pr/\beta : Y \to X$ is normal with Galois group isomorphic to the quotient $\pi_1(X, x)/\pi_1(Y, y)$. Therefore, the Galois group $G_{pr/\beta}$ is generated by the generators of $\pi_1(X, x)$ modulo $\pi_1(Y, y)$.

There is a bijection from the set of conjugacy classes of nonidentity elements in the fundamental group $\pi_1(X, x)$ of X to the set of equivalence classes of geodesic cycles in X such that the conjugacy classes of primitive elements in $\pi_1(X, x)$ correspond to the equivalence classes of primitive geodesic cycles in X. Recall that an element of a group G is called *primitive* if it generates its centralizer in G.

6.4 Graph-Theoretical Chebotarëv Density Theorem

6.4.1 Prime Decomposition

Let $\alpha : Y \to X$ be a cover. Let $\mathfrak{P}$ be a prime of Y. Then the projection $\alpha(\mathfrak{P})$ to X is the kth power of a unique prime $\mathfrak{p}$ of X for some positive integer k. We say that $\mathfrak{P}$ lies *above* $\mathfrak{p}$ and call k the *degree* of $\mathfrak{P}$ over $\mathfrak{p}$. Note that the lengths of the two primes are related by $\ell(\mathfrak{P}) = k\ell(\mathfrak{p})$. We say $\mathfrak{p}$ *splits completely* in Y if all primes $\mathfrak{P}$ of Y above $\mathfrak{p}$ have degree 1. Conversely, given a prime $\mathfrak{p}$ of X represented by a primitive geodesic cycle C in X starting at x, we can lift C to a path C_1 starting at a vertex y_1 in the fiber $\alpha^{-1}(x)$. If the ending vertex y_2 of C_1, which also lies in $\alpha^{-1}(x)$, is not y_1, then we lift C in Y to a path C_2 starting at y_2, repeating this procedure to get a sequence of paths $C_1, C_2, \ldots$ with distinct ending vertices $y_1, y_2, \ldots$ in the fiber $\alpha^{-1}(x)$. If α is a finite cover, then after finitely many, say k, liftings, this procedure will stop and we obtain a primitive geodesic cycle C' that is C_1 followed by $C_2, \ldots, C_k$, which represents a prime $\mathfrak{P}$ of Y above $\mathfrak{p}$ of degree k. If there are vertices in $\alpha^{-1}(x)$ not occurring in C', then we repeat this process to find a new prime of Y above $\mathfrak{p}$. As α has finite degree, this procedure finally stops and we obtain finitely many primes of Y lying above $\mathfrak{p}$, and their total degree is equal to the degree of α. If, in addition, α is normal, then all primes above $\mathfrak{p}$ have the same degree (or length) because the Galois group G_α of α acts transitively on the primes above $\mathfrak{p}$ and is length-preserving. In this case observe that the subgroup of elements in G_α stabilizing the prime $\mathfrak{P}$ represented by C' is cyclic of order k generated by the *Frobenius element*, denoted by $\mathrm{Frob}(\mathfrak{P}/\mathfrak{p})$, which maps y_1 to y_2; in other words, it sends C' to an equivalent cycle whose starting vertex is distance $\ell(\mathfrak{p})$ ahead of that of C'. In this case, all primes above $\mathfrak{p}$ are given by $\tau(\mathfrak{P})$ for some $\tau \in G_\alpha$, and the Frobenius element $\mathrm{Frob}(\tau(\mathfrak{P})/\mathfrak{p})$ is conjugate to $\mathrm{Frob}(\mathfrak{P}/\mathfrak{p})$. Denote by $\mathrm{Frob}(\mathfrak{p})$ the conjugacy class of $\mathrm{Frob}(\mathfrak{P}/\mathfrak{p})$; it is independent of the choice of the prime $\mathfrak{P}$ above $\mathfrak{p}$, and depends only of the prime $\mathfrak{p}$ of X.

6.4.2 Chebotarëv Density Theorem in Dirichlet Density

The Chebotarëv density theorem in number theory is a generalization of Dirichlet's theorem on the uniform distributions of primes in arithmetic progressions of a fixed modulus. More precisely, it concerns the uniform distribution of the set of primes whose Frobenius conjugacy classes are the given conjugacy class of a finite Galois group. The result is expressed in terms of the Dirichlet density of the set. It has a corresponding graph-theoretical version as stated in the text that follows. Let $\alpha : Y \to X$ be a finite normal unramified cover and let $\mathcal{C}$ be a conjugacy class of the Galois group G_α. Then the set

$$P(\alpha; \mathcal{C}) = \{\mathfrak{p} \text{ is a prime of } X \mid \mathrm{Frob}(\mathfrak{p}) = \mathcal{C}\}$$

has Dirichlet density $\frac{|\mathcal{C}|}{|G_\alpha|}$, provided that the fundamental group of X has rank $r(X) \geq 2$. This was first proved in the book by Terras [17, Theorem 22.3].

Here the Dirichlet density of a subest P of primes of X is defined as follows. Recall that the radius of convergence around 0 of the Ihara zeta function $Z(X, u)$ is $R_X = \lambda_X^{-1}$. The set P is said to have *Dirichlet density* δ if the limit

$$\lim_{u \to R_X} \frac{\sum_{\mathfrak{p} \in P} u^{\ell(\mathfrak{p})}}{\sum_{\mathfrak{p} \text{ prime of } X} u^{\ell(\mathfrak{p})}}$$

exists and is equal to δ. We remark that the set of all primes of X has Dirichlet density 1. Hence the Dirichlet density of P, if exists, lies in the interval $[0, 1]$.

6.4.3 Chebotarëv Density Theorem in Natural Density

The *natural density* of a set P of primes of X is defined as

$$\lim_{r \to \infty} \frac{|\{\mathfrak{p} \in P \mid \ell(\mathfrak{p}) < r\}|}{|\{\text{primes } \mathfrak{p} \text{ of } X \mid \ell(\mathfrak{p}) < r\}|}$$

if the limit exists. It is a general fact that if the natural density of a set P of primes exists, then so does the Dirichlet density of P and they are equal. Since the Chebotarëv density theorem for finite normal graph covers is shown to exist in Dirichlet density, it is natural to ask whether it also holds in the stronger natural density. For number fields, this holds, while for function fields, this does not always hold.

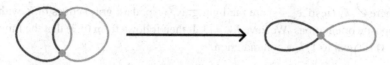

Fig. 6.1. A normal cover of degree 2

This question for graph covers was considered in [6, Theorem 4(i)], where the graph-theoretical Chebotarëv density theorem in natural density was claimed. It turns out that the result is in fact conditional. We shall first exhibit an example pinpointing the failure of the natural density, then present a necessary and sufficient condition for the Chebotarëv density theorem to hold in natural density.

Example 6.1. Consider the degree 2 normal cover $\alpha : Y \to X$, as displayed in Fig. 6.1, which preserves the coloring of edges.

It is clear that the fundamental group of X has rank 2, $\Delta_X = 1$ and $\Delta_Y = 2$. The Galois group $G_\alpha = \{\pm id\}$ has two conjugacy classes $C_+ = \{id\}$ and $C_- = \{-id\}$. It is clear that $P(\alpha; C_+)$ (resp. $P(\alpha; C_-)$) consists of primes of X with even (resp. odd) length, and each set has Dirichlet density $1/2$. We claim that neither $P(\alpha; C_+)$ nor $P(\alpha; C_-)$ has natural density. Since the argument is similar, we shall assume that the natural density of $P(\alpha; C_+)$ exists and hence is equal to $1/2$, and derive a contradiction.

First, the following two ratios have the same limit:

$$\lim_{\ell \to \infty} \frac{|\{\mathfrak{p} \in P(\alpha; C_+) \mid \ell(\mathfrak{p}) \le 2\ell\}|}{|\{\mathfrak{p} \text{ is a prime of } X \mid \ell(\mathfrak{p}) \le 2\ell\}|}$$

$$= \lim_{\ell \to \infty} \frac{|\{\mathfrak{p} \in P(\alpha; C_+) \mid \ell(\mathfrak{p}) \le 2\ell + 1\}|}{|\{\mathfrak{p} \text{ is a prime of } X \mid \ell(\mathfrak{p}) \le 2\ell + 1\}|} = \frac{1}{2}.$$

Since all primes in $P(\alpha; C_+)$ have even length, the subset with length $\le 2\ell$ agrees with the subset with length $\le 2\ell + 1$, hence the foregoing equality yields

$$\lim_{\ell \to \infty} \frac{|\{\mathfrak{p} \text{ is a prime of } X \mid \ell(\mathfrak{p}) \le 2\ell\}|}{|\{\mathfrak{p} \text{ is a prime of } X \mid \ell(\mathfrak{p}) \le 2\ell + 1\}|} = 1. \tag{6.4}$$

On the other hand, the matrix T_X is equal to

$$\begin{array}{cc} & \begin{array}{cccc} e_1 & \bar{e}_1 & e_2 & \bar{e}_2 \end{array} \\ \begin{array}{c} e_1 \\ \bar{e}_1 \\ e_2 \\ \bar{e}_2 \end{array} & \left(\begin{array}{cccc} 1 & 0 & 1 & 1 \\ 0 & 1 & 1 & 1 \\ 1 & 1 & 1 & 0 \\ 1 & 1 & 0 & 1 \end{array} \right) \end{array},$$

where $e_1, \bar{e}_1$ (resp. $e_2, \bar{e}_2$) are the light gray (resp. dark gray) edges of X with opposite orientations. We have $\lambda_X = 3$. It then follows from (6.3) that the limit (6.4) is equal to $1/3$, a contradiction.

We end this section by proving the following criterion for the graph-theoretical Chebotarëv density theorem to hold in natural density.

Theorem 6.4. *Assume that $r(X) \geq 2$. Let $\alpha : Y \to X$ be a finite normal cover with Galois group G_α. Then the natural density of $P(\alpha; C)$ exists, and hence is equal to $\frac{|C|}{|G_\alpha|}$, for any conjugacy class C of G_α if and only if $\Delta_X = \Delta_Y$.*

Stark and Terras [14, Corollary 1] showed that either $\Delta_Y = \Delta_X$ or $\Delta_Y = 2\Delta_X$ for any cover $Y \to X$ of finite degree, provided that $r(X) \geq 2$. The proof that follows will not use this fact. The foregoing example is indeed the case $\Delta_Y = 2\Delta_X$.

The first step toward proving the theorem is the following observation.

Lemma 6.2. *Let $\alpha : Y \to X$ be a finite normal cover. Then $\lambda_Y = \lambda_X > 1$.*

Proof. Recall the assumption that all vertices of X have degree at least 2. By [4, Lemma 5.13], we have

$$\det T_X = -(-1)^{r(X)} \prod_{x \text{ vertex of } X} (\deg x - 1).$$

The assumption that $r(X) \geq 2$ forces the existence of at least one vertex of X having degree ≥ 3. Therefore $|\det T_X| \geq 2$ and thus

$$\lambda_X > 1.$$

Denote by m the number of edges in X. To a d-dimensional unitary irreducible representation ρ of the Galois group G_α of α, Hashimoto [6] associated an Artin L-function $L(X, \rho, u)$ and proved the existence of a $2md \times 2md$ matrix $T_X(\rho)$ so that

$$L(X, \rho, u) = \frac{1}{\det(I - T_X(\rho)u)}$$

is the reciprocal of a polynomial in u of degree $2md$. When ρ is the trivial representation, the associated L-function agrees with the zeta function $Z(X, u)$. He also proved the decomposition of the zeta function of Y as a product of the Artin L-functions, just like what happens in number theory:

$$Z(Y, u) = Z(X, u) \prod_{\rho \in \widehat{G_\alpha}'} L(X, \rho, u)^{\deg \rho},$$

or equivalently,

$$\det(I - T_Y u) = \det(I - T_X u) \prod_{\rho \in \widehat{G_\alpha}'} \det(I - T_X(\rho) u)^{\deg \rho}.$$

Here $\widehat{G_\alpha}$ denotes the set of finite-dimensional unitary irreducible representations of G_α over $\mathbb{C}$, and $\widehat{G_\alpha}'$ consists of the nontrivial representations in $\widehat{G_\alpha}$. This shows that the eigenvalues of T_Y are those of $T_X(\rho)$ with $\rho \in \widehat{G_\alpha}$. Therefore $\lambda_Y \geq \lambda_X$.

The reverse inequality is an immediate consequence of the following proposition on the eigenvalues of $T_X(\rho)$ proved by Hashimoto. Hence the lemma is proved. □

Proposition 6.3 (Lemma 5.1, [6]). *Let $\rho \in \widehat{G_\alpha}'$. The eigenvalues λ of $T_X(\rho)$ satisfy $|\lambda| \leq \lambda_X$ and the equality holds only if $\deg \rho = 1$, $\lambda = \lambda_X e^{2\pi i \theta}$, and*

$$\rho(\mathrm{Frob}(\mathfrak{P}/\mathfrak{p})) = e^{2\pi i \theta \ell(\mathfrak{p})} \tag{6.5}$$

for all primes $\mathfrak{p}$ of X and all primes $\mathfrak{P}$ of Y over $\mathfrak{p}$.

Remark. Proposition 6.3 is the graph-theoretical analogue of the same statement for manifolds proved by Adachi and Sunada in [1]. The behavior of the Artin L-functions for graphs was also discussed in Sunada [16].

Consequently, if $T_X(\rho)$ for a representation $\rho \in \widehat{G_\alpha}'$ contains an eigenvalue λ of the form $\lambda_X e^{2\pi i \theta}$, then ρ has degree 1 and all eigenvalues of $T_X(\rho)$ with absolute value λ_X are $\lambda_X e^{2\pi i \left(\theta + \frac{h}{\Delta(X)}\right)}$ for $h = 1, \ldots, \Delta(X)$ because $\det(I - T_X(\rho) u)$ is a polynomial in u^{Δ_X} by definition. Hence ρ depends only on the conjugacy classes of G_α. As observed in Section 6.3.1, for a normal cover α, the Frobenius elements $\mathrm{Frob}(\mathfrak{P}/\mathfrak{p})$ for the primes $\mathfrak{P}$ of Y over a prime $\mathfrak{p}$ of X are conjugate to each other. So Eq. (6.5) can be restated in terms of the Frobenius conjugacy class as

$$\rho(\mathrm{Frob}(\mathfrak{p})) = e^{2\pi i \theta \ell(\mathfrak{p})}$$

for all primes $\mathfrak{p}$ of X. Hence for this ρ, $\lambda_X e^{2\pi i \theta}$ is an eigenvalue of T_Y with multiplicity 1. In view of Lemma 6.1, the eigenvalues of T_Y with absolute value λ_Y are $\lambda_Y e^{2\pi i h/\Delta_Y}$ for $1 \leq h \leq \Delta_Y$, which are eigenvalues of the degree 1 representations ρ_h, $1 \leq h \leq \Delta_Y/\Delta_X$, of G_α given by

$$\rho_h(\mathrm{Frob}(\mathfrak{p})) = e^{2\pi i h \ell(\mathfrak{p})/\Delta_Y} \tag{6.6}$$

for all primes $\mathfrak{p}$ of X. In particular, if $\Delta_Y = \Delta_X$, then all ρ_h are trivial representations. This proves

Proposition 6.4. *Let* $\alpha : Y \to X$ *be a finite normal cover with Galois group* G_α.

(1) *If* $\Delta_Y = \Delta_X$, *then for all nontrivial irreducible unitary representations* ρ *of* G_α, *the associated Artin L-function* $L(X, \rho, u)$ *is holomorphic on the closed disc* $|u| \leq (\lambda_X)^{-1}$. *Equivalently, the eigenvalues of* $T_X(\rho)$ *have absolute value* $< \lambda_X$.

(2) *If* $\Delta_Y > \Delta_X$, *then* ρ_h, $1 \leq h < \Delta_Y / \Delta_X$, *are the distinct nontrivial irreducible unitary representations of* G_α *whose associated Artin L-functions have* Δ_X *poles on the circle* $|u| = (\lambda_X)^{-1}$.

Proof of Theorem 6.4. (Necessity) Since a prime of Y projects to a geodesic cycle of X, we have $\Delta_X | \Delta_Y$. Suppose $\Delta_Y \neq \Delta_X$. Then $\Delta_Y = k\Delta_X$ for an integer $k \geq 2$. Let $\mathcal{C}$ be a conjugacy class of G_α. We claim that the set

$$P(\alpha; \mathcal{C}) = \{\text{primes } \mathfrak{p} \text{ of } X : \text{Frob}(\mathfrak{p}) = \mathcal{C}\}$$

has no natural density. Suppose otherwise. Then its natural density is equal to its Dirichlet density $|\mathcal{C}|/|G_\alpha|$, which is nonzero. We shall derive a contradiction.

By (6.6), for all $\mathfrak{p} \in P(\alpha; \mathcal{C})$, we have

$$\rho_1(\text{Frob}(\mathfrak{p})) = e^{2\pi i \ell(\mathfrak{p})/\Delta_Y} = \rho_1(\mathcal{C}),$$

which in turn implies that the lengths $\ell(\mathfrak{p})$ for all $\mathfrak{p} \in P(\alpha; \mathcal{C})$ are congruent modulo Δ_Y to $m\Delta_X$ for some integer m satisfying $1 \leq m \leq k$ since the length $\ell(\mathfrak{p})$ is a multiple of Δ_X. As a consequence of the existence of natural density, we have

$$\lim_{\ell \to \infty} \frac{|\{\mathfrak{p} \in P(\alpha; \mathcal{C}) \mid \ell(\mathfrak{p}) \leq (m + \ell k)\Delta_X\}|}{|\{\text{primes } \mathfrak{p} \text{ of } X \mid \ell(\mathfrak{p}) \leq (m + \ell k)\Delta_X\}|}$$

$$= \lim_{\ell \to \infty} \frac{|\{\mathfrak{p} \in P(\alpha; \mathcal{C}) \mid \ell(\mathfrak{p}) \leq (m + 1 + \ell k)\Delta_X\}|}{|\{\text{primes } \mathfrak{p} \text{ of } X \mid \ell(\mathfrak{p}) \leq (m + 1 + \ell k)\Delta_X\}|}.$$

Since there are no primes $\mathfrak{p} \in P(\alpha; \mathcal{C})$ with length satisfying $(m + \ell k)\Delta_X < \ell(\mathfrak{p}) \leq (m + 1 + \ell k)\Delta_X$, the two numerators agree. The foregoing equality then implies

$$\lim_{\ell \to \infty} \frac{|\{\text{primes } \mathfrak{p} \text{ of } X \mid \ell(\mathfrak{p}) \leq (m + k\ell)\Delta_X\}|}{|\{\text{primes } \mathfrak{p} \text{ of } X \mid \ell(\mathfrak{p}) \leq (m + 1 + k\ell)\Delta_X\}|} = 1.$$

On the other hand, computing the foregoing limit using (6.3) yields the result $1/(\lambda_X)^{\Delta_X}$. This implies that $\lambda_X = 1$, contradicting Lemma 6.2.

(Sufficiency) Assume $\Delta_Y = \Delta_X$. The proof of this part results from the holomorphy of the Artin L-functions $L(X, \rho, u)$ on the closed disk $|u| \leq (\lambda_X)^{-1}$ for nontrivial unitary irreducible representations ρ of G_α (Proposition 6.4, (1)), as sketched in [16, Proposition B] and carefully carried out in [6, Proof of Theorem 3].

6.5 Finite Criteria for Equivalent and Sunada Equivalent Subcovers

The final section of this chapter is devoted to developing finite combinatorial characterizations for equivalent subcovers and Sunada equivalent subcovers of a finite normal cover. They are graph theoretical analogues of number theoretic statements, but we improve the number theoretic results by finding finite necessary and sufficient conditions.

In number theory, it is well known that finite Galois extensions of a number field K are determined by the set of prime ideals of K splitting completely in the extension. In other words, two finite Galois extensions F_1 and F_2 of K in an algebraic closure of K are equal if and only if the sets of prime ideals of K splitting completely in F_1 and F_2, respectively, coincide. We will see that the same holds for finite covers of a graph. In fact, we shall prove a stronger statement that only the agreement of completely splitting primes with bounded length is needed.

In what follows we fix a finite connected undirected graph X on n vertices and m edges. Then $r(X) = m - n + 1$ is the rank of its fundamental group $\pi_1(X, x)$. Fix a spanning tree $\mathcal{T}_X$ of X. Denote by $e_1, e_2, \ldots, e_{r(X)}$ the edges of X not in $\mathcal{T}_X$. For each $1 \leq i \leq r(X)$, the graph $\mathcal{T}_X \cup \{e_i\}$ contains a geodesic cycle C_i' by the maximality of the spanning tree. Let T_i be the shortest path in $\mathcal{T}_X$ from x to C_i', which meets C_i' at the vertex x_i. It is called a *tail*. Choose an orientation on C_i' so that C_i' is a geodesic cycle starting at x_i. Then T_i followed by C_i' followed by T_i in reverse direction yields a backtrackless cycle C_i starting at x and passing through exactly one edge e_i not in $\mathcal{T}$. Further C_i has length $\ell(C_i) \leq 2n - 1$. It is well known that the homotopy classes $\langle C_1 \rangle$, $\langle C_2 \rangle, \ldots, \langle C_{r(X)} \rangle$ generate $\pi_1(X, x)$ as a free group; they are called the *canonical* generators of $\pi_1(X, x)$ with respect to the tree $\mathcal{T}_X$.

Fix a finite normal cover $\alpha : Y \to X$. Let y be a vertex of Y in the fiber $\alpha^{-1}(x)$. Then the Galois group G_α is generated by $\langle C_1 \rangle, \ldots, \langle C_{r(X)} \rangle$ modulo $\pi_1(Y, y)$.

We give two criteria for two subcovers of α to be equivalent. The first one is in terms of conicidence of cycles with bounded length.

Theorem 6.5. *For a subcover β of α and an integer $\ell > 0$ let $C_\ell(\beta)$ be the collection of all cycles of X starting at x with lengths $\leq \ell$ which lift to cycles in*

$\beta(Y)$ *starting at* $\beta(y)$ *via* α/β. *Then two subcovers* β *and* β' *of* α *are equivalent if and only if* $C_{2nd-1}(\beta) = C_{2nd-1}(\beta')$, *where* $d = \max(\deg(\alpha/\beta), \deg(\alpha/\beta'))$.

Proof. ($\Rightarrow$) The equivalence of β and β' implies that α/β and α/β' are isomorphic. Hence $C_\ell(\beta) = C_\ell(\beta')$ for all $\ell > 0$.

($\Leftarrow$) By Theorem 6.3, it suffices to show that $G_\beta = G_{\beta'}$. Since G_α, G_β, and $G_{\beta'}$ are respectively $\pi_1(X, x)$, $\pi_1(\beta(Y), \beta(y))$ and $\pi_1(\beta'(Y), \beta'(y))$ modulo $\pi_1(Y, y)$, this amounts to proving that $\pi_1(\beta(Y), \beta(y))$, and $\pi_1(\beta'(Y), \beta'(y))$ are identified with the same subgroup of $\pi_1(X, x)$. Let $d = \max(\deg(\alpha/\beta), \deg(\alpha/\beta'))$. Then both $\beta(Y)$ and $\beta'(Y)$ are graphs with at most nd vertices. As observed before, all canonical generators $\langle D \rangle \in \pi_1(\beta(Y), \beta(y))$ with respect to a spanning tree $\mathcal{T}_{\beta(Y)}$ of $\beta(Y)$ are represented by cycles D satisfying $\ell(D) \le 2nd - 1$. The cycle D projects to a cycle C in X via α/β. Since $\ell(C) = \ell(D) \le 2nd - 1$, it follows that $C \in C_{2nd-1}(\beta)$. The assumption $C_{2nd-1}(\beta) = C_{2nd-1}(\beta')$ implies $C \in C_{2nd-1}(\beta')$. Hence C lifts to a cycle D' in $\beta'(Y)$ starting at $\beta'(y)$ via α/β'. This shows that when identifying $\pi_1(\beta(Y), \beta(y))$ and $\pi_1(\beta'(Y), \beta'(y))$ as subgroups of $\pi_1(X, x)$, both $\langle D \rangle$ and $\langle D' \rangle$ are identified with $\langle C \rangle$. Since the $\langle D \rangle$'s generate $\pi_1(\beta(Y), \beta(y))$, we conclude that $\pi_1(\beta(Y), \beta(y))$ is contained in $\pi_1(\beta'(Y), \beta'(y))$ as subgroups of $\pi_1(X, x)$. By symmetry, the reverse inclusion also holds. This proves the theorem. $\square$

The second criterion is for normal subcovers, in terms of agreement of completely splitting primes with bounded length. This criterion is parallel to, but stronger than, the result in number theory.

Theorem 6.6. *Assume that* $r(X) \ge 2$. *For a normal subcover* β *of* α *and an integer* $\ell > 0$ *denote by* $P_\ell(\beta)$ *the set of all primes* $\mathfrak{p}$ *of* X *with* $\ell(\mathfrak{p}) \le \ell$ *that split completely in* $\beta(Y)$ *via* α/β. *Then two normal subcovers* β *and* β' *of* α *are equivalent if and only if* $P_{(4n-1)d-1}(\beta) = P_{(4n-1)d-1}(\beta')$, *where* $d = l.c.m.(\deg(\alpha/\beta), \deg(\alpha/\beta'))$.

Proof. ($\Rightarrow$) It is obvious.

($\Leftarrow$) We follow a similar line of ideas and use the same notation as in the proof of Theorem 6.5. By symmetry, it suffices to prove that, as subgroups of $\pi_1(X, x)$, $\pi_1(\beta(Y), \beta(y))$ is contained in $\pi_1(\beta'(Y), \beta'(y))$. With $d = l.c.m.(\deg(\alpha/\beta), \deg(\alpha/\beta'))$ we note that both $\beta(Y)$ and $\beta'(Y)$ have at most nd vertices. Let $\langle D \rangle$ be a canonical generator of $\pi_1(\beta(Y), \beta(y))$ with respect to the tree $\mathcal{T}_{\beta(Y)}$ represented by a cycle D of length $\ell(D) \le 2nd - 1$ starting at $\beta(y)$. Let C be the image of D in X.

To proceed, we claim that there exists a canonical generator $\langle C' \rangle$ of $\pi_1(X, x)$ with respect to $\mathcal{T}_X$ of $\ell(C') \le 2n - 1$ such that $\langle C' \rangle^d \langle C \rangle$ is not a positive power of an element of $\pi_1(X, x)$ other than itself. To see this, first express $\langle C \rangle$ as a reduced word in the canonical generators of $\pi_1(X, x)$ with respect to the tree $\mathcal{T}_X$. Since $r(X) \geq 2$ by assumption, choose $\langle C' \rangle$ to be a canonical generator of $\pi_1(X, x)$ different from the leading generator in $\langle C \rangle$. Suppose $\langle C' \rangle^d \langle C \rangle = \langle Q \rangle^k$ for some $\langle Q \rangle \in \pi_1(X, x)$ and some integer $k \geq 2$. Then the reduced word expression for $\langle Q \rangle$ in the same canonical generators of $\pi_1(X, x)$ starts with $\langle C' \rangle^d$, and $\langle C' \rangle^d$ appears in the reduced word expression of $\langle C \rangle$. Since $\langle C \rangle$ lifts to the canonical generator $\langle D \rangle$ in $\pi_1(\beta(Y), \beta(y))$ and $\langle C' \rangle^d$ also lifts to an element in $\pi_1(\beta(Y), \beta(y))$, which is a product in canonical generators of $\pi_1(\beta(Y), \beta(y))$, we conclude that the two liftings agree. Therefore $\langle C \rangle = \langle C' \rangle^d$, which contradicts the choice of $\langle C' \rangle$. This proves the claim.

Denote by R the backtrackless and tailless cycle obtained from $(C')^d \cdot C$ by removing the backtracks and the tail T from x to R that meets R at x'. Then R is a primitive geodesic cycle of length $\ell(R) \leq \ell((C')^d \cdot C) \leq (4n - 1)d - 1$.

Since β is a normal subcover of α, by Theorem 6.3 the cover $\alpha/\beta : \beta(Y) \to X$ is normal and its Galois group $G_{\alpha/\beta}$ has order $\deg(\alpha/\beta)$ dividing d. Hence $\langle C' \rangle^d$ is identified with an element $\langle S \rangle$ in $\pi_1(\beta(Y), \beta(y))$. Together with $\langle C \rangle$ identified with $\langle D \rangle$ in $\pi_1(\beta(Y), \beta(y))$, we conclude that $\langle C' \rangle^d \langle C \rangle$ is identified with $\langle S \rangle \langle D \rangle$ in $\pi_1(\beta(Y), \beta(y))$. This shows that one and hence all liftings of the cycle R in $\beta(Y)$ via α/β are again primitive geodesic cycles since α/β is normal. In other words, the prime $[R]$ splits completely in $\beta(Y)$. As the length of $[R]$ is $\leq (4n - 1)d - 1$, the prime $[R]$ is in $P_{(4n-1)d-1}(\beta)$.

Since $P_{(4n-1)d-1}(\beta) = P_{(4n-1)d-1}(\beta')$ by assumption, we also have $[R] \in P_{(4n-1)d-1}(\beta')$. Lift the path T to a path T' in $\beta'(Y)$ starting at $\beta'(y)$ via α/β'. It ends at a vertex z' that projects to x' in X. Lift R to a path R' in $\beta'(Y)$ starting at z'. Since R splits completely in $\beta'(Y)$, the path R' is a primitive geodesic cycle in $\beta'(Y)$. Let D' be the cycle in $\beta'(Y)$ that is T' followed by R' followed by T' in reverse direction. Then $\langle D' \rangle \in \pi_1(\beta'(Y), \beta'(y))$ is identified with $\langle C' \rangle^d \langle C \rangle$ in $\pi_1(X, x)$, as is $\langle S \rangle \langle D \rangle$ in $\pi_1(\beta(Y), \beta(y))$. This shows that, viewed in $\pi_1(X, x)$, $\langle S \rangle \langle D \rangle$ lies in $\pi_1(\beta'(Y), \beta'(y))$.

On the other hand, since β' is a normal subcover of α and the Galois croup $G_{\alpha/\beta'}$ has order dividing d, $\langle C' \rangle^d$ is identified with an element $\langle S' \rangle$ in $\pi_1(\beta'(Y), \beta'(y))$. Thus $\langle S \rangle \in \pi_1(\beta(Y), \beta(y))$ and $\langle S' \rangle \in \pi_1(\beta'(Y), \beta'(y))$ are identified with the same element in $\pi_1(X, x)$. Consequently $\langle D \rangle$ lies in $\pi_1(\beta'(Y), \beta'(y))$. This proves that, as subgroups of $\pi_1(X, x)$, $\pi_1(\beta(Y), \beta(y))$ is contained in $\pi_1(\beta'(Y), \beta'(y))$, as desired. The theorem follows. $\square$

Two subgroups H, H' of a finite group G are said to be *locally conjugate* in G if each conjugacy class of G meets H and H' in the same number of elements. The concept of local conjugacy first appeared in number theory in the context that $G = Gal(K/F)$ is the Galois group of a finite Galois extension K of a number field F; two subgroups H and H' of G are locally conjugate if and only if their fixed fields have the same Dedekind zeta function. This criterion was extended by Sunada in [15], where G is the Galois group of a finite normal cover $M \to G\backslash M$ of compact Riemannian manifolds. Sunada showed that two subgroups H and H' of G are locally conjugate if and only if the two manifolds $H\backslash M$ and $H'\backslash M$ are isospectral. This result was then used to construct many pairs of isospectral, but nonisometric, compact Riemannian manifolds, giving explicit counterexamples to the question "Can you hear the shape of a drum?" As discussed in [16], Sunada expected a similar statement for graph covers, which was carried out by Somodi in [11].

Two subcovers β and β' of a finite normal cover α are said to be *Sunada equivalent* if G_β and $G_{\beta'}$ are locally conjugate in G_α. Sunada equivalence is weaker than equivalence.

Theorem 6.7 (Theorem 5.1, [11]). *Two subcovers β and β' of α are Sunada equivalent if and only if for every prime $\mathfrak{p}$ of X there is a length-preserving bijection between the primes of $\beta(Y)$ above $\mathfrak{p}$ and those of $\beta'(Y)$.*

Consequently if β and β' are Sunada equivalent, then $\beta(Y)$ and $\beta'(Y)$ have the same Ihara zeta functions. Moreover, $\beta(Y)$ and $\beta'(Y)$ have the same number of vertices, edges, and spanning trees, respectively, and the same Laplacian spectrum by [3, Theorem 1]. Hence $\beta(Y)$ and $\beta'(Y)$ are isospectral. Somodi [11] also proved a slightly stronger criterion for Sunada equivalence:

Theorem 6.8 (Corollary 5.2, [11]). *Two subcovers β and β' of α are Sunada equivalent if and only if for every prime $\mathfrak{p}$ of X the number of primes in $\beta(Y)$ above $\mathfrak{p}$ with length equal to $\ell(\mathfrak{p})$ agrees with that in $\beta'(Y)$.*

We end this chapter by further improving the foregoing criterion with bounded length.

Theorem 6.9. *Assume $r(X) \geq 2$. Two subcovers β, β' of α are Sunada equivalent if and only if for each prime $\mathfrak{p}$ of X with length $\ell(\mathfrak{p}) \leq 2n \deg \alpha$, the number of primes in $\beta(Y)$ above $\mathfrak{p}$ with length equal to $\ell(\mathfrak{p})$ agrees with that in $\beta'(Y)$.*

Proof. It was shown in [15] that the Galois subgroups G_β and $G_{\beta'}$ of G_α are locally conjugate if and only if the trivial representations on G_β and $G_{\beta'}$ induce isomorphic representations ρ_β and $\rho_{\beta'}$ of G_α. In other words, β and β' are Sunada equivalent if and only if tr $\rho_\beta =$ tr $\rho_{\beta'}$. Since trace is a class function, the value depends only on the conjugacy classes of G_α.

As observed before, G_α is $\pi_1(X, x)$ modulo the subgroup $\pi_1(Y, y)$. Write d for $\deg \alpha$. Recall that the conjugacy classes of nonidentity elements in $\pi_1(X, x)$

correspond to the equivalence classes of geodesic cycles in X. Furthermore, by the Chebotarëv density theorem (in Dirichlet density), each conjugacy class of G_α is Frob($\mathfrak{p}$) for infinitely many primes $\mathfrak{p}$ of X. We claim that for each $\sigma \in G_\alpha$, the conjugacy class of σ is equal to the Frobenius conjugacy class at a prime $\mathfrak{p}_\sigma$ of X with length at most $2nd$.

To prove the claim, first choose a spanning tree $\mathcal{T}_Y$ of Y. Elements in G_α are uniquely determined by their images at y, which belong to the fiber $\alpha^{-1}(x)$. Let $y' \in \alpha^{-1}(x)$ not be equal to y. In the tree $\mathcal{T}_Y$ there is a unique backtrackless path from y to y' of length $\le nd$, which is sent (via α) to a backtrackless cycle $C_{y'}$ in X starting and ending at x. Its homotopy class $\langle C_{y'} \rangle \in \pi_1(X, x)$ modulo $\pi_1(Y, y)$ is the element $\sigma \in G_\alpha$ satisfying $\sigma(y) = y'$. After removing the tail in the cycle $C_{y'}$, we obtain a geodesic cycle C_σ whose equivalence class $[C_\sigma]$ corresponds to the conjugacy class of σ in G_α, and it has length at most nd. The trivial element in G_α is represented by the equivalence class of any geodesic cycle in Y. There is a prime $\mathfrak{P}$ of Y with length at most nd, for instance, a prime using exactly one edge of Y outside the tree $\mathcal{T}_Y$. We may take $[C_{id}]$ to be $\alpha(\mathfrak{P})$. Now for $\sigma \in G_\alpha$, if C_σ is primitive, choose $\mathfrak{p}_\sigma$ to be the class $[C_\sigma]$. In case that C_σ is not primitive, there is a geodesic cycle D_σ of length $\le n$ such that $(D_\sigma)^d C_\sigma$ is a primitive geodesic cycle of length $\le 2nd$, as shown in the proof of Theorem 6.6. Since G_α has order d, $(D_\sigma)^d$ lifts to a geodesic cycle in Y so that, changing the base point if necessary, we may assume that $\langle (D_\sigma)^d \rangle$ lies in $\pi_1(Y, y)$. Thus we may choose $\mathfrak{p}_\sigma$ to be the class $[(D_\sigma)^d C_\sigma]$, which has length $\le 2nd$, as desired.

The representation ρ_β is the regular representation on the coset space $G_\beta \backslash G_\alpha$ via right translation. Given $\sigma \in G_\alpha$, $\operatorname{tr} \rho_\beta(\sigma) = \operatorname{tr} \rho_\beta(\operatorname{Frob}(\mathfrak{p}_\sigma))$ counts the number of G_β-cosets fixed by σ, which in turn is equal to the number of liftings of $\mathfrak{p}_\sigma$ in $\beta(Y)$ that are closed. In other words, this is the number of primes of $\beta(Y)$ above $\mathfrak{p}_\sigma$ with length equal to $\ell(\mathfrak{p}_\sigma)$. Hence the theorem follows. $\square$

Acknowledgments

The research of Hau-Wen Huang is partially supported by the Ministry of Science and Technology (MOST) grant 105-2115-M-008-013. Part of the research was done when he was supported by the National Center for Theoretical Sciences of Taiwan and the Council for Higher Education of Israel.

The research of Wen-Ching Winnie Li is partially supported by the National Science Foundation grant DMS-1101368 and by a grant from the Simons Foundation #355798. Part of the research was done when she was visiting the National Tsing Hua University and the National Center for Theoretical Sciences in Hsinchu, Taiwan. She would like to thank the Mathematics Department of NTHU and NCTS for their hospitality.

References

1. T. Adachi and T. Sunada. Twisted Perron-Frobenius theorem and L-functions. *J. Funct. Anal.* **71** (1987), 1–46.
2. H. Bass. The Ihara-Selberg zeta function of a tree lattice. *Int. J. Math.* **3** (1992), 717–797.
3. L. Halbeisen and N. Hungerbuhler. Generation of isospectral graphs. *J. Graph Theory* **31** (1999), 255–265.
4. K. Hashimoto. Zeta functions of finite graphs and representations of p-adic groups. *Adv. Stud. Pure Math.* **15** (1989), 211–280.
5. K. Hashimoto. On zeta and L-functions of finite graphs. *Int. J. Math.* **1** (1990), 381–396.
6. K. Hashimoto. Artin type L-functions and the density theorem for prime cycles on finite graphs. *Int. J. Math.* **3** (1992), 809–826.
7. H-W. Huang and W-C. W. Li. A unified approach to the Galois closure problem. *J. Number Theory.* **180** (2017), 251–279.
8. Y. Ihara. On discrete subgroups of the two by two projective linear group over p-adic fields. *J. Math. Soc. Japan* **18** (1966), 219–235.
9. J.C. Lagarias and A.M. Odlyzko. Effective versions of the Chebotarev density theorem in *Algebraic Number Fields*, ed. A Fröhlich, Academic Press, London, 1977, pp. 409–464.
10. J-P. Serre. *Trees.* Translated from the French by John Stillwell, Springer-Verlag, Berlin, 1986.
11. M. Somodi. On Sunada equivalent graph coverings. *J. Comb. Number Theory* **7**, no. 2, (2015) 79–94.
12. H. Stark and A. Terras. Zeta functions of finite graphs and coverings. *Adv. Math.* **121** (1996), 124–165.
13. H. Stark and A. Terras. Zeta functions of finite graphs and coverings Part II. *Adv. Math.* **154** (2000), 132–195.
14. H. Stark and A. Terras. Zeta functions of finite graphs and coverings III. *Adv. Math.* **208** (2007), 467–489.
15. T. Sunada. Riemannian coverings and isospectral manifolds. *Ann. Math.* **121** (1985), 169–186.
16. T. Sunada. L-functions in geometry and some applications. Curvature and topology of Riemannian manifolds (Katata, 1985), pp. 266–284. Lecture Notes in Mathematics, Vol. 1201. Springer-Verlag, Berlin, 1986.
17. A. Terras. *Zeta Functions of Graphs: A Stroll through the Garden.* Cambridge Studies in Advanced Mathematics, Vol. 128. Cambridge University Press, Cambridge (2011).

7

The First Function and Its Iterates

Carl Pomerance

Abstract

Let $s(n)$ denote the sum of the positive divisors of n except for n itself. Discussed since Pythagoras, s may be the first function of mathematics. Pythagoras also suggested iterating s, so perhaps considering the first dynamical system. The historical legacy has left us with some colorful and attractive problems, mostly still unsolved. Yet the efforts have been productive in the development of elementary, computational, and probabilistic number theory. In the context of the Catalan–Dickson conjecture and the Guy–Selfridge counter-conjecture, we discuss the geometric mean of the numbers $s(s(2n))/s(2n)$, thus extending recent work of Bosma and Kane. We also discuss the number of integers m with $s(m) = n$.

7.1 Introduction

Let $\sigma(n)$ denote the sum of the natural divisors of the positive integer n. Let $s(n)$ be the sum of only the proper divisors of n, so that $s(n) = \sigma(n) - n$. "Perfect" and "amicable" numbers are attributed to Pythagoras. A perfect number is one, like 6, where $s(n) = n$, and an amicable number, like 220, is not perfect, but satisfies $s(s(n)) = n$. (The name "amicable" stems from the pair of numbers n and $s(n) = m$, where s of one is the other.) Euclid found the formula $2^{p-1}(2^p - 1)$ that gives perfect numbers whenever the second factor is prime, and Euler proved that all even perfect numbers are given by this formula. No odd perfect numbers are known. The search for even perfect numbers spurred theoretical developments, such as the Lucas–Lehmer primality test, a forerunner of all of modern primality testing (see [24]).

The ancient Greeks also distinguished two types of nonperfect numbers, the "deficient" ones, where $s(n) < n$, and the "abundant" ones, where $s(n) > n$.

125

This concept spurred the development of probabilistic number theory, when Davenport showed that the sets of deficient and abundant numbers each have an asymptotic density, and more generally showed that $s(n)/n$ has a distribution function.

Prominent among the many unsolved problems about s is the century-old Catalan–Dickson conjecture [4, 5]. This asserts that starting from any positive integer n and iterating s, one arrives eventually at 1, then stopping at 0, or one enters a cycle, such as a 1-cycle (perfect numbers), a 2-cycle (amicable pairs), or a higher order cycle. That is, the conjecture asserts that every orbit is bounded. This conjecture has helped to spur on modern factorization algorithms (since one needs the prime factorization of n to compute $s(n)$). The first number in doubt is 276, where thousands of iterates have been computed, reaching beyond 200 decimal digits. Although we know of no unbounded sequences of this type, Guy and Selfridge [11] came up with a "counter" conjecture, namely that for almost all[1] even seeds, the sequence is unbounded, while for almost all odd seeds it is bounded.

We extend the definition of s to include $s(0) = 0$, and for $n \geq 0$, we let $s_k(n)$ denote the kth iterate of s at n. The sequence $n, s(n), s_2(n), \ldots$ is known as the "aliquot" sequence with seed n. Numbers in a cycle under the s-iteration are called "sociable." It's known (see [14]) that the set of even sociable numbers has asymptotic density 0 and the set of odd sociable numbers has upper asymptotic density at most the density of the set of odd abundant numbers (which is ≈ 0.002). It is conjectured that the set of odd sociable numbers has density 0. It is not known if there are infinitely many sociable numbers, though it is conjectured that this is the case.

The evidence either for the Catalan–Dickson conjecture or the Guy–Selfridge counter-conjecture is mixed. Perhaps pointing toward Guy–Selfridge is the theorem of Lenstra [15] that there exist arbitrarily long strictly increasing aliquot sequences, and the strengthening of Erdős [8] that for each k and almost all abundant numbers n, the aliquot sequence with seed n strictly increases for k steps. (The asymptotic density of the abundant numbers is known to be ≈ 0.2476 [13].) These theoretical results are borne out in practice, where it is found not uncommonly that certain divisors (such as 24), known as "drivers," persist for long stretches in the sequence, and when these divisors are abundant, the sequence grows geometrically at least as long as this persistence.

However, the persistence of an abundant driver is not absolute, and its dominance can be broken, sometimes by a "down driver," such as 2 (where the number is not divisible by 4 or 3), which can also persist and tends to drive

---

[1] Whenever we say "almost all" we mean except for a set of asymptotic density 0.

the sequence lower geometrically. It is rare for a sequence to switch parity; this occurs if and only if one hits a square or its double. Since the even numbers are where the principal disagreement in the two conjectures lies, Bosma and Kane [3] considered the geometric mean, on average, for $s(2n)/2n$. Namely, they showed that there is a real number β such that

$$\frac{1}{x} \sum_{n \leq x} \log(s(2n)/2n) \to \beta \quad \text{as } x \to \infty,$$

and that $\beta \approx -0.03$ is *negative*. This may be interpreted as evidence in favor of Catalan–Dickson and against Guy–Selfridge.

However, very little is known about the set of numbers $s(n)$, with even less known about $s_k(n)$ when $k \geq 2$, so statistical results about seeds n already seem less relevant when one proceeds a single step into the aliquot sequence. We do know that almost all odd numbers are of the form $s(n)$, a result whose proof depends on approximations to Goldbach's conjecture that even numbers are the sum of two primes. In fact, almost all odd numbers are in the image of every s_k, see [9, Theorem 5.3]. More mysterious are the even numbers in the range of s. Erdős [7] showed that a positive proportion of even numbers are not of the form $s(n)$, while it was only very recently shown in [17] that a positive proportion of even numbers are of the form $s(n)$. In [22] a heuristic argument is given with some numerical evidence that the asymptotic density of the even numbers of the form $s(n)$ is about $1/3$; also see [26].

Our principal result is an extension of the Bosma–Kane theorem to the next iterate.

Theorem 7.1. *The average value of* $\log(s_2(2n)/s(2n))$ *is asymptotically equal to the average value of* $\log(s(2n)/2n)$. *That is,*

$$\frac{1}{x} \sum_{2 \leq n \leq x} \log(s_2(2n)/s(2n)) \sim \frac{1}{x} \sum_{1 \leq n \leq x} \log(s(2n)/2n) \sim \beta, \quad \text{as } x \to \infty.$$

(The reason the first sum excludes $n = 1$ is that $s_2(2) = 0$.) The proof of Theorem 7.1 uses some ideas from [17] and [18].

We also consider the sets $s^{-1}(n)$, obtaining what is likely to be an asymptotic formula for its size when $n > 1$ is odd. The set $s^{-1}(1)$ is the set of primes, but for $n > 1$, $s^{-1}(n)$ is finite (in fact every preimage of n under s is smaller than n^2).

We mention some other recent work on s. In Bosma [2] the aliquot sequence is computed for each seed to 10^6 until it terminates, cycles, or surpasses 10^{99}. He found that about one-third of the even seeds are in this last category, perhaps lending some support to Guy–Selfridge. In Troupe [27] it is shown that the normal number of prime factors of $s(n)$ is $\log \log n$, thus extending the theorem of Hardy and Ramanujan about the normal number of prime factors of n. In [25] it is shown that for all large x the number of amicable numbers $n \leq x$ is smaller than $x/ \exp(\sqrt{\log x})$.

By way of notation, we have $\omega(n)$ as the number of distinct prime divisors of the natural number n, $\Omega(n)$ as the total number of prime factors of n with multiplicity, and $\tau(n)$ as the number of positive divisors of n. We let $P^+(n)$ denote the largest prime factor of $n > 1$, with $P^+(1) = 1$. We let $\text{rad}(n)$ denote the largest squarefree divisor of n. We write $a \parallel b$ if $a \mid b$ and $(a, b/a) = 1$. We reserve the letters p, q for primes. We let $\log_k x$ denote the k-fold iteration of the natural logarithm at x, and when we use this notation we assume that the argument is large enough for the expression to be defined. We write $f(x) \ll g(x)$ if $f(x) = O(g(x))$, and we write $f(x) \asymp g(x)$ if $f(x) \ll g(x) \ll f(x)$.

7.2 The Double Iterate

In this section we prove Theorem 7.1.

Proof. It follows from [8] and [9] (also see [21] and [18]) that $s_2(n)/s(n) \sim s(n)/n$ as $n \to \infty$ on a set of asymptotic density 1. Thus,

$$\log(s_2(n)/s(n)) = \log(s(n)/n) + o(1)$$

on this same set. However, some terms here are unbounded, both on the positive side and the negative side, and it is conceivable that a set of asymptotic density 0 could be of consequence when averaging. This event commonly occurs. For example, when computing the average of $\tau(n)$, the number of divisors of n, one finds the normal size of $\tau(n)$ is considerably smaller than the average size.

By the theorem of Bosma and Kane,

$$\sum_{n \leq x} \log(s(2n)/2n) \sim \beta x, \quad \text{as } x \to \infty,$$

where $\beta \approx -0.03$. Thus, we need to show that for each fixed $\epsilon > 0$, there is some number B such that the contribution to the two sums in our theorem from terms that have absolute value $> B$ has absolute value at most ϵx.

Note that no term in the Bosma–Kane sum is smaller than $-\log 2$. However, the terms $\log(s_2(2n)/s(2n))$ can be smaller than this; for example, when $n = 2$ we have $-\log 4$. However, using $s(n) \ll n \log_2 n$, we have

$$\log(s_2(2n)/s(2n)) \geq \log(1/s(2n)) \geq -\log n - \log_3 n + O(1).$$

Now if $s_2(2n)/s(2n) < 1/2$, we have $s(2n)$ odd, which implies that n or $2n$ is a square. Thus,

$$\frac{1}{x} \sum_{\substack{1 < n \leq x \\ s_2(2n)/s(2n) < 1/2}} \log(s_2(2n)/s(2n)) \geq -2 \frac{\log x + \log_3 x + O(1)}{x^{1/2}}.$$

So it remains to show that large positive values are of small consequence. A useful result is the following.

Theorem 7.2. *Uniformly for every positive number x, the number of positive integers $n \leq x$ with $s(n)/n > y$ is at most*

$$x / \exp(\exp((e^{-\gamma} + o(1))y)), \quad as \ y \to \infty.$$

This result is essentially due to Erdős, see [14, Theorem B].

We can use Theorem 7.2 as follows. Let y_0 be so large that the count in Theorem 7.2 is smaller than $x/\exp(\exp(y/2))$ for all $y \geq y_0$. Let $B \geq y_0$ be a large fixed number. The contribution to the Bosma–Kane sum from those $n \leq x$ with $B^{2^{j-1}} < s(2n)/2n \leq B^{2^j}$ is at most

$$\frac{2x \log(B^{2^j})}{\exp(\exp(B^{2^{j-1}}/2))}.$$

Summing this for $j \geq 1$ we get a quantity that is $\ll x \log B / \exp(\exp(B/2)))$. Since B may be fixed as arbitrarily large, the numbers $n \leq x$ with $s(2n)/2n > B$ give a contribution of vanishing importance.

We need to show the same result for $s_2(2n)/s(2n)$, and this is the heart of the argument.

Let $y = y(x) = (\log_2 x)/(\log_3 x)^2$.

Proposition 7.1. *But for $O(x/y^{4/3})$ integers $n \leq x$ we have*

$$\left| \frac{s_2(n)}{s(n)} - \frac{s(n)}{n} \right| \ll \frac{\log_4 x}{\log_3 x} \cdot \frac{\sigma(n)}{n}.$$

Proof. We first show that we may assume that for each integer $m \leq y$ there are five distinct primes $p_1, \ldots, p_5$ with each $p_i \| n$ and $p_i \equiv -1 \pmod{m}$. To see this, let P_m denote the set of primes $p \in (y^2, x^{1/5})$ with $p \equiv -1 \pmod{m}$. Using [23] we have

$$S_m := \sum_{p \in P_m} \frac{1}{p} = \frac{\log(\log(x^{1/5})/\log y)}{\varphi(m)} + O(1),$$

uniformly for each $m \leq y$. Thus, for x sufficiently large, we have each

$$S_m \geq \frac{\log_2 x}{2m}.$$

The number of $n \leq x$ divisible by some p^2 with $p > y^2$ is $O(x/y^2)$, so we may assume that the numbers n we are considering are never divisible by p^2 for p in some P_m. The number of $n \leq x$ not divisible by five different members of P_m is, via the sieve,

$$\ll \left(1 + S_m + \frac{1}{2}S_m^2 + \frac{1}{6}S_m^3 + \frac{1}{24}S_m^4 \right) \frac{x}{\exp(S_m)} \ll \frac{x(\log_2 x)^4}{\exp\left(\frac{\log_2 x}{2m} \right)}$$

$$\leq \frac{x(\log_2 x)^4}{\exp\left(\frac{1}{2}(\log_3 x)^2 \right)}.$$

Summing for $m \leq y$, we get an estimate that is $\ll x(\log_2 x)^5 / \exp(\frac{1}{2}(\log_3 x)^2)$. Thus, but for a negligible set of $n \leq x$, $m^5 \mid \sigma(n)$ for every $m \leq y$.

For $n \leq x$, let $m = m(n)$ be the largest divisor of n supported on the primes $\leq y$. If $p^a \mid m(n)$ with $p^a > y^2$, then $a \geq 3$. The number of such n is at most

$$\sum_{\substack{p \leq y \\ p^a > y^2}} \frac{x}{p^a} \ll \frac{x}{y^{4/3}},$$

so we may assume that each $p^a \mid m(n)$ has $p^a \leq y^2$. We claim that if n has not so far been excluded, then $m(n) = m(s(n))$. If $p \leq y$, then $p^5 \mid \sigma(n)$, so that $p \mid n$ if and only if $p \mid s(n)$. Suppose $p^a \| n$ with $p^a \leq y$. Then $p^{5a} \mid \sigma(n)$, so $p^a \| s(n)$. It remains to consider the case of $p^a \| m(n)$ with $y < p^a \leq y^2$. Let b be the largest integer with $p^b \leq y$. Then $p^b \geq y^{1/2}$, so that $5b > a$. But, as we have seen, $p^{5b} \mid \sigma(n)$. Thus $p^a \| s(n)$. We conclude that $m(n) = m(s(n))$.

To complete the proof of the proposition, we must show that primes $p > y$ do not overly influence the values $s(n)/n$ and $s_2(n)/s(n)$. The number of integers $n \leq x$ with $\omega(n) > 3\log_2 x$ is $\ll x/\log x$ by a well-known result of Hardy and Ramanujan. So, we may assume that $\omega(n) \leq 3\log_2 x$. The sum of reciprocals of the first $\lfloor 3\log_2 x \rfloor$ primes larger than y is $\ll \log_4 x / \log_3 x$. Thus,

$$\frac{s(n/m(n))}{n/m(n)} \ll \frac{\log_4 x}{\log_3 x},$$

and so

$$\left| \frac{s(n)}{n} - \frac{s(m(n))}{m(n)} \right| = \frac{s(n/m(n))}{n/m(n)} \cdot \frac{\sigma(m(n))}{m(n)} \ll \frac{\log_4 x}{\log_3 x} \cdot \frac{\sigma(m(n))}{m(n)}$$

$$\leq \frac{\log_4 x}{\log_3 x} \cdot \frac{\sigma(n)}{n}.$$

It thus suffices to prove a similar result for $s_2(n)/s(n)$. By a result of [6], the number of $n \leq x$ with $P^+(n) \leq x^{1/\log_3 x}$ is at most $x/(\log_2 x)^{(1+o(1))\log_4 x}$, as

$x \to \infty$. Thus, we may assume that $P^+(n) > x^{1/\log_3 x}$. Suppose that $\omega(s(n)) > 7\log_2 x \log_3 x$. Write $n = pk$ where $p = P^+(n)$. By a foregoing result we may assume that $p \nmid k$. Thus,

$$s(n) = ps(k) + \sigma(k). \tag{7.1}$$

Since $\omega(s(n)) > 7\log_2 x \log_3 x$, there is a divisor u of $s(n)$ with $u < x^{1/\log_3 x}$ and $\omega(u) > 7\log_2 x$. Let u_1 be the largest divisor of u that is coprime to n. From (7.1) we have u_1 coprime to $s(k)$. Since we may assume that $\omega(n) \le 3\log_2 x$, we have $\omega(u_1) > 4\log_2 x$. Reading (7.1) as a congruence mod u_1, we see that for a given choice of k and u_1, p is determined mod u_1. Since $k < x^{1-1/\log_3 x}$ and $u_1 < x^{1/\log_3 x}$, it follows that the number of choices for p, and thus for n is

$$\le \sum_{k < x^{1-1/\log_3 x}} \sum_{\substack{u_1 < x^{1/\log_3 x} \\ \omega(u_1) > 4\log_2 x}} \frac{x}{ku_1} \ll \frac{x}{\log x},$$

again using the result of Hardy and Ramanujan mentioned earlier. Thus, we may assume that $\omega(s(n)) \le 7\log_2 x \log_3 x$. Since the reciprocal sum of the first $7\log_2 x \log_3 x$ primes $> y$ is $\ll \log_4 x / \log_3 x$ and $m(s(n)) = m(n)$, we have

$$\left| \frac{s_2(n)}{s(n)} - \frac{s(m(n))}{m(n)} \right| \ll \frac{\log_4 x}{\log_3 x} \cdot \frac{\sigma(m(n))}{m(n)} \le \frac{\log_4 x}{\log_3 x} \cdot \frac{\sigma(n)}{n}.$$

This completes the proof of the proposition. $\qquad\square$

We now complete the proof of the theorem. We have seen that the negative terms in the sums with large absolute values are negligible, and that large positive values of $\log(s(n)/n)$ are also negligible. Since $s(n)/n \ll \log_2 n$ for $n \ge 3$, it follows that for $n \le x$ with $s_2(2n)/s(2n) \ge 1$, we have $\log(s_2(2n)/s(2n)) \ll \log_3 x$. Thus, the contribution to the sum from those terms not satisfying the inequality in Proposition 7.1 is $\ll x(\log_3 x)/y^{4/3} = o(x)$. Since $\sum_{n \le x} \sigma(n)/n \ll x$, it follows that the difference of the two sums for those $n \le x$ that satisfy the inequality of Proposition 7.1 is $\ll x \log_4 x / \log_3 x = o(x)$. This completes the proof of the theorem. $\qquad\square$

Corollary 7.1. *We have*

$$\sum_{n \le x} \frac{s_2(n)}{s(n)} \sim \sum_{n \le x} \frac{s(n)}{n} \quad \text{as } x \to \infty$$

and

$$\sum_{n \le x} \frac{s_2(n)}{n} \sim \sum_{n \le x} \left(\frac{s(n)}{n} \right)^2 \quad \text{as } x \to \infty.$$

These results follow from the tools used to prove Theorem 7.1. Note that, where ζ is the Riemann zeta function,

$$\sum_{n\le x}\frac{s(n)}{n} \sim (\zeta(2)-1)x, \quad \sum_{n\le x}\left(\frac{s(n)}{n}\right)^2 \sim \left(\frac{\zeta(2)^2\zeta(3)}{\zeta(4)}-2\zeta(2)+1\right)x$$

as $x \to \infty$.

In [9] the following conjecture is proposed.

Conjecture 7.1. *If A is a set of natural numbers of asymptotic density 0, then $s^{-1}(A)$ has asymptotic density 0.*

Theorem 7.3. *Assuming Conjecture 7.1, then for each integer $k \ge 2$ there is a set A_k of asymptotic density 1 such that*

$$\frac{1}{x}\sum_{\substack{n\le x \\ n\in A_k}}\log(s_k(n)/s_{k-1}(n)) \to \beta, \quad as\ x \to \infty.$$

Proof. We first note that a consequence of the conjecture is that if A has density 0, then $s_k^{-1}(A) = \{n : s_k(n) \in A\}$ has density 0. This is clear for $k = 1$. For $k \ge 2$, by the case $k = 1$ and induction, $s_k^{-1}(A) = s_{k-1}^{-1}(s^{-1}(A))$ has density 0.

Let A be the set of integers n such that for each $m \le \log_2 n/(\log_3 n)^2$ there are at least five distinct primes $p \parallel n$ with $p \equiv -1 \pmod{n}$. We have seen in the proof of Proposition 7.1 that A has asymptotic density 1. If we also insist that members n of A satisfy $\omega(n) \le 3\log_2 n$, then A still has asymptotic density 1. Letting B denote the complement of A, the conjecture implies that $s_j^{-1}(B)$ has density 0 for each $j < k$. Let A_k be the part of A lying outside of each of these sets $s_j^{-1}(B)$ so that A_k has density 1. We have seen in the proof of Proposition 7.1 that if $n, s(n) \in A$, then $s_2(n)/s(n) \sim s(n)/n$. However, if $n \in A_k$ then all of $n, s(n), \dots, s_{k-1}(n)$ are in A, so all of the ratios $s_{j+1}(n)/s_j(n)$ are asymptotic to each other. This proves the theorem. $\qquad\square$

In [3] the authors also study the full sum $\sum_{1<n\le x}\log(s(n)/n)$, showing that it is asymptotically $-e^{-\gamma}x\log_2 x$. We can prove this for $\log(s_2(n)/s(n))$, with the proviso that n runs over composite numbers to avoid undefined summands. The sum of $\log(s(n)/n)$ is analyzed by singling out those n with the same smallest prime factor q. The terms when $q < \exp((\log x)^\epsilon)$ account for a vanishingly small portion of the sum when ϵ is small (and so this is another case where the asymptotics are dominated by those terms corresponding to a set of density 0). To prove the result for $\log(s_2(n)/s(n))$, one reduces to the case when $p = P^+(n) > x^{1/\log_2 x}$, writes $n = pm$, and assumes that m has no prime factors up to

$(\log_2 x)^2$. Fixing a prime q in the range $\exp((\log x)^\epsilon) < q < \exp((\log x)^{1-\epsilon})$ and a value for m, one counts primes $p \le x/m$ such that $s(pm)$ has least prime factor q. This can be done asymptotically correctly using the fundamental lemma of the sieve; see [12].

We do not have an analogue of Theorem 7.3 for $\log(s_k(n)/s_{k-1}(n))$ since presumably the sum of these terms is principally supported on a set of n of density 0.

7.3 The Inverse Image

For a positive integer n, let $G(n)$ denote the number of pairs of primes $p > q$ with $n = p + q$.

Lemma 7.1. *Suppose that n, v, D are given positive integers with $n > 1$, $(v, nD) = 1$, and $\mathrm{rad}(D) \mid n$. The number of prime powers p^a coprime to vD with $s(p^a vD) = n$ is $O(\log n)$, while there is at most one number u coprime to nvD with $u < v$ and $s(uvD) = n$.*

Proof. For the first assertion, using $s(p^a vD) = n$ and $(p^a, vD) = 1$, we have

$$p^a s(vD) + s(p^a) \sigma(vD) = n.$$

Consider the polynomial $f(x) = s(vD)x^a + \sigma(vD)(x^{a-1} + x^{a-2} + \cdots + 1)$, so that f is increasing for $x > 0$. This implies that for a, v, D, n fixed, there is at most one prime p with $f(p) = n$. Since $p^a < n^2$, there are at most $O(\log n)$ choices for a, that is, at most $O(\log n)$ polynomials. This proves the first assertion.

Now we consider the case when $u < v$. We have $n = s(uvD) = \sigma(u)\sigma(vD) - uvD$, so that

$$n \equiv -uvD \pmod{\sigma(vD)}.$$

Let $d = (n, \sigma(vD))$, so that $d \mid uvD$. Since n is coprime to uv, we have $d \mid D$. Thus,

$$\frac{n}{d} \equiv -uv\frac{D}{d} \ \left(\mathrm{mod}\ \frac{\sigma(vD)}{d}\right)$$

and this shows that given n, D, d, v, we have u determined modulo $\sigma(vD)/d$. But

$$u < v < \sigma(vD)/d,$$

so u is determined. This proves the second assertion. $\qquad\square$

Lemma 7.2. *Suppose that m is a number with $\omega(m) \geq 2$, with p^a the greatest prime power dividing m, and with q^b the greatest prime power dividing m/p^a. If $p^a < m^{1/2}$, then m has a factorization as uv where u, v are positive coprime integers and $u < v < (mq^b)^{1/2}$.*

Proof. Let $1 \leq u < v$ be coprime with $uv = m$ and v minimal. Assuming $p^a < m^{1/2}$, then $v \neq p^a$. Let r^c be the least prime power with $r^c \| v$, so that $r^c \leq q^b$, and let $u' = ur^c$, $v' = v/r^c$. By the minimality of v, we have $u' > v'$ and $u' \geq v$. But $u'v = uvr^c \leq mq^b$, so that $v = \min\{u', v\} \leq (mq^b)^{1/2}$. $\qquad\square$

Theorem 7.4. *For a fixed integer $n > 1$, the number of integers m with $s(m) = n$ and $(m, n) > 1$ is $O_\epsilon(n^{2/3+\epsilon})$ for each $\epsilon > 0$.*

Proof. If $s(m) = n$, we have $m < n^2$. Let $1 < D < n^2$ run over numbers with $\mathrm{rad}(D) \mid n$. Every m with $s(m) = n$ and $(m, n) > 1$ may be written as $m_0 D$ for some such D, where $(m_0, Dn) = 1$. Since m_0 is a proper divisor of m, we have $m_0 < s(m) = n$. If $m_0 = 1$, this is 1 possibility for m_0. If $m_0 = p^a$, Lemma 7.1 implies there are at most $O(\log n)$ possibilities for m_0. If $m_0 = p^a q^b$ with $p^a > q^b$, then $q^b < n^{1/2}$ and each choice of q^b gives $O(\log n)$ possibilities for m_0. So, there are $O(n^{1/2})$ possibilities for m_0 in this case.

Now assume that $\omega(m_0) = k \geq 3$. If p^a is the largest prime power dividing m_0, we may assume that $p^a < m_0^{1/3}$, since otherwise, m_0/p^a is an integer smaller than $n^{2/3}$ which determines p^a in at most $O(\log n)$ ways. Then Lemma 7.2 implies that there are coprime integers u, v with $m_0 = uv$ and $u < v < n^{2/3}$. Thus, the second part of Lemma 7.1 implies there are at most $n^{2/3}$ possibilities for m_0.

Given n, there are at most $n^{o(1)}$ choices for $D < n^2$ with $\mathrm{rad}(D) \mid n$, as $n \to \infty$ (see the proof of Theorem 11 in [10] or [20, Lemma 4.2]). Thus, there are at most $n^{2/3+o(1)}$ choices for m with $s(m) = n$ and $(m, n) > 1$. $\qquad\square$

Theorem 7.5. *For $n > 1$, the number of integers m with $(m, n) = 1$ and $s(m) = n$ is*

$$G(n - 1) + O(n^{3/4} \log n).$$

Proof. Using $n > 1$, if $\omega(m) \leq 1$ there are $O(\log n)$ choices for m. If $\omega(m) = 2$ and $m = pq$ is squarefree, then a solution to $s(pq) = n$ is equivalent to a solution to $p + q + 1 = n$ with $p > q$. Thus, the number of choices for m is $G(n - 1)$. Now suppose that $\omega(m) = 2$ and m is not squarefree, so that $\Omega(m) \geq 3$. Let $m = p^a q^b$ with $p > q$ and note that

$$q^2 < p^a q^{b-1} < s(m) = n,$$

so that $q < n^{1/2}$. For q fixed, each of the $O(\log n)$ choices of b gives rise to $O(\log n)$ choices for p^a, so that with q running over primes, there are $O(n^{1/2} \log n)$ choices for m in this case.

Now suppose that $\omega(m) = 3$ and m is squarefree. Write $m = pqr$ where $p > q > r$. Since $pq < s(m) = n$, we have $q < n^{1/2}$. Let $l = s(qr) = q + r + 1$, so that $l \ll n^{1/2}$. With x a polynomial variable, we have

$$(x - q)(x - r) \equiv x^2 + x + n \pmod{l}.$$

Indeed $-q - r \equiv 1 \pmod{l}$, and using $\sigma(qr) = qr + l$, we have

$$n = s(m) = ps(qr) + \sigma(qr) = pl + qr + l \equiv qr \pmod{l}.$$

For a given choice of l, the polynomial $x^2 + x + n$ has at most $O(l^{1/2})$ roots modulo l. Thus, as $l \ll n^{1/2}$, there are at most $O(n^{3/4})$ choices for m in this case. If m is not squarefree, write $m = p^a q^b r^c$, with $a \geq 2$ and $q^b > r^c$. We have $p^a r^c < n$ and $q^b r^c < n$. The latter implies that $r^c < n^{1/2}$ and the former implies that $p^a < n/r^c$. The number of prime powers p^a with $a \geq 2$ corresponding to r^c is thus at most $O((n/r^c)^{1/2}/\log n)$. Now summing over prime powers $r^c < n^{1/2}$, we get $O(n^{3/4}/(\log n)^2)$ pairs p^a, r^c, so by Lemma 7.1, there are at most $O(n^{3/4}/\log n)$ choices for m in this case.

The rest of the proof will follow from Proposition 7.2 together with Lemma 7.1 in the case $D = 1$. ∎

Proposition 7.2. *If $n > 1$, $s(m) = n$, and $\omega(m) \geq 4$, then there are positive integers u, v with $(u, v) = 1$, $m = uv$, $v \leq n^{3/4}$, and either $u < v$ or $\omega(u) = 1$.*

Proof. We write $m = p_1^{a_1} \ldots p_k^{a_k}$ where $p_1^{a_1} > \cdots > p_k^{a_k}$. We also write $p_i^{a_i} = n^{\theta_i}$, so that $\theta_1 > \cdots > \theta_k > 0$. Since $m/p_k^{a_k} < n$, we have

$$\theta_1 + \cdots + \theta_{k-1} < 1. \tag{7.2}$$

Consider the case $k = 4$. By way of contradiction, we may assume that the lesser of $\theta_1 + \theta_3$ and $\theta_2 + \theta_3 + \theta_4$ exceeds $\frac{3}{4}$; say it is $\frac{3}{4} + \epsilon$, where $\epsilon > 0$. Then $\theta_1 + \theta_2 > \frac{3}{4} + \epsilon$, so by (7.2), we have $\theta_3 < \frac{1}{4} - \epsilon$, so that $\theta_4 < \frac{1}{4} - \epsilon$. This then implies that $\theta_2 > \frac{1}{4} + 3\epsilon$. Since $\theta_1 + \theta_3 \geq \frac{3}{4} + \epsilon$, we have $\theta_1 > \frac{1}{2} + 2\epsilon$. Thus, $\theta_1 + \theta_2 > \frac{3}{4} + 5\epsilon$. We continue, starting with this inequality, getting $\theta_4 < \theta_3 < \frac{1}{4} - 5\epsilon$, $\theta_2 > \frac{1}{4} + 11\epsilon$, $\theta_1 > \frac{1}{2} + 6\epsilon$, and $\theta_1 + \theta_2 > \frac{3}{4} + 17\epsilon$. Continuing the process j times starting from $\theta_1 + \theta_2 > \frac{3}{4} + \epsilon$, we get $\theta_1 + \theta_2 > \frac{3}{4} + (2 \cdot 3^j - 1)\epsilon$. If j is large enough, we have $\theta_1 + \theta_2 > 1$, contradicting (7.2).

Now suppose that $k \geq 5$. Let $\alpha = \sum_1^k \theta_i$. Note that (7.2) implies that $\alpha < \frac{5}{4}$. Since we may assume that $\sum_2^k \theta_i > \frac{3}{4}$, it follows that $\theta_1 < \frac{1}{2}$. By Lemma 7.2 we thus may assume that $\theta_2 > \frac{3}{2} - \alpha$, so that $\theta_1 + \theta_2 > 3 - 2\alpha$. By (7.2), we have $\sum_3^{k-1} \theta_i < 1 - (\theta_1 + \theta_2)$, so that using $k \geq 5$,

$$\sum_{i=3}^{k} \theta_i < \frac{3}{2}(1 - (\theta_1 + \theta_2)).$$

Then, since $\theta_1 + \theta_2 = \alpha - \sum_3^k \theta_i$, we have

$$\theta_1 + \theta_2 > \alpha - \frac{3}{2}(1 - (\theta_1 + \theta_2)),$$

which implies that $\theta_1 + \theta_2 < 3 - 2\alpha$, a contradiction.   □

Corollary 7.2. *We have*

$$\#s^{-1}(n) = \begin{cases} G(n-1) + O(n^{3/4}\log n), & n > 1, \ odd, \\ O_\epsilon(n^{2/3+\epsilon}), & n > 0, \ even. \end{cases}$$

Proof. The first assertion follows immediately from Theorems 7.4 and 7.5; in fact, it is not necessary that n be odd. The second assertion will follow from Theorem 7.4 if we show that when n is even, there are not too many integers m coprime to n with $s(m) = n$. Such an integer m must be an odd square, say it is $p^2 l^2$, where $p = P^+(m)$. Then $pl^2 < n$ and if $p \nmid l$,

$$n = s(m) = (p+1)\sigma(l^2) + p^2 s(l^2).$$

Given l, the right side is increasing with p, so there is at most 1 choice for p. So in all, there are at most 2 choices for p: the 1 determined from the foregoing equation and $P^+(l)$. Hence l is an integer smaller than $\sqrt{n}$ and it determines at most two choices for p. Hence there are at most $O(\sqrt{n})$ choices for m. This completes the proof.   □

Remarks. A very recent result of Booker [1, Corollary 2.3] improves our estimate when n is even to $O_\epsilon(n^{1/2+\epsilon})$. The formula given for $\#s^{-1}(n)$ when n is odd is likely to be an asymptotic formula in that $G(n-1)$ is likely to be larger than $n^{1-\epsilon}$ for all sufficiently large odd n. In fact, a strong form of Goldbach's conjecture asserts that for k even,

$$G(k) \sim \frac{k^2}{2\varphi(k)(\log k)^2} \prod_{p \nmid k} \left(1 - \frac{1}{(p-1)^2}\right),$$

and it is known that this asymptotic holds for almost all even numbers k (see, e.g., [19, Section 8]). We conjecture that the error term exponent for odd n in Corollary 7.2 is $1/2 + \epsilon$. An averaging argument shows that it cannot be improved to $1/2 - \epsilon$. In the case of n even, the exponent may be $o(1)$. It is not hard to show that for each positive integer k the upper density of the set of even numbers n with $\#s^{-1}(n) \geq k$ is $O(1/k)$. On the other hand, it seems difficult to show there are infinitely many even n with $\#s^{-1}(n) \geq 3$. In this regard see the forthcoming paper "Divisor-sum fibers" with P. Pollack and L. Thompson in Mathematika.

Acknowledgment

I am grateful to Noah Lebowitz-Lockard and Paul Pollack for some very helpful comments.

References

1. A. R. Booker. Finite connected components of the aliquot graph. Math. Comp., to appear, DOI:https://doi.org/10.1090/mcom3299.
2. W. Bosma. Aliquot sequences with small starting values. arXiv:1604.03004 [math.NT].
3. W. Bosma and B. Kane. The aliquot constant. *Quart J. Math.* **63** (2012), 309–323.
4. E. Catalan. Propositions et questions diverses. *Bull. Soc. Math. France* **16** (1888), 128–129.
5. L. E. Dickson. Theorems and tables on the sum of the divisors of a number. *Quart J. Math.* **44** (1913), 264–296.
6. N. G. de Bruijn. On the number of positive integers $\leq x$ and free of prime factors $> y$. *Nederl. Acad. Wetensch. Proc. Ser. A* **54** (1951), 50–60.
7. P. Erdős. Über die Zahlen der Form $\sigma(n) - n$ und $n - \varphi(n)$. *Elem. Math.* **28** (1973), 83–86.
8. P. Erdős. On asymptotic properties of aliquot sequences. *Math. Comp.* **30** (1976), 641–645.
9. P. Erdős, A. Granville, C. Pomerance, and C. Spiro. On the normal behavior of the iterates of some arithmetic functions. Analytic Number Theory (Allerton Park, IL, 1989), *Progr. Math.*, Vol. 85, Birkhäuser Boston, Boston, MA, 1990, pp. 165–204.
10. P. Erdős, F. Luca, and C. Pomerance. *On the proportion of numbers coprime to a given integer. Anatomy of Integers.* CRM Proc. Lecture Notes, Vol. 46. Amer. Math. Soc., Providence, RI, 2008, pp. 47–64.
11. R. K. Guy and J. L. Selfridge. What drives an aliquot sequence? *Math. Comp.* **29** (1975), 101–107.
12. H. Halberstam and H.-E. Richert. *Sieve Methods.* Academic Press, London, 1974.
13. M. Kobayashi. *On the Density of the Abundant Numbers.* PhD dissertation, Dartmouth College, 2010.
14. M. Kobayashi, P. Pollack, and C. Pomerance. On the distribution of sociable numbers. *J. Number Theory* **129** (2009), 1990–2009.
15. H. W. Lenstra, Jr. Problem 6064. *Amer. Math. Monthly* **82** (1975), 1016. Solution by the proposer, op. cit. **84** (1977), 580.
16. F. Luca and C. Pomerance. *Irreducible radical extensions and Euler function chains. Combinatorial Number Theory.* de Gruyter, Berlin, 2007, pp. 351–361; *Integers* **7** (2007), no. 2, #A25.

17. F. Luca and C. Pomerance. The range of the sum-of-proper-divisors function. *Acta Arith.* **168** (2015), 187–199.
18. F. Luca and C. Pomerance. *Local behavior of the composition of the aliquot and cototient functions*. In *Number Theory: In Honor of Krishna Alladi's 60th Birthday*. G. Andrews and F. Garvan, eds. Spinger Science + Business Media, New York, forthcoming.
19. H. L. Montgomery and R. C. Vaugan. The exceptional set in Goldbach's problem, *Acta Arith.* **27** (1975), 353–370.
20. P. Pollack. On the greatest common divisor of a number and its sum of divisors. *Michigan Math. J.* **60** (2011), 199–214.
21. P. Pollack. Some arithmetic properties of the sum of proper divisors and the sum of prime divisors. *Illinois J. Math.* **58** (2014), 125–147.
22. P. Pollack and C. Pomerance. Some problems of Erdős on the sum-of-divisors function. *Trans. Amer. Math. Soc. Ser. B* **3** (2016), 1–26.
23. C. Pomerance. On the distribution of amicable numbers. *J. Reine Angew. Math.* **293/294** (1977), 217–222.
24. C. Pomerance. Primality testing: Variations on a theme of Lucas. *Proceedings of the 13th Meeting of the Fibonacci Association*, Congressus Numerantium **201** (2010), 301–312.
25. C. Pomerance. On amicable numbers. In *Analytic Number Theory: In Honor of Helmut Maier's 60th birthday*, M. Rassias and C. Pomerance, eds., Springer, Cham, Switzerland, 2015, pp. 321–327.
26. C. Pomerance and H.-S. Yang. Variant of a theorem of Erdős on the sum-of-proper-divisors function. *Math. Comp.* **83** (2014), no. 288, 1903–1913.
27. L. Troupe. On the number of prime factors of values of the sum-of-proper-divisors function. *J. Number Theory*, **150** (2015), 120–135.

8

Erdős, Klarner, and the $3x + 1$ Problem

Jeffrey C. Lagarias

Abstract

This chapter describes work of Erdős, Klarner, and Rado on semigroups of integer affine maps and on sets of integers they generate. It gives the history of problems they studied, some solutions, and new unsolved problems that arose from them.

8.1 Introduction

This chapter describes the history of an Erdős problem on iteration of integer affine functions, gives its solution, and tours some related work. An *integer affine function* of one variable is a function of the form $f(x) = mx + b$ for integers m, b. The Erdős problem concerns the structure of integer orbits of a particular finitely generated semigroup of integer affine functions, with the semigroup operation being composition of maps.

In the early 1970s David Klarner and Richard Rado studied integer orbits of semigroups of such affine functions in an arbitrary number of variables, motivated by the work of Crampin and Hilton on self-orthogonal Latin squares described in the text that follows. In response to a question they posed about a particular example, Paul Erdős proved a theorem on the size of an orbit for certain semigroups of univariate functions, upper bounding the number of integers below a given cutoff T occurring in such orbits, cf. [37, Theorem 8]. Erdős's interest in this orbit problem led him to offer a reward for a particular semigroup iteration problem. This problem was solved by Crampin and Hilton in 1972, but their solution was never published. We supply a reconstructed solution here.

We also present a history of selected later developments, including work of Mike Fredman, Don Knuth, David Klarner, and Don Coppersmith. Their work addresses the structure of particular affine integer semigroups, and sufficient conditions for an integer affine semigroup to be freely generated. The latter topic led Klarner to pose in 1982 several easy-to-state problems in the spirit of Erdős's prize problem, given at the end of the chapter, which remain unsolved.

In the last section we explain how the $3x + 1$ problem may be formulated in terms of orbits of an affine semigroup. To do this requires allowing affine functions having rational coefficients. We show that its simpler cousin, the $x + 1$ problem, a solved problem, is expressible in terms of orbits of integer affine semigroups in exactly the formulation studied by Klarner and Rado.

8.2 Self-orthogonal Latin Squares

A motivation for the study of iterates of affine functions arose from the work of Joan Crampin and Anthony Hilton on a combinatorial problem concerning the orders of self-orthogonal Latin squares (SOLS), which we define in the text that follows.

Recall that a $n \times n$ Latin square can be viewed as an $n \times n$ matrix having entries taking values 1 to n, with each value appearing once in each row and column. It is a special kind of magic square. A pair (M, N) of two $n \times n$ Latin squares are *(mutually) orthogonal* (called an MOLS, or a *Greco-Latin square*) if their vector of ordered pairs of entries (M_{ij}, N_{ij}) takes all n^2 values $\{(k, \ell) : 1 \le k, \ell \le n\}$ exactly once.

In a 1782 paper Euler [19] made an extensive study of magic squares. This paper, written in French, is his only paper published in a Dutch journal. He opens it by saying that a curious question is whether there exists an MOLS with $n = 6$ (the 36 officers problem). He says: "But after spending much effort to resolve the problem, we must acknowledge that such an arrangement is absolutely impossible, though we cannot give a rigorous proof." ([20, paragraph 1]). His paper presents constructions showing that MOLS exist for all odd n and for all n divisible by 4. In [19, paragraph 17] Euler showed that MOLS do not exist for $n = 2$. A large part of the paper treats attempts to resolve the case $n = 6$. In paragraph 140 he says that all his systematic methods failed to find one and that it remains to examine all possible cases. In paragraph 144 he says that he cannot produce a complete Latin square with 36 entries and that he does not hesitate to conclude the same impossibility extends to all cases $n = 4k + 2$. A history of work on this problem is given in Klyve and Stemkowski [38]. The first complete proof of impossibility for $n = 6$ was given in 1900 by Tarry [43]. In 1959 and 1960 Bose, Shrikande, and Parker ([2], [3], [4]) disproved the belief of Euler, showing that there exist MOLS of order $4n + 2$ for all $n \geq 2$. Their proof used various constructions of larger MOLS from smaller ones of specific types.

A refinement of this Euler problem concerns orthogonal Latin squares of a more restricted form, *self-orthogonal Latin squares* (SOLS). These are Latin squares M such that (M, M^T) form a pair of orthogonal Latin squares. By permuting the rows and columns one may always reduce such a square to the case that the main diagonal of M is $(1, 2, \ldots, n)$. An example for $n = 4$ is

$$M = \begin{bmatrix} 1 & 3 & 4 & 2 \\ 4 & 2 & 1 & 3 \\ 2 & 4 & 3 & 1 \\ 3 & 1 & 2 & 4 \end{bmatrix} \quad M^T = \begin{bmatrix} 1 & 4 & 2 & 3 \\ 3 & 2 & 4 & 1 \\ 4 & 1 & 3 & 2 \\ 2 & 3 & 1 & 4 \end{bmatrix} \quad (M, M^T) = \begin{bmatrix} 11 & 34 & 42 & 23 \\ 43 & 22 & 14 & 31 \\ 24 & 41 & 33 & 12 \\ 32 & 13 & 21 & 44 \end{bmatrix}.$$

We have abbreviated ordered pair entries (k, ℓ) to $k\ell$ in the MOLS on the right.

Problem *For which integers n do there exist SOLS?*

In the period 1970–1971 Anthony J. W. Hilton, along with D. Joan Crampin, then both at the University of Reading, approached this problem by using combinatorial constructions starting from some neglected constructions of Sade [41, 42]. An example of these constructions starts with a triple of Latin squares (P, Q_1, Q_2) in which P is a $q \times q$ SOLS containing a smaller $p \times p$ SOLS in its upper left corner, together with a pair (Q_1, Q_2) of MOLS of size $(p - q) \times (p - q)$. This triple can be used to produce from any SOLS of side v a larger SOLS of side $f(v) := (q - p)v + p$. In this construction $f(x) = (q - p)x + p$ is an integer affine function. Starting from a set of such constructions, with associated functions $f_1(x), \ldots, f_k(x)$ plus a known set of initial SOLS of sizes $c_1, c_2, \ldots, c_m$ found by direct construction, one can produce SOLS of all sizes in the set

$$\langle R : A \rangle := \langle f_1, \ldots, f_k : c_1, \ldots, c_m \rangle,$$

which denotes the closure of the initial set $A = \{c_1, \ldots, c_m\}$ under the action of all elements of the integer semigroup $\mathcal{S} = \mathcal{S}(R)$ generated by the set $R = \{f_1, \ldots, f_s\}$ of affine maps. Their object was to obtain a positive proportion of all integers in such an orbit. Using this approach they eventually obtained the following results (Crampin and Hilton [15, Theorems 3 and 5]), which relied on extensive computer calculations to complete an induction step.

Theorem 8.1. (Crampin and Hilton (1972, 1975)

(1) For every $n \geq 36372$ there exists SOLS of sidelength n that contains in its top left corner an SOLS of sidelength 22.
(2) For every $n > 482$ there is an SOLS of sidelength n.

Their result was announced at the British Mathematical Colloquium in 1972. However, their detailed paper was not ready in time to appear in the conference proceedings, because the computer calculations were incomplete. In the meantime, the following research announcement [5] appeared.

Theorem 8.2. (Brayton, Coppersmith, and Hoffman (1974)) *There exists an SOLS of every sidelength n except $n = 2$, 3, and 6.*

Two of the authors were at IBM Yorktown Heights, where the available computer resources were extensive. The paper of Crampin and Hilton [15] appeared in 1975, after revision in light of the results of Brayton, Coppersmith, and Hoffman. The detailed paper of Brayton et al. appeared in 1976 [6].

8.3 Klarner and Rado (1971–1972)

In 1970–1971 David Klarner visited the University of Reading as a postdoctoral fellow, where Richard Rado was a professor. Motivated by discussions with A. J. W. Hilton on iteration of affine functions coming from the problem in the preceding section, Klarner began a collaboration with Rado on the study of integer sequences generated by iteration of affine functions. They formulated the problem as the study of orbits of semigroups S of affine functions $f : \mathbb{N} \to \mathbb{N}$ of the form $f(x) = \beta x + \alpha$, with α, β nonnegative integers. More generally, they considered orbits of affine maps $f_i : \mathbb{N}^n \to \mathbb{N}^n$ of the form $f_i(x_1, x_2, \ldots, x_n) = \beta_{i,1}x_1 + \cdots + \beta_{i,n}x_n + \alpha_i$, in which all $\beta_{i,j}$ and α_i are nonnegative integers. Given a set R of generating functions f_i and an initial set of nonnegative integers A, the *orbit* $\langle R : A \rangle$ consists of all possible integers that can be obtained under iteration of the operation of substituting into any of the functions f_i for each of its variables separately any integer already obtained, i.e., $f_i(a_1, a_2, \ldots, a_n)$. Rado formulated the problem using the terminology of universal algebra, taking the viewpoint that the orbit $\langle R : A \rangle$ is characterized by the property that it is a completed whole: the smallest set of integers that contains A and is closed under the operations of substituting any of its values into any function f_i. A contrasting local viewpoint taken by Klarner is to build up the orbit recursively, starting from the initial seed A and applying the functions a given number of times. Their results appeared first as a Stanford technical report in 1972 ([35]), later published in 1974 ([37]).

Klarner and Rado obtained strong results on the structure of possible orbits when the functions in the semigroup had two or more variables, giving conditions when the sets were of positive density in the integers, including infinite arithmetic progressions. For example, for $f_1(x, y) = 2x + 3y$ the orbit $\langle 2x + 3y : 1 \rangle$ consists of all integers congruent to 1 or 5 modulo 12. They proved under some conditions that if an orbit contains an infinite arithmetic progression then it is a finite union of arithmetic progressions ([37, Theorem 4]). For the univariate case $n = 1$ they noted a restricted family of semigroups giving near arithmetic progressions ([37, Theorem 7]). However, they also noted that other examples had a more complicated structure.

In particular, they mentioned the minimal set of integers $S = S_{KR}$ containing 1 and closed under the action of the two affine functions $f_2(x) := 2x + 1$ and $f_3(x) := 3x + 1$. This sequence begins

$$S_{KR} = 1, 3, 4, 7, 9, 10, 13, 15, 19, 21, 22, 27, 28, 31, 39, 40, 43, \ldots.$$

We will call it the *Klarner–Rado sequence*.[1] It can be recursively generated in a tree fashion

$$S = \bigcup_{r=0}^{\infty} S_r,$$

in which $S_0 = \{1\}$ and

$$S_r := (2S_{r-1} + 1) \cup (3S_{r-1} + 1).$$

We can picture the elements in the set S generated in a rooted binary tree, starting from the root node assigned a value $S_0 = 1$, with the two edges of tree from a given node labeled the functions f_2, f_3, respectively, and by each node in the tree being assigned a value given by the evaluation of the composed function at the value $x = 1$ of the root node. The initial part of this tree is pictured in Fig. 8.1.

Each node of the binary tree is assigned a *level* (or *rank*) r that counts the number of functions occurring in the composition, with the root node assigned level 0, and there are 2^r nodes at level r. The values of nodes at level r give the integers in S_r, and the integers in S are the complete set of values that occur in the entire tree. In this tree integers may appear more than once, and this leads

---

[1] In [37, p. 455] Klarner and Rado say: "For example, we have studied the set $S = \langle 2x + 1, 3x + 1 : 1 \rangle$ which seems to be fairly complicated."

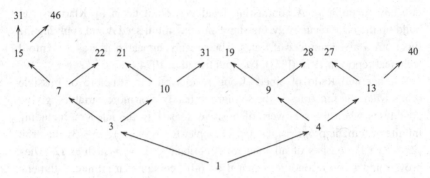

Fig. 8.1. Iteration binary tree for $\mathcal{S} = \langle f_1, f_2 \rangle$.

us to also consider *multisets*, which are sets with elements assigned positive integer multiplicities. We define the multiset

$$S^\# := \coprod_{r=0}^{\infty} S_r^\#,$$

which is defined by a similar recursion

$$S_r^\# := \left(2S_{r-1}^\# + 1\right) \coprod \left(3S_{r-1}^\# + 1\right).$$

and $\coprod$ means "disjoint union." The multiset $S^\#$ contains the same integers as S, but also records their multiplicity of occurrence in the tree. The multiplicity of each integer is finite, because all integers n occurring at level r satisfy $n \geq 2^r$.

The multiset $S^\#$ differs from S, because integers may be generated by this procedure in more than one way. The smallest integer of multiplicity exceeding 1 is $n = 31 \in S_3 \cap S_4$ having multiplicity two, with

$$f_3 \circ f_3 \circ f_2(1) = f_2 \circ f_2 \circ f_2 \circ f_2(1) = 31.$$

Klarner raised several questions about this set:

(1) What is the size of the Klarner–Rado set S? (The size may be measured by an appropriate notion of density of a set of integers.) (Treated in [35, Sect. 2]).
(2) Does the Klarner–Rado set S contain an infinite arithmetic progression (also treated in [35, Sect. 2]) considered more generally in [31]?
(3) Can the complement $\mathbb{N} \setminus S$ of the Klarner–Rado set be covered by infinite arithmetic progressions? (Raised as [9, Problem 1].)

These three questions have all been answered, as follows:

(1) S has natural density zero.
(2) S does not contain an infinite arithmetic progression.

(3) The complement of S can be partitioned into a countable union of infinite arithmetic progressions.

Paul Erdős answered question (1). Erdős had a long-term collaboration with Richard Rado, starting with correspondence in 1933, which resulted in 18 joint papers as of 1987, which he recalls in [16]. He often visited the University of Reading. Klarner and Rado included a proof of answer (1) in their paper [37, Theorem 8], crediting it to Erdős. We present this proof in Section 8.4. Question (2) is answered as a direct corollary of the answer to question (1), given as [37, Corollary to Theorem 8]. Question (3) is more difficult, and it is resolved as a special case of results of Coppersmith [13] in 1975 (see Section 8.8).

The answers to these questions led to further research questions. The first of these was a prize problem posed by Erdős in 1972, presented in Section 8.7, solved afterwards by Crampin and Hilton. Its resolution led to the later formulation of various new problems by Klarner, stated in Section 8.9, which today remain unsolved.

8.4 Semigroups of Affine Maps

We now recall some details of the Klarner and Rado framework for orbits of semigroups of affine maps. We restrict attention to the one-dimensional case. It proves useful here to extend their framework to allow multisets of integers, given in the following definitions.

Definition 8.1. *(1) Let $R = \{f_i : i \in \mathcal{I}\}$ denote a collection of affine maps (finite or countably infinite) having real coefficients*

$$f_i(x) = m_i x + b_i, \quad (i \in \mathcal{I}),$$

in which all the b_i are nonnegative and the multipliers satisfy

$$\inf_{i \in \mathcal{I}} \{m_i\} > 1. \tag{8.1}$$

Let $S(R)$ denote the semigroup of all (distinct) affine maps generated by the maps in R under composition. (In general this semigroup does not include an identity element.)

(2) The labeled semigroup $S(R)^{\#}$ is

$$S(R)^{\#} := \{f_{i_1, i_2, \ldots, i_r}(x) := f_{i_1} \circ f_{i_2} \circ \cdots \circ f_{i_r}(x) : \ r \geq 1, \text{ with each } i_k \in \mathcal{I}\}.$$

That is, we treat elements with different labels $I = (i_1, i_2, \ldots, i_r)$ as distinct, denoting them $f_I(x)$, even if they represent the same affine function. Thus the labeled semigroup $S(R)^{\#}$ is a free semigroup.

Next we consider the action of this semigroup on a given initial multiset of nonnegative integers A.

Definition 8.2. *(1) Suppose that $A \subset \mathbb{N} = \{0, 1, 2, ...\}$ is a discrete multiset (possibly countably infinite) having no finite limit point. For a set R of integer affine maps as earlier, we define the set*

$$\langle R : A \rangle$$

to be the smallest subset of $\mathbb{N}$ that contains the elements of A, and is closed under iteration by the maps in R.

(2) With a multiset A satisfying the conditions in (i) we define the multiset

$$\langle R : A \rangle^{\#}$$

to be the multiset obtained by applying to each element of A (with multiplicity) all the elements of the labeled semigroup

$$\mathcal{S}(R)^{\#} := \{f_{i_1, i_2, \dots, i_r} = f_{i_1} \circ f_{i_2} \circ \cdots \circ f_{i_r} : \ r \geq 1, \ with \ each \ i_k \in \mathcal{I}\},$$

in which elements with different labels $(i_1, i_2, \dots, i_r)$ are treated as distinct, even if they represent the same function. Condition (8.1) ensures that each element in $\langle R : A \rangle^{\#}$ occurs with finite multiplicity.

To measure the size of sets of natural numbers, we introduce some notions of density of such a set; here we extend the standard definitions to apply to multisets.

Definition 8.3. *(1) The* upper asymptotic density *$\bar{d}(S^{\#})$ of a multiset of non-negative integers $S^{\#}$ is*

$$\bar{d}(S^{\#}) := \limsup_{x \to \infty} \frac{1}{x} \# \left\{ S^{\#} \bigcap [0, x] \right\}.$$

Here we count integers with multiplicity, and we allow the value $\bar{d}(S^{\#}) = +\infty$.

(2) The lower asymptotic density *$\underline{d}(S^{\#})$ of a multiset of nonnegative real numbers $S^{\#}$ is*

$$\underline{d}(S^{\#}) := \liminf_{x \to \infty} \frac{1}{x} \# \left\{ S^{\#} \bigcap [0, x] \right\}.$$

(3) The (natural) density *$d(S^{\#})$ of a multiset of nonnegative integers $S^{\#}$ exists if and only if its upper and lower asymptotic densities are equal. In that case, one sets $d(S^{\#}) := \underline{d}(S^{\#}) = \bar{d}(S^{\#})$.*

These densities can take any value in $[0, \infty)$, including $+\infty$. If the multiset $S^{\#}$ has all multiplicities 1, so that it is a set S, then we have the standard density bounds

$$0 \leq \underline{d}(S) \leq \bar{d}(S) \leq 1.$$

In this framework we obtain the Klarner–Rado sequence as

$$S := \langle R : A \rangle = \langle f_2(x) = 2x + 1, f_3(x) = 3x + 1 : 1 \rangle$$

(choosing the index set $\mathcal{I} = \{2, 3\}$ as a mnemonic). We may also interpret S as the forward orbit of the value $x = 1$ under the action of the functions in the semigroup $\mathcal{S}^* = \langle f_2, f_3 \rangle$ (not counting multiplicities), which we call the *Klarner–Rado semigroup*. We also have a multiset version $\langle 2x + 1, 3x + 1 : a \rangle^\#$ of the Klarner–Rado sequence, which we call the *Klarner–Rado multisequence*.

One can further extend the foregoing definitions to real-valued functions and sets. The various definitions of density given earlier extend to arbitrary discrete (multi)sets of nonnegative real numbers. For the functions $f_i(x)$ we may allow m_i, b_i to be real numbers, and similarly for the initial value (multi)set A. For integers m_i, b_i the discreteness property is built in. For the general case of real numbers, some restrictions on the values $\{(m_i, b_i) : i \in \mathcal{I}\}$ are needed to get discrete orbits; the conditions that all $b_i \geq 0$, all $m_i > 1$ and that for each $T > 0$ there are finitely many $m_i \leq T$ are sufficient for discreteness. For affine functions with rational coefficients appearing in connection with the $3x + 1$ problem treated in Section 8.10 orbits are not discrete.

8.5 Erdős's Upper Bound (1972)

Erdős obtained a general result giving an upper bound on the size of the associated set. His result is reported in a 1974 paper of Klarner and Rado [37, Theorem 8], who said:

> P Erdős has kindly communicated to us the essentials of a result which shows that for certain sets R of unary linear operations and certain sets A the set $\langle R : A \rangle$ has density zero and is therefore neither a per-set or a near per-set[2]. This applies for example, to $\langle 2x + 1, 3x + 1 : 1 \rangle$.

We present here a proof that applies to multisets.

Theorem 8.3. (Erdős) *Let $R = \{f_i(x) = m_i x + b_i : i \in \mathcal{I}\}$ be a (possibly infinite) set of affine functions with real coefficients m_i, b_i having all $m_i \geq 1$ and all $b_i \geq 0$. Suppose there is given a positive real number σ such that*

$$\alpha := \sum_{i \in \mathcal{I}} \frac{1}{m_i^\sigma} < 1. \tag{8.2}$$

---

[2] A *per-set* is a union of infinite arithmetic progressions, and a *near per-set* is one that differs from a per-set at a finite number of integers.

Let $A \subset \mathbb{R}_{>0}$ be a (possibly infinite) set of generators having no finite limit point. Then for any $T \geq 1$,

$$|\langle R : A \rangle^{\#} \bigcap [0, T]| \leq \frac{1}{1 - \alpha} \left(\sum_{\{a \in A : 0 \leq a \leq T\}} \frac{1}{a^{\sigma}} \right) T^{\sigma}.$$

Proof of Theorem 8.3. We first note that under the assumption all $m_i \geq 1$, the hypothesis (8.2) on the (possibly infinite) set of f_i cannot be satisfied unless the quantity

$$\delta := \inf_{i \in \mathcal{I}} \{m_i\} \tag{8.3}$$

has $\delta > 1$.

Each affine function in $\mathcal{S}(R)$ necessarily has the form

$$f_I(x) = f_{i_1, \ldots, i_r}(x) := f_{i_1} \circ f_{i_2} \circ \cdots \circ f_{i_r}(x) = m_{i_1} m_{i_2} \cdots m_{i_r} x + n_{i_1, \ldots, i_r},$$
$$\tag{8.4}$$

in which $n_I := n_{i_1, i_2, \ldots, i_r} \geq 0$. We treat functions as specified by the set of indices, so the same function may potentially appear more than once in the list if the semigroup is not free. To the empty index set we assign the function $f_\emptyset(x) = x$ and $m_\emptyset = 1$, $n_\emptyset = 0$. For a real value $T \geq 0$, set

$$N(T) := \{I = (i_1, \ldots, i_r) : r \geq 0, \ m_{i_1, \ldots, i_r} \leq T\}.$$

This set has cardinality $|N(T)| = 0$ for $0 \leq T < 1$ and $|N(1)| = 1$, with our convention on the empty set.

Claim 1. *For all $T \geq 1$*

$$|N(T)| \leq \frac{1}{1 - \alpha} T^{\sigma}.$$

We prove Claim 1 holds on the interval $0 \leq T \leq \delta^n$ by induction on $n \geq 0$. The base case corresponds to $0 \leq T \leq 1$, where it holds by the bounds above. For the induction step for the interval $\delta^n < T \leq \delta^{n+1}$, and any $(i_1, \ldots, i_r)$ with $r \geq 1$ with $m_{i_1} m_{i_2} \cdots m_{i_r} \leq T \leq \delta^{n+1}$ we necessarily have $m_{i_2} m_{i_3} \cdots m_{i_r} \leq \frac{T}{m_{i_1}}$, whence

$$|N(T)| \leq 1 + \sum_{i \in \mathcal{I}} \left| N \left(\frac{T}{m_i} \right) \right|.$$

Now all $\frac{T}{m_i} \le \frac{T}{\delta} \le \delta^n$, so the induction hypothesis together with $T \ge 1$ applies to give

$$|N(T)| \le 1 + \frac{1}{1-\alpha} \left(\sum_{i \in \mathcal{I}} \left(\frac{T}{m_i} \right)^\sigma \right)$$

$$= 1 + \frac{1}{1-\alpha} \left(\sum_{i \in \mathcal{I}} \frac{1}{m_i^\sigma} \right) T^\sigma$$

$$= \frac{1-\alpha}{1-\alpha} + \frac{\alpha}{1-\alpha} T^\sigma \le \frac{1}{1-\alpha} T^\sigma,$$

proving the claim.

Now we count elements for each $a \in A$ separately, using the fact that

$$\langle R : A \rangle^{\#} = \bigcup_{a \in A} \langle R : \{a\} \rangle^{\#}.$$

Claim 2. *For all* $T \ge 1$

$$|\langle R : \{a\} \rangle^{\#} \bigcap [0, T]| \le \frac{1}{1-\alpha} \left(\frac{T}{a} \right)^\sigma.$$

To prove Claim 2 we have from (8.4) that

$$f_I(a) := m_{i_1} m_{i_2} \cdots m_{i_r} a + n_{i_1, \ldots, i_r} \ge m_{i_1} m_{i_2} \cdots m_{i_r} a.$$

Assuming $f_{i_1, \ldots, i_r}(a) \le T$ gives

$$|\langle R : \{a\} \rangle^{\#} \bigcap [0, T]| \le \left| N \left(\frac{T}{a} \right) \right| \le \frac{1}{1-\alpha} \left(\frac{T}{a} \right)^\sigma$$

by Claim 1, so Claim 2 follows.

Finally, since $|N(T)| = 0$ for $T < 1$ we have

$$|\langle R : A \rangle^{\#} \bigcap [0, T]| \le \sum_{\{a \in A : a \le T\}} |\langle R : \{a\} \rangle^{\#} \bigcap [0, T]|$$

$$\le \frac{1}{1-\alpha} \left(\sum_{\{a \in A : 0 \le a \le T\}} \left(\frac{T}{a} \right)^\sigma \right)$$

$$\le \frac{1}{1-\alpha} \left(\sum_{\{a \in A : 0 \le a \le T\}} \frac{1}{a^\sigma} \right) T^\sigma,$$

completing the proof. $\qquad \square$

A useful feature of Theorem 8.3 is that it applies to some cases where the set R of functions is infinite; see Section 8.7. As a special case, we immediately obtain for Klarner's original functions $R = \{2x + 1, 3x + 1\}$ the following corollary.

Corollary 8.1. (Erdős) Set $R = \{2x + 1, 3x + 1\}$. Let τ be the unique real solution to

$$\frac{1}{2^\tau} + \frac{1}{3^\tau} = 1, \tag{8.5}$$

with $\tau \approx 0.78788$, and for any $\sigma = \tau + \epsilon$, with $\epsilon > 0$, set

$$\alpha_\sigma := \frac{1}{2^\sigma} + \frac{1}{3^\sigma},$$

which has $\alpha_\sigma < 1$. The Klarner–Rado multiset $S^\# := \langle R : \{1\}\rangle^\#$ satisfies the bounds

$$|S^\# \bigcap [0, T]| \leq \frac{1}{1 - \alpha_\sigma} T^\sigma. \tag{8.6}$$

It follows that the set $S(R)$ has zero density.

Here (8.6) says that for each $\epsilon > 0$ there is a positive constant $C(\epsilon)$ such that

$$|S^\# \bigcap [0, T]| \leq C(\epsilon) T^{\tau + \epsilon}.$$

Remarks. (1) The 1972 Stanford computer science PhD thesis of Mike Fredman [21] acknowledges input from David Klarner. On [21, p. 79] Fredman applies Theorem 2.3.1 of his thesis to Klarner's problem given earlier to obtain the improved upper bound

$$|S^\# \bigcap [0, T]| \leq C_1 T^\tau.$$

Theorem 2.3.1 reappears as Lemma 4.1 in Fredman and Knuth [22]. Further results in the Fredman and Knuth paper ([22, Theorem 4.3]) apply to this problem (since $\frac{\log 2}{\log 3}$ is irrational) and yield the existence of an asymptotic formula

$$|S^\# \bigcap [0, T]| = c\, T^\tau + o(T^\tau),$$

for a particular $c > 0$.

(2) In 1981 Klarner [30] gave an effective decision procedure to determine whether orbits of certain highly structured integer affine semigroups all have density 0. It applies to semigroups having the property that all the multipliers a_i of the affine functions generating the semigroup are powers of a fixed element $m > 1$.

8.6 Structure of the Klarner–Rado Sequence

We digress to establish further properties of the internal structure of the Klarner–Rado sequence $S = \langle f_2(x), f_3(x) : 1 \rangle$, by studying the associated multiset $S^{\#} := \langle f_2(x), f_3(x) : 1 \rangle^{\#}$. We show that its maps generate a free semigroup, and bound the multiplicity with which a given integer n can appear in an orbit.

Theorem 8.4. *Let $R = \{f_2(x) = 2x + 1, f_3(x) = 3x + 1\}$. Then the following hold.*

(1) The semigroup $\mathcal{S}(R)$ is a free semigroup. That is, all the affine functions $f_{i_1,\ldots,i_r}(x)$ are distinct, for all $r \geq 1$ and all $(i_1, \ldots, i_r) \in \{2, 3\}$.

(2) For each $r \geq 1$, all elements in the multiset $S_r^{\#}$ have multiplicity 1, so that $S_r^{\#} = S_r$.

We will deduce this result from the next theorem, which requires some additional notation to state. Let $\mathcal{S}_r^{\#}$ denote the set of all vectors $\mathbf{I} = (i_1, i_2, \ldots, i_r)$ having all $i_j = 2$ or 3. Then for each $0 \leq k \leq r$ let $\mathcal{S}_{r,k}^{\#}$ denote the set of vectors $\mathbf{I} = (i_1, i_2, \ldots, i_r)$ having all $i_j = 2$ or 3, with exactly k values equal to 3. This set of vectors has size $\binom{r}{k}$. We let $|\mathbf{I}|$ denote the *weight* of a vector, given by

$$|\mathbf{I}| := i_1 + i_2 + \cdots + i_r = 2(r - k) + 3k = k + 2r, \tag{8.7}$$

which takes a constant value on the vectors in $\mathcal{S}_{r,k}^{\#}$. We define a total order $>_R$ on these vectors by lexicographic ordering, read from right to left, with 2 being viewed as larger than 3 in this ordering. That is, letting $\mathbf{J} = (j_1, j_2, \ldots, j_r) \in S_{r,k}$, we set $\mathbf{I} >_R \mathbf{J}$ if and only if there is some s with $1 \leq s \leq r$ with $i_t = j_t$ if $s < t \leq r$ and $i_s < j_s$ (i.e., $i_s = 2, j_s = 3$).

The following result and its proof follow ideas of Klarner [31] given in Section 8.9.

Theorem 8.5. *Let $R = \{f_2(x) = 2x + 1, f_3(x) = 3x + 1\}$. Then for $r \geq 1$ the following properties (P1) and (P2) hold.*

(P1) For $0 \leq k \leq r$ and all elements $\mathbf{I}, \mathbf{J} \in \mathcal{S}_{r,k}^{\#}$,

$$f_I(1) > f_J(1) \iff \mathbf{I} >_R \mathbf{J}.$$

That is, the natural ordering of $f_I(1) \in S_r^{\#}$ as real numbers agrees with that of the lexicographic ordering $>_R$ on vectors $\mathbf{I}$ in $\mathcal{S}_{r,k}^{\#}$.

(P2) For fixed r and $\mathbf{I} \in \mathcal{S}_{r,k_1}^{\#}$ and $\mathbf{J} \in \mathcal{S}_{r,k_2}^{\#}$,

$$k_1 > k_2 \implies f_I(1) > f_J(1).$$

Remarks. (1) As an example for property (P1), for $\mathcal{S}_{4,2}^{\#}$ this ordering is

$$(3,3,2,2) >_R (3,2,3,2) >_R (2,3,3,2) >_R (3,2,2,3) >_R (2,3,2,3)$$
$$>_R (2,2,3,3).$$

Letting $f_{i_1 i_2 i_3 i_4} := f_{i_1 i_2 i_3 i_4}(1)$, we have

$$f_{3322} = 67 > f_{3232} = 64 > f_{2332} = 63 > f_{3223} = 58 > f_{2323} = 57 > f_{2233} = 55.$$

(2) As an example for property (P2), for $\mathcal{S}_{4,3}^{\#}$ we have

$$f_{3332} = 94 > f_{3323} = 85 > f_{3233} = 82 > f_{2333} = 81,$$

and these values exceed all foregoing values $f_{\mathbf{I}}(1)$ for $\mathbf{I} \in \mathcal{S}_{4,2}^{\#}$.

Proof. We prove (P1) and (P2) together by induction on $r \geq 1$. The base case $r = 1$ is easily checked; both $\mathcal{S}_{1,0}^{\#}$ and $\mathcal{S}_{1,1}^{\#}$ have one element, and (P2) holds since $f_3(1) = 4 > f_2(1) = 3$.

Now suppose the induction hypotheses (P1), (P2) hold for $r - 1$, we must treat the value r. We first address part (P1). Let $I = (i_1, \ldots, i_r)$, $J = (j_1, \ldots, j_r)$, suppose both in $\mathcal{S}_{r,k}^{\#}$. The function determined by $\mathbf{I} \in \mathcal{S}_{r,k}^{\#}$ has the form

$$f_{\mathbf{I}}(x) := f_{i_1,\ldots,i_r}(x) = 2^{r-k}3^k x + n_{\mathbf{I}}, \tag{8.8}$$

for some positive integer $n_{\mathbf{I}}$. Thus

$$f_{\mathbf{I}}(1) = 2^{r-k}3^k + n_{\mathbf{I}}.$$

Similarly for $f_{\mathbf{J}}(1)$, so we will have $f_{\mathbf{I}}(1) = f_{\mathbf{J}}(1)$ if and only if $n_{\mathbf{I}} = n_{\mathbf{J}}$. Now let $\mathbf{I}' = (i_1, \ldots, i_{r-1})$ with $\mathbf{I}' \in \mathcal{S}_{r-1,k'}^{\#}$ where $k' = k$ if $i_r = 2$ and $k' = k - 1$ if $i_r = 3$; define $\mathbf{J}' = (j_1, \ldots, j_{r-1})$ similarly, with associated k'', so that $\mathbf{J}' \in \mathcal{S}_{r-1,k''}^{\#}$. Now we have

$$f_{\mathbf{I}'}(x) = 2^{r-k'}3^{k'} x + n_{\mathbf{I}'}$$

and

$$f_{\mathbf{I}}(x) = f_{\mathbf{I}'}(f_{i_r}(x)) = f_{\mathbf{I}'}(i_r x + 1)$$
$$= 2^{r-k}3^k x + 2^{r-k'}3^{k'} + n_{\mathbf{I}'}.$$

Substituting $x = 0$ yields

$$n_{\mathbf{I}} = f_{\mathbf{I}}(0) = f_{\mathbf{I}'}(1) = 2^{r-k'}3^{k'} + n_{\mathbf{I}'}$$

and we have similarly

$$n_{\mathbf{J}} = f_{\mathbf{J}}(0) = f_{\mathbf{J}'}(1) = 2^{r-k''}3^{k''} + n_{\mathbf{J}'}.$$

We deduce for $\mathbf{I}, \mathbf{J} \in \mathcal{S}_{r,k}^{\#}$ that

$$f_{\mathbf{I}}(1) = f_{\mathbf{J}}(1) \iff f_{\mathbf{I}'}(1) = f_{\mathbf{J}'}(1). \tag{8.9}$$

We now treat three cases for $\mathbf{I}, \mathbf{J} \in \mathcal{S}_{r,k}^{\#}$: $i_r = j_r = 2$, $i_r = j_r = 3$, and $\{i_r, j_r\} \equiv \{2, 3\}$.

Suppose $i_r = j_r = 2$. Then $\mathbf{I}', \mathbf{J}' \in \mathcal{S}_{r-1,k}^{\#}$. By the induction hypothesis (P1) for $r - 1$ we have

$$f_{\mathbf{I}'}(1) > f_{\mathbf{J}'}(1) \iff \mathbf{I}' >_R \mathbf{J}'.$$

However, we also have

$$\mathbf{I}' >_R \mathbf{J}' \iff \mathbf{I} = (\mathbf{I}', 2) >_R \mathbf{J} = (\mathbf{J}', 2).$$

We deduce from these two equivalences and from (8.9) that if $i_r = j_r = 2$, then

$$f_{\mathbf{I}}(1) > f_{\mathbf{J}}(1) \iff \mathbf{I} >_R \mathbf{J}.$$

Suppose $i_r = j_r = 3$. This case is done the same way, using the induction hypothesis (P1) for $r - 1$ for $\mathcal{S}_{r-1,k-1}^{\#}$, and we deduce that if $i_r = j_r = 3$, then

$$f_{\mathbf{I}}(1) > f_{\mathbf{J}}(1) \iff \mathbf{I} >_R \mathbf{J}.$$

Suppose $i_r = 2$, $j_r = 3$. Now by definition of the lexicographic order we have $\mathbf{I} >_R \mathbf{J}$. But we also have $\mathbf{I}' \in \mathcal{S}_{r-1,k}^{\#}$ while $\mathbf{J}' \in \mathcal{S}_{r-1,k-1}^{\#}$. Now (P2) of the induction hypothesis for $r - 1$ applies to show that $f_{\mathbf{I}'}(1) > f_{\mathbf{J}'}(1)$, and (8.9) yields $f_{\mathbf{I}}(1) > f_{\mathbf{J}}(1)$. A similar argument holds exchanging $\mathbf{I}$ and $\mathbf{J}$, so we conclude that if $\{i_r, j_r\} \equiv \{2, 3\}$, then

$$f_{\mathbf{I}}(1) > f_{\mathbf{J}}(1) \iff \mathbf{I} >_R \mathbf{J}.$$

This covers all cases, and induction hypothesis (P1) for r is verified.

To verify induction hypothesis (P2) for r, we first note it suffices to verify for $0 \leq k \leq r - 1$ that for all $\mathbf{I} \in \mathcal{S}_{r,k+1}^{\#}$ and $\mathbf{J} \in \mathcal{S}_{r,k}^{\#}$ there holds $f_{\mathbf{I}}(1) > f_{\mathbf{J}}(1)$. Additionally it suffices to verify this inequality for the smallest value of $f_{\mathbf{I}}(1)$ compared with the largest value of $f_{\mathbf{J}}(1)$. By the already verified induction hypothesis (P1) for r these values occur for $\mathbf{I} = (\underbrace{2, \ldots, 2}_{r-k-1}, \underbrace{3, \ldots, 3}_{k+1})$ and $\mathbf{J} = (\underbrace{3, \ldots, 3}_{k}, \underbrace{2, \ldots, 2}_{r-k})$. We recursively compute that

$$f_{\mathbf{I}}(1) = 2^{r-k-1}3^{k+1} + 2^{r-k-1}3^k + \cdots + 2^{r-k-1}3 + 2^{r-k-1} + 2^{r-k-2} + \cdots + 1$$

$$= 2^{r-k-1}\left(\frac{3^{k+2} - 1}{2}\right) + \left(2^{r-k-1} - 1\right)$$

$$= 9 \cdot 2^{r-k-2}3^k + 2^{r-k-2} - 1$$

and

$$f_{\mathbf{J}}(1) = 2^{r-k}3^k + 2^{r-k-1}3^k + \cdots + 2\cdot3^k + 3^k + 3^{k-1} + \cdots + 1$$
$$= \left(2^{r-k+1} - 1\right)3^k + \left(\frac{3^k - 1}{2}\right)$$
$$= 8 \cdot 2^{r-k-2}3^k - \left(\frac{3^k + 1}{2}\right).$$

It follows that

$$f_{\mathbf{I}}(1) - f_{\mathbf{J}}(1) = 2^{r-k-2}3^k + 2^{r-k-2} - 1 + \left(\frac{3^k + 1}{2}\right) > 0,$$

which verifies (P2) and completes the induction step.   □

Proof of Theorem 8.4. To verify assertion (1), asserting freeness of the semigroup $\mathcal{S}(R)$, we must show that all functions $f_{\mathbf{I}}(x)$ are distinct. The function $f_{\mathbf{I}}(x) = 2^{r-k}3^k x + n_{\mathbf{I}}$ in (8.8) uniquely determines its values of r and k, hence if $f_{\mathbf{J}}(x) \equiv f_{\mathbf{I}}(x)$, then necessarily $\mathbf{I}, \mathbf{J} \in \mathcal{S}_{r,k}^{\#}$. But now Theorem 8.5 (P1) asserts that if $\mathbf{I} \neq \mathbf{J}$, then $f_{\mathbf{I}}(1) \neq f_{\mathbf{J}}(1)$, since $>_R$ is a total ordering. Thus the two functions cannot be identical, and the semigroup is free.

In view of Theorem 8.5 the total ordering $>_R$ on each $\mathcal{S}_{r,k}^{\#}$ separately can be extended to a total ordering $>_{RR}$ on $\mathcal{S}_r^{\#}$ by setting $\mathbf{I} >_{RR} \mathbf{J}$ whenever the weights $|\mathbf{I}| > |\mathbf{J}|$. and having the ordering agree with $>_R$ whenever $|\mathbf{I}| = |\mathbf{J}|$.

Now Theorem 8.5 (P1), (P2) says that the ordering $f_{\mathbf{I}}(1)$ on $\mathcal{S}_r^{\#}$ matches this total order $>_{RR}$ on $\mathcal{S}_r^{\#}$, and in particular, implies that all such values $f_{\mathbf{I}}(1)$ are distinct. Since $\mathcal{S}_r^{\#} := \{f_{\mathbf{I}}(1) : \mathbf{I} \in \mathcal{S}_r^{\#}\}$, this verifies assertion (2).   □

We use this result to bound the multiplicity of occurrence of elements in the Klarner–Rado semigroup.

Corollary 8.2. The multiplicity $m_{KR}(n)$ of occurrence of a positive integer n in the Klarner–Rado multisequence $S_{KR} = \langle 2x + 1, 3x + 1; 1 \rangle^{\#}$ is finite, with

$$m_{KR}(n) \leq \log_2 n + 1.$$

Proof. Theorem 8.5 shows that each integer n occurs at most once on each level of the tree, because the elements at that level are totally ordered. It is easy to see that the smallest element at the rth level of the tree is of size at least 2^r, so it follows that the value n can occur only on the first $\log_2 n + 1$ levels of the tree of iterates.   □

We note that $m_{KR}(31) = 2$, with $n = 31$ occurring at levels 4 and 5 of the tree, viewing the base node $a = 1$ as occupying level 1. How large can the multiplicity function get?

Question. *Does the Klarner–Rado sequence have bounded multiplicities? That is, does $m_{KR}(n) = O(1)$ hold? If not, can the upper bound in Corollary 8.2 be improved?*

A reviewer noted that the function $m_{KR}(x)$ for $x \geq 2$ is easily calculable using the recursion

$$m_{KR}(x) = m_{KR}\left(\frac{x-1}{2}\right) + m_{KR}\left(\frac{x-1}{3}\right),$$

where one sets $m_{KR}(1) = 1$ and $m_{KR}(x) = 0$ whenever x is not an integer. Using the recursion, the reviewer found that $m_{KR}(20479) = 3$, $m_{KR}(3988094143) = 4$, and $m_{KR}(7238266879) = 5$.

8.7 Erdős's Density Problem (1972)

After proving his result given earlier (Theorem 8.3), Erdős offered a prize for a solution to the following problem[3].

Erdős Positive Density Problem. *Let $R = \{f_2(x) = 2x + 1, f_3(x) = 3x + 1, f_6(x) = 6x + 1\}$ and set $A = \{1\}$. Does the set $S = \langle R : A \rangle$ have a positive density? More precisely, does S have a positive lower asymptotic density $\underline{d}(S) > 0$?*

This problem is of interest because Theorem 8.3 does not apply to give any nontrivial upper bound on the number of elements in S, since $\frac{1}{2} + \frac{1}{3} + \frac{1}{6} = 1$. Erdős offered a prize of £10 for its solution in 1972.

The problem was answered in the negative soon afterwards by Joan R. Crampin and Anthony J. W. Hilton (see [31, p. 140], also [25]). They collected £5 each from Erdős. A. J. W. Hilton made a copy of his check; see Figure 8.2.

The key observation underlying their solution is that the semigroup of affine maps $\mathcal{S}(R)$ is not a free semigroup; it has the nontrivial relation

$$f_2 \circ f_2 \circ f_3(x) = f_6 \circ f_2(x) = 12x + 7.$$

This relation implies density 0 for integers in an orbit counted without multiplicity.

Theorem 8.6. (Crampin and Hilton) *Suppose $R = \langle f_2(x) = 2x + 1, f_3(x) = 3x + 1, f_6(x) = 6x + 1 \rangle$ and $A = \{1\}$, and let $S_1 = \langle R : A \rangle$. Then, for each $\epsilon > 0$, there is a positive constant $C(\epsilon)$ such that*

$$|S_1 \cap [0, T]| \leq C(\epsilon)T^{\tau_1 + \epsilon},$$

---

[3] A. J. W. Hilton's [25] recollection is that this problem may have been formulated by Klarner, and that Erdős liked it and offered a prize for its solution.

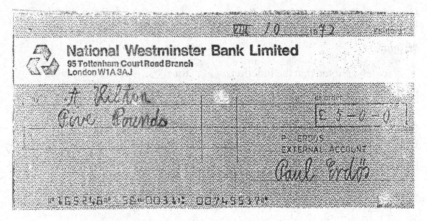

Fig. 8.2. Erdős's cheque to A. J. W. Hilton.

in which τ_1 is the unique positive root of

$$\left(\frac{1}{6}\right)^{\tau_1} + \sum_{k=0}^{\infty} \left(\frac{1}{3 \cdot 2^k}\right)^{\tau_1} = 1.$$

Here $\tau_1 \approx 0.900526 < 1$.

Remark. Orbits that count integers with multiplicity have a different growth rate. Letting $S_1^{\#} = \langle 2x + 1, 3x + 1, 6x + 1 : 1 \rangle^{\#}$, one can show instead that, for any $\epsilon > 0$,

$$|S_1^{\#} \bigcap [0, T]| \geq C(\epsilon) T^{1-\epsilon}.$$

Proof. Label the three functions by the symbols $2 (= 2x + 1)$, 3, $(= 3x + 1)$, $6 (= 6x + 1)$. The semigroup $S^{\#}$ generated by $2, 3, 6$ under composition of functions has a relation: $223 = 62$, that is

$$f_{223}(x) := f_2 \circ f_2 \circ f_3(x) = f_6 \circ f_2(x) =: f_{62}(x) = 12x + 7. \qquad (8.10)$$

We construct a new semigroup that eliminates this relation. First take any word W in the semigroup and replace all patterns 62 with 223, getting an equivalent word W^* in S^*, representing the same affine function, but omitting the pattern 62. Call $\mathcal{T}$ the set of all words of the new form. It is a subset of $S^{\#}$ (but is not a semigroup), consisting of all symbol sequences in $2, 3, 6$ that omit the pattern 62. The words in $\mathcal{T}$ list all possible different functions occurring is $S^{\#}$ (possibly with repetition, if there are further relations in the semigroup $S^{\#}$ that we don't know about).

Second, define the new semigroup

$$S^* = \langle g_0, g_1, g_2, \cdots \rangle \subset S(R)^{\#}$$

having generators g_i corresponding to each of the words **6, 3, 32, 322, 3222** . . . , taking $g_0 = f_6$, $g_k = f_3 \circ f_2^{\circ(k-1)}$, for all $k \geq 1$. This semigroup $S^* \subset S(R)^{\#}$ avoids all compositions of functions containing the pattern **62**. It has infinitely many generators. No word in S^* begins with a **2**.

Claim 1. *For each word W^* in T, either $W^* \in S^*$ if it does not start with a **2**; otherwise the word $3W^* \in S^*$ (i.e., add a prefix **3** on front of W^*).*

We prove the claim by factoring the word W^* from right to left into a product of generators of S^*. We remove any rightmost symbol that is **3** or **6**; otherwise we remove a block of consecutive **2**'s (reading right to left) plus the following non-**2** symbol to the left, which will give a generator in S^*. After this factor is removed, the smaller word is still in T, so we may continue. This factorization procedure works except possibly for the leftmost block in W^*, where it fails if the leftmost block consists entirely of **2**'s. In that case we glue on an extra prefix **3** to obtain a generator of S^*. This proves the claim.

Now let the *dilation factor* $w(W)$ of a word W be the product of its symbol labels **2, 3** or **6**; that is, the affine function $f_W(x)$ corresponding to the word W has the form

$$f_W(x) := w(W)x + c(W)$$

with $w(W)$ and $c(W)$ positive integers.

Claim 2. *Let J be the smallest set of integers containing 1 which is closed under applying functions in the semigroup $S^{\#}$. Then the number of integers in J below a bound T is at most the number of words in T having dilation factor less than T, which is at most the number of words in the generators of S^* having dilation factor less than $3T$.*

This claim follows on noting that each element of J is of the form $f_W(1) = f_{W^*}(1)$ for some word $W^* \in T$ obtained by replacing **62** with **223**, and that the dilation factor of W^* is the same as that of W. Now we can complete the proof. The dilation factors of generators g_i of S^* satisfy the bound

$$\sum_{i=0}^{\infty} \frac{1}{w(g_i)} = \frac{1}{6} + \sum_{k=1}^{\infty} \frac{1}{3 \cdot 2^{k-1}} = \frac{5}{6} < 1.$$

In consequence the constant τ_1 for which $\sum_{i=0}^{\infty} \frac{1}{w(g_i)^{\tau_1}} = 1$ satisfies $0 < \tau_1 < 1$, and one finds numerically $\tau_1 \approx 0.900526$. Theorem 8.3 of Erdős applies to

Problem 1.

Consider the set $<2x+1, 3x+1: 1)$ defined to be the smallest set of natural numbers which contains 1 and is closed under the operations $x \rightarrow 2x+1$ or $3x+1$. The set can be constructed by iterating these operations as indicated in the following tree.

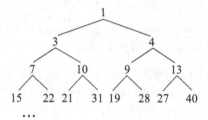

Michael Fredman showed in his thesis that this set has density 0 in the set of all natural numbers; hence, $S = <2x+1, 3x+1; 1)$ does not contain an infinite arithmetic progression. Let N denote the set of all natural numbers. Is it true that $N \backslash S$ may be expressed as a disjoint union of infinite arithmetic progressions?

Fig. 8.3. Selected combinatorial research problem 1.

this infinitely generated semigroup S^* (since (8.2) is satisfied) and shows that the number of words in S^* of weight less than $3T$ is at most $\frac{1}{\epsilon}(3T)^{\tau_1+\epsilon}$, for $0 < \epsilon < 1 - \tau_1$. Theorem 8.6 follows on combining this bound with Claim 2.   □

8.8 Coppersmith's Complement Covering Criterion (1975)

The complement covering problem for the Klarner–Rado sequence was posed as problem number #1 on a list of research problems compiled by V. Chvatal, D. Klarner, and D. Knuth [9] in a 1972 Stanford Research Report; see Fig. 8.3.

In 1975 Don Coppersmith [13] solved this problem. He considered integer semigroups S generated by a finite set of affine maps $f_i(x) = a_i x + b_i$ with $a_i \geq 2$ and $b_i \geq 0$. He defined an integer semigroup to have the property of being *"good"* if the forward orbit of every positive integer has its complement expressible as a finite disjoint union of infinite arithmetic progressions $P(d, s) := \{dn + s : n \in \mathbb{N}\}$, with $d \geq 1, s \geq 1$. He gave a sufficient condition for a semigroup to be "good," which applies to the Klarner–Rado sequence.

To state his criterion we need some terminology. A *feedback element* of a semigroup S is any integer m (positive or negative) that is a fixed point of some element $f(x) \in S$. A feedback element m *propagates into the positive region* if some element $g(x) \in S$ has $g(m) > m$. Iterating $g(x)$ will eventually map this feedback element into the positive region. If such an element $g(x)$ exists, he shows that one can always choose it to be an element of the generating set.

Theorem 8.7. (Coppersmith (1975)) *Suppose that a given semigroup S as in the foregoing satisfies one of the following.*

(1) *No feedback elements exist, i.e., the fixed points of all $f(x) \in S$ are nonin-teger rational numbers.*

(2) *Feedback elements exist, but none of them propagate into the region $x \geq 1$.*

Then the semigroup S is "good."

Coppersmith shows that the properties (1) and (2) are finitely checkable for such a semigroup S. His result applies to the Klarner–Rado semigroup.

Theorem 8.8. *For each integer $a \geq 1$ the complement $\mathbb{N} \smallsetminus S_a$ of the Klarner–Rado orbit $S_a = \langle 2x + 1, 3x + 1 : a \rangle$ is a finite disjoint union of infinite arithmetic progressions In particular, $S_1 = S_{KR}$ gives the original Klarner–Rado sequence.*

Proof. The semigroup $S := \langle 2x + 1, 3x + 1 \rangle$ is proved to be "good" by the criterion of Theorem 8.7. One may prove by an induction argument that all elements of S have the form $f(x) = ax + b$, where $1 \leq b < a$. The fixed point $f(x) = x$ has $m = -\frac{b}{a-1}$, which gives $-1 \leq m < 0$. A feedback element $m = -1$ does occur for $f(x) = 2x + 1$. However, it does not propagate to the positive region since the other generator has $f_3(-1) = -2$. $\qquad\square$

We sketch a direct proof of this result for the special case of the orbit $S_1 = \langle 2x + 1, 3x + 1 : \{1\} \rangle$. If we look at the tree of iterates (mod 6) we find that all elements at the third level appear at earlier levels, with only three distinct values $1, 3, 4 \pmod 6$ taken. The complement then contains the infinite arithmetic progressions $0, 2, 5 \pmod 6$, a set of density $\frac{1}{2}$. Now we consider $S_1 \pmod{36}$. One checks that all iterates at the fifth level (mod 36) repeat values at earlier levels. There are at most $1 + 2 + 4 + 8 = 15$ residue classes (mod 36) hit at the first four levels, so at least $\frac{21}{36}$ residue classes (mod 36) fall in $\mathbb{N} \smallsetminus S_1$. Similar properties hold $\pmod{6^n}$, where the $(2n + 1)$st level of the tree repeats values at earlier levels. Consequently a fraction of at least $1 - \frac{2^n - 1}{6^n}$ of all residue classes $\pmod{6^n}$ is wholly contained in the complement $\mathbb{N} \smallsetminus S_1$. Letting $n \to \infty$ we conclude the complement contains a set of arithmetic progressions that cover a set of density 1, and hence S_1 has natural density 0. A refinement of this approach allows one to show that every element in $\mathbb{N} \smallsetminus S_1$

in the complement can be covered by arithmetic progressions. Finally a lemma of Coppersmith [13, Lemma 1] shows that if the complement can be covered by arithmetic progressions, then this covering can be refined to give a covering with a countable union of *pairwise disjoint* arithmetic progressions.

Coppersmith [13, Theorem 2] also presented a (very complicated to state) sufficient condition for such a semigroup to be "bad," meaning that at least one of its orbits is a "bad" orbit. A "bad" orbit is one having a complement not expressible as a disjoint union of arithmetic progressions. His result does not rule out some orbits being disjoint unions of arithmetic progressions, but his results identified some specific "bad" orbits as well. We give a simple example, coming from the semigroup generated by $2x$ and $2x + 1$.

Theorem 8.9. *For $a \geq 2$ the complement $\mathbb{N} \setminus \tilde{S}_a$ of the orbit $\tilde{S}_a = \langle 2x, 2x + 1 : a \rangle$ cannot be covered by a finite set of infinite arithmetic progressions.*

Proof. This result is a special case of [13, Theorem 2]. It can also be proved by a simple direct argument, which we exhibit for $a = 2$. The first level of the tree is $\{4, 5\}$. The second level of the tree is $\{8, 9, 10, 11\}$. By induction on n the nth level of the tree consists of 2^n consecutive integers. In particular, every arithmetic progression of period P will hit the nth level elements as soon as $2^n > P$. So no infinite arithmetic progression lies in $\mathbb{N} \setminus \tilde{S}_2$. $\square$

Remarks. (1) The sets $\tilde{S}_a = \cup_{n=0}^{\infty} [2^n a, 2^n a + 2^n - 1]$ have a positive density for all $a \geq 1$. For $a \geq 2$ the set $\tilde{S}_a$ does not have a limiting natural density; instead its density in $[0, x]$ oscillates as $x \to \infty$. The lower asymptotic density of $\tilde{S}_a$ is $\frac{1}{a}$ and its upper asymptotic density is $\frac{2}{a+1}$.

(2) These sets $\tilde{S}_a$ are well structured in the sense that their base 2 expansions can be described as the output of a finite automaton. That is, they form an *automatic sequence* to base 2 in the sense of Allouche and Shallit [1, Chapter 5]. This fact follows from later work of Klarner [32], see also Klarner and Post [34].

8.9 Klarner's Criterion for Free Semigroups (1982)

David Klarner continued to investigate properties of affine semigroups and their orbits for much of his career. His student Dean Hoffman wrote a PhD thesis on this topic in 1976 ([26]), which resulted in several joint papers ([27, 28]) on multivariable orbits. In the 1980s Klarner studied special classes of integer orbits that were recognizable by a finite automaton ([32]), thus providing interesting examples of automatic sequences (see Allouche and Shallit [1]). In the 1990s he continued to raise problems (Klarner and Post [34]).

The result of Crampin and Hilton on Erdős's density problem showed the importance of the requirement of freeness of the semigroup. In 1982 Klarner [31] obtained a more general criterion for freeness of a semigroup of integer affine maps. His criterion for freeness is a generalization of properties (P1) and (P2) in Theorem 8.5. Consider the semigroup $\mathcal{S} = \langle \alpha_1, \ldots, \alpha_k \rangle$ with $\alpha_i(x) = m_i x + a_i$ for $i = 1, 2, \ldots, k$ with integers $A = \{a_1, a_2, \ldots, a_k\}$, with each $m_i \geq 2$ and all $a_i \geq 0$. A necessary condition for freeness of the semigroup is

$$\frac{1}{m_1} + \frac{1}{m_2} + \cdots + \frac{1}{m_k} \leq 1.$$

The case of equality is the *critical case*. Set $p_i = \frac{a_i}{m_i - 1}$, noting that $p_i \geq 0$. We may reduce to the case that the maps are ordered so that

$$p_1 < p_2 < \cdots < p_k.$$

This may be done because the equality $p_i = \frac{a_i}{m_i-1} = \frac{a_{i+1}}{m_{i+1}-1} = p_{i+1}$ implies that the maps α_i and α_{i+1} commute, in which case $\mathcal{S}$ would not be free. The criterion is as follows [31, Theorem 2.3].

Theorem 8.10. (Klarner (1982)) *Consider the integer semigroup* $\mathcal{S} = \langle \alpha_1, \ldots, \alpha_k \rangle$ *with* $\alpha_i(x) = m_i x + a_i$ *with each* $m_i \geq 2$, *each* $a_i \geq 0$, *such that* $p_1 < p_2 < \cdots < p_k$ *with* $p_i = \frac{a_i}{m_i - 1}$. *Then* $\mathcal{S}$ *is a free semigroup whenever*

$$\frac{p_k + a_i}{m_i} \leq \frac{p_1 + a_{i+1}}{m_{i+1}} \quad \text{holds for} \quad 1 \leq i \leq k - 1.$$

The proof shows that the compositions have a coefficient ordering that agrees with a certain lexicographic ordering by levels of subscripts. This lexicographic ordering is that exhibited in property (P1) in Theorem 8.5.

In other work Klarner continued to explore the difficulty of recognizing where a semigroup of integer affine maps is a free semigroup. Composition of such maps can be encoded as multiplication of integer matrices. In 1991 Klarner, along with Birget and Satterfield [33], showed the undecidability of such questions in general by a reduction of the Post correspondence problem; the reduction used 3×3 matrices. The affine maps $f(x) = ax + b$ in the foregoing can be encoded as 2×2 matrices $\begin{bmatrix} a & b \\ 0 & 1 \end{bmatrix}$ with nonnegative entries, and are not covered by this result. This problem in this case is discussed in more detail in Cassaigne, Harju, and Karhumäki [7, Section 4]. It remains an open problem and they describe it as "very challenging." The current state of knowledge is surveyed in Cassaigne and Nicolas [8].

Klarner's motivation for his freeness criterion in Theorem 8.10 was to apply it to establish freeness in certain cases where a set of affine maps

$R = \{\alpha_i = a_i x + b_i : 1 \le i \le n\}$ satisfies the threshold condition

$$\sum_{i=1}^{n} \frac{1}{a_i} = 1.$$

He applied his criterion to the simplest nontrivial threshold parameters $(a_1, a_2, a_3) = (2, 3, 6)$ and identified several semigroups that were free ([31, pp. 147–148]).

Theorem 8.11. (Klarner (1982)) *The following six sets of affine transformations each generate a free semigroup.*

(1) $\mathcal{S}_1 = \langle 2x, 3x + 3, 6x + 10 \rangle$.
(2) $\mathcal{S}_2 = \langle 2x, 3x + 2, 6x + 3 \rangle$.
(3) $\mathcal{S}_3 = \langle 2x + 2, 3x, 6x + 15 \rangle$.
(4) $\mathcal{S}_4 = \langle 2x + 1, 3x, 6x + 2 \rangle$.
(5) $\mathcal{S}_5 = \langle 2x + 2, 3x + 1, 6x \rangle$.
(6) $\mathcal{S}_6 = \langle 2x + 1, 3x + 6, 6x \rangle$.

These free semigroups give rise to (corrected) variants of the problem raised by Erdős.

Klarner's Question. *Do any of the orbits of 0 of the foregoing six sets of affine transforma, given by $\langle \alpha_1, \alpha_2, \alpha_3 : 0 \rangle$, have positive density?*

Richard K. Guy [23] included the foregoing semigroup $\mathcal{S}_2$, where

$$f_2(x) = 2x, \quad f_3(x) = 3x + 2, \quad f_6(x) = 6x + 3,$$

in his 1983 paper "Don't try to solve these problems!". As an application of earlier results, we note the following special property of orbits of Guy's particular semigroup.

Theorem 8.12. *For infinitely many integers a, the complement $\mathbb{N} \setminus S_a$ of the orbit $S_a^* := \langle 2x, 3x + 2, 6x + 3 : a \rangle$ cannot be covered by a finite set of infinite arithmetic progressions. In particular the initial value $a = 4$ has this property.*

Proof. One can show that this map satisfies the conditions of Coppersmith [13, Theorem 2]. Coppersmith explicitly mentions the related semigroup $\mathcal{S} = \langle 2x, 3x + 2, 3x + 3 \rangle$ as satisfying the hypotheses of this theorem.   □

Guy repeated the problem in his influential problem book [24, Problem E36]. Klarner's question has not been answered for any of these six sequences.

8.10 Affine Semigroups and the $3x + 1$ Problem

The $3x + 1$ problem concerns the forward orbits of iterating the function

$$T(x) = \begin{cases} \dfrac{3x + 1}{2} & \text{if } x \equiv 1 \ (\text{mod } 2) \\ \dfrac{x}{2} & \text{if } x \equiv 0 \ (\text{mod } 2) \end{cases}$$

for starting value a positive integer. The $3x + 1$ *conjecture* asserts that the iterates of any integer $m \geq 1$ eventually reach the value $n = 1$ and then repeat in the periodic orbit $\{1, 2\}$. Entry to the extensive literature on it can be found in [39, 40].

The $3x + 1$ problem, also called the *Collatz problem*, seems to be one of the most difficult problems in mathematics. Lothar Collatz has stated that he invented the problem before 1952, in the form of iterating the function

$$C(x) = \begin{cases} 3x + 1 & \text{if } x \equiv 1 \ (\text{mod } 2) \\ \dfrac{x}{2} & \text{if } x \equiv 0 \ (\text{mod } 2) \end{cases}$$

and circulated it to Helmut Hasse in 1952; see [10]. However it was first stated in the literature (to my knowledge) in 1971 by H. S. M. Coxeter [14], in the published version of a lecture he gave in Australia in 1970. Coxeter offered 50 dollars[4] for a proof, and 100 dollars for a counterexample. Significant work on the $3x + 1$ problem was done soon after by John H. Conway [11], who in 1972 showed that a generalization of the problem is undecidable. His work was done about the same time as the work of Klarner, Rado, and Erdős, so one may say that iteration problems of this sort were "in the air" around this time. For Conway's more recent views on its unsettleability, see [12, Appendix A].

The $3x + 1$ problem can be rephrased in terms of orbits of affine maps; this is done by considering the semigroup generated by the inverse maps $g_2(x) = 2x$, $g_3(x) = \frac{2x-1}{3}$ associated to the function $T(x)$. We will denote this semigroup

$$S_3 := \left\langle 2x, \frac{2x - 1}{3} \right\rangle.$$

The backwards orbit of the $3x + 1$ iteration from a given value x_0 generates a tree of inverse iterates using these maps; however, it restricts the map $g_3(x)$ to be applied only to an x whose image $g_3(x)$ is an integer.

---

[4] Presumably Canadian dollars, but maybe Australian dollars!

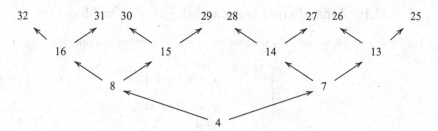

Fig. 8.4. Initial orbit of $a = 4$ for the inverse $x + 1$ map.

As a warmup problem, consider the simpler map giving the inverse iterates for the $(x + 1)$ problem. Recall that the $(x + 1)$ problem iterates the map

$$T_1(x) = \begin{cases} \dfrac{x+1}{2} & \text{if } x \equiv 1 \ (\mathrm{mod}\ 2) \\ \dfrac{x}{2} & \text{if } x \equiv 0 \ (\mathrm{mod}\ 2). \end{cases}$$

For the map T_1 applied to integers $m \geq 1$, it is easy to prove each iterate strictly decreases at each step until $a = 1$ is reached, and that $a = 1$ is a fixed point of the map. The corresponding semigroup of inverse iteration maps is easy to describe: it is

$$S := \langle 2x, 2x - 1 \rangle.$$

The argument just given establishes that the orbit of 1 under S is

$$\langle 2x, 2x - 1 : 1 \rangle = \mathbb{N} \setminus \{0\}.$$

This semigroup is nearly of the Klarner–Rado form considered in Section 8.4, except that one of the maps contains a negative coefficient. Moreover, it is conjugate to the semigroup $\tilde{S}$ of Klarner–Rado type treated in Theorem 8.9 by the transformation $U(x) = x - 1$, namely $\tilde{f}_1(x) = U \circ f_1 \circ U^{-1}(x) = 2x + 1$, $\tilde{f}_2(x) = U \circ f_2 \circ U^{-1}(x) = 2x$.

The tree of allowed inverse iterates of T_1 starting from value $a = 1$ is a full binary tree, which has all the integers $2^n \leq m \leq 2^{n+1} - 1$ on its nth level. Its other orbits also have lots of structure. As an example, Fig. 8.4 pictures the inverse orbit of the $x + 1$ problem starting from $a = 4$. The complement of this orbit is infinite, but it contains no infinite arithmetic progression, which one can see directly because the inverse orbit contains arbitrarily long blocks of consecutive integers. It is given by

$$\langle 2x, 2x - 1 : 4 \rangle = \bigcup_{k=0}^{\infty} [2^{k+2} - 2^k + 1, 2^{k+2}].$$

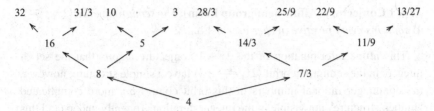

Fig. 8.5. Orbit of $a = 4$ for the inverse $3x + 1$ map.

A result analogous to Theorem 8.9 holds: the integers at the kth level of the inverse tree form a consecutive block of integers. It is also evident that this orbit has positive lower asymptotic density in the set $\mathbb{N}$, and the associated semigroup S is free on two generators.

We now return to the $3x + 1$ problem. One may describe its tree of inverse iterates from a given starting point $x = a$ as (part of) the orbit of the affine semigroup $\langle 2x, \frac{2x-1}{3} : a \rangle$. Figure 8.5 pictures the start of the orbit of this semigroup for $a = 4$. This forward orbit contains the inverse orbit of the $3x + 1$ map applied to $a = 4$, which the $3x + 1$ conjecture asserts will be all positive integers excluding 1 and 2.

The mappings in this affine semigroup S_3 differ from the ones treated by Klarner and Erdős in two important ways:

(1) Some mappings contain negative coefficients. Consequently elements of S_3 can have positive fixed points.
(2) Some mappings have rational coefficients, not all of which are integers. In particular, every orbit under this semigroup that starts at an integer contains noninteger rational numbers.

The $3x + 1$ problem concerns the subtree of inverse iterates that take integer values. It does not apply the map $\frac{2x+1}{3}$ to a value x unless the output is an integer. The semigroup formulation allows all inverse iterates. A nice feature of this particular semigroup action is that: *once an iterate becomes noninteger, it stays noninteger forever after, under application of the semigroup maps.* That is, when applied to any rational number the power of 3 appearing in its denominator is nondecreasing under application of either of the two maps in S_3. An iterate of an integer becomes fractional under these maps only through acquiring a denominator of 3. It follows that the set of integer values of an orbit that starts at an integer a forms a connected subtree of the backwards iteration tree.

We may restate the $3x + 1$ conjecture in the following form, using starting value $a = 4$, which avoids issues with its cycle $\{1, 2\}$. The inverse iterates of $a = 4$ contain no cycles.

$3x + 1$ **Conjecture (Affine Semigroup Form).** *The semigroup orbit* $\langle 2x, \frac{2x-1}{3} : 4 \rangle$ *contains every positive integer except* 1 *and* 2.

This affine reformulation of the $3x + 1$ conjecture asserts that the set of integers in the semigroup orbit $\langle 2x, \frac{2x-1}{3} : 4 \rangle$ have a simple structure; however, the noninteger rational numbers in this orbit comprise a more complicated "hidden structure" not visible in the integer iteration. To really understand this problem, should we not understand these rational numbers as well? One may easily show the rationals in the orbit take arbitrary large positive and negative values. Additionally, each semigroup orbit $\langle 2x, \frac{2x-1}{3} : a \rangle$ is not discrete for any real value of a.

Paul Erdős worked extensively in combinatorial number theory, formulating problems and conjectures concerning the interactions of special types of integers in arbitrarily long arithmetic progressions, including Ramsey theory (see Erdős and Graham [17, 18]). However, he did not publish any result on the $3x + 1$ problem, which he characterized in conversation as "hopeless." The upper bound on orbit size given in Theorem 8.3 seems to be the closest he ever came to working on problems like the $3x + 1$ problem.

Acknowledgments

I am grateful to Anthony J. W. Hilton for describing to me his work on this problem and for providing a copy of his Erdős cheque. I thank Mike Zieve for some initial discussions related to this work and Corey Everlove and Lara Du for corrections. I thank the reviewers for very helpful comments and computations.

References

1. J.-P. Allouche and J. Shallit. *Automatic Sequences. Theory, Applications, Generalizations.* Cambridge University Press, Cambridge, 2003.
2. R. C. Bose and S. S. Shrikande. On the falsity of Euler's conjecture about the nonexistence of two orthogonal Latin squares of order $4t + 2$. *Proc. Natl. Acad. Sci. USA* **45** (1959) 734–737.
3. R. C. Bose and S. S. Shrikande. On the construction of sets of mutually orthogonal Latin squares and the falsity of a conjecture of Euler, Latin squares of order $4t + 2$. *Trans. Amer. Math. Soc.* **95** (1960) 191–209.
4. R. C. Bose, S. S. Shrikande, and E. T. Parker. Further results on the construction of mutually orthogonal Latin squares and the falsity of Euler's conjecture. *Canad. J. Math.* **12** (1960) 189–203.
5. R. K. Brayton, D. Coppersmith, and A. J. Hoffman. Self-orthogonal Latin squares of all orders $n \neq 2, 3, 6$. *Bull. Amer. Math. Soc.* **80** (1974) 116–118.

6. R. K. Brayton, D. Coppersmith, and A. J. Hoffman. Self-orthogonal Latin squares. In *Colloquio Internazionale sulle Teorie Combinatorie (Rome, 1973)*, Tomo II, pp. 509–517. Atti dei Convegni Lincei No. 17, Accad. Naz. Lincei, Rome 1976.

7. J. Cassaigne, T. Harju, and J. Karhumäki. On the undecidability of freeness of matrix semigroups. *Int. J. Algebra Comput.* **9**, no. 3–4 (1999) 295–305.

8. J. Cassaigne and F. Nicolas. On the decidability of semigroup freeness. *RAIRO Theor. Inform. Appl.* **46**, no. 3 (2012) 355–399.

9. V. Chvatal, D. Klarner, and D. E. Knuth. Selected combinatorial research problems. Stanford Computer Science Dept. Technical Report STAN-CS-72-292, June 1972.

10. L. Collatz. On the motivation and origin of the $3x + 1$ problem (Chinese). *J. Qufu Normal Univ., Natural Sci. Ed. [Qufu shi fan da xue xue bao. Zi ran ke xue ban]*, **12**, no. 3 (1986) 9–11. (English translation in [40, 241–247].)

11. J. H. Conway. Unpredictable iterations. In Proc. 1972 Number Theory Conference. University of Colorado, Boulder, CO. 1972. 49–52. [Reprinted in [40].]

12. J. H. Conway. On unsettleable arithmetical problems. *Amer. Math. Monthly* **120**, no. 3 (2013) 192–198.

13. D. Coppersmith. The complement of certain recursively defined sets. *J. Combin. Theory Ser. A* **18**, no. 3 (1975) 243–251. (MR 51 #5477).

14. H. S. M. Coxeter. Cyclic Sequences and Frieze Patterns, (The Fourth Felix Behrend Memorial Lecture). *Vinculum* **8** (1971) 4–7. [Reprinted in [40].]

15. D. J. Crampin and A. J. W. Hilton. Remarks on Sade's disproof of the Euler conjecture with an application to Latin squares orthogonal to their transpose. *J. Comb. Theory Ser. A* **18** (1975), 47–59.

16. P. Erdős. My joint work with Richard Rado. In *Surveys in combinatorics 1987 (New Cross 1987)*, London Math. Soc. Lecture Notes No. 123. Cambridge University Press, Cambridge, 1987. 53–80.

17. P. Erdős and R. L. Graham. Old and new problems and results in combinatorial number theory: van der Waerden's theorem and related topics. *Enseign. Math.* **25**, no. 3–4 (1979) 325–344. (MR 81f:10005).

18. P. Erdős and R. L. Graham. *Old and new problems and results in combinatorial number theory*. Monographie No. 28 de L'Enseignement Mathématique. Kundig, Geneva 1980.

19. L. Euler. Recherches sur une nouvelle espace de quarres magiques. *Verhandelingern uitgegeven door het zeeuwch Genootschap der Wetenschappen te Vissingen* **9** Middelburg 1782, 85–239. (E530 in Enëstrom index of Euler's works) (Opera Omnia: Series 1, Vol. 7, 291–392.)

20. L. Euler. Investigations of a new type of magic square, trans. of [19], by Audie Ho and Dominic Klyve. (Available at Euler archive.)

21. M. L. Fredman. *Growth properties of a class of recursively defined functions*. Ph.D. thesis, Stanford Computer Science Department, June 1972. (Also issued as Stanford Computer Science Dept. Technical Report STAN-CS-72-296.)

22. M. L. Fredman and D. E. Knuth. Recurrence relations based on minimization. *J. Math. Anal. Appl.* **48** (1974) 534–559.

23. R. K. Guy. Don't try to solve these problems! *Amer. Math. Monthly* **90** (1983) 35–41.

24. R. K. Guy. *Unsolved Problems in Number Theory*, 3rd edn. Problem Books in Mathematics. Springer-Verlag, New York, 2004.

25. A. J. W. Hilton. Private communications, 2010 and 2014.

26. D. G. Hoffman. *Sets of integers closed under affine operators*. Ph.D. thesis, University of Waterloo, 1976.

27. D. G. Hoffman and D. A. Klarner. Sets of integers closed under affine operators—The closure of finite sets. *Pacific J. Math.* **78**, no. 2 (1978) 337–344.

28. D. G. Hoffman and D. A. Klarner. Sets of integers closed under affine operators—The finite basis theorem. *Pacific J. Math.* **83**, no. 1 (1979) 135–144.

29. D. A. Klarner. Sets generated by iteration of a linear operation. Stanford Computer Science Department report STAN-CS-72-275, March 1972.

30. D. A. Klarner. An algorithm to determine when certain sets have 0 density. *J. Algorithms* **2** (1981) 31–43.

31. D. A. Klarner. A sufficient condition for certain semigroups to be free. *J. Algebra* **74** (1982) 140–148.

32. D. A. Klarner. *m*-recognizability of sets closed under certain affine functions. *Discrete Appl. Math.* **21**, no. 3 (1988) 207–214.

33. D. A. Klarner, J.-C. Birget, and W. Satterfield. On the undecidability of the freeness of integer matrix semigroups. *Int. J. Algebra Comput.* **1**, no. 2 (1991) 223–226.

34. D. A. Klarner and K. Post. Some fascinating integer sequences. *Discrete Math.* **106/107** (1992) 303–309.

35. D. A. Klarner and R. Rado. Arithmetic properties of certain recursively defined sets. Stanford Computer Science Dept. Technical Report, STAN-CS-72-269, Stanford University, March 1972. [Published as [37].]

36. D. A. Klarner and R. Rado. Linear combinations of sets of consecutive integers. *Amer. Math. Monthly* **80**, no. 9 (1973) 985–989.

37. D. A. Klarner and R. Rado. Arithmetic properties of certain recursively defined sets. *Pacific J. Math.* **53**, no. 2 (1974) 445–463.

38. D. Klyve and L. Stemkowski. Graeco-Latin squares and a mistaken conjecture of Euler. *College Math. J.* **37**, no. 1 (2006) 2–15.

39. J. C. Lagarias. The $3x + 1$ problem and its generalizations. *Amer. Math. Monthly* **92** (1985) 3–23. [Reprinted with corrections in [40].]

40. J. C. Lagarias (Editor). *The Ultimate Challenge: The $3x + 1$ Problem.* American Mathematical Society, Providence, RI, 2010.

41. A. Sade. Contribution à la théorie des quasi-groupes: Diviseurs singuliers. *C. R. Acad. Sci. Paris* **237** (1953) 272–274.

42. A. Sade. Produit direct-singulier de quasigroupes orthogonaux et anti-abéliens. *Ann. Soc. Sci. Bruxelles Sér. I* **74** (1960) 91–99.

43. G. Tarry. Le probléme des 36 officiers. *C. R. Assoc. France Av. Sci.* **29**, part 2 (1900) 170–203.

9

A Short Proof for an Extension of the Erdős–Ko–Rado Theorem

Peter Frankl and Andrey Kupavskii

Abstract

A proof with almost no computation is given for the following inequality due to Pyber. If $A, B \subset \binom{[n]}{k}$ satisfy $A \cap B \neq \emptyset$ for all $A \in \mathcal{A}, B \in \mathcal{B}$ and $n \geq 2k$ then $|\mathcal{A}||\mathcal{B}| \leq \binom{n-1}{k-1}^2$ holds.

9.1 Introduction

Let n, k be positive integers with $n \geq 2k$ and let $\binom{[n]}{k}$ chapter the collection of all k-subsets of $[n] = \{1, \ldots, n\}$. Two families $\mathcal{A}, \mathcal{B} \subset \binom{[n]}{k}$ are said to be *cross-intersecting* if $A \cap B \neq \emptyset$ holds for all $A \in \mathcal{A}$ and $B \in \mathcal{B}$.

The aim of this short Chapter is to give a short, simple proof of the following theorem of Pyber.

Theorem 9.1 (Pyber [10]). *If $\mathcal{A}, \mathcal{B} \subset \binom{[n]}{k}$ are cross-intersecting and $n \geq 2k$, then one has*

$$|\mathcal{A}||\mathcal{B}| \leq \binom{n-1}{k-1}^2. \tag{9.1}$$

Let us mention that in the case $\mathcal{A} = \mathcal{B}$ (9.1) implies $|\mathcal{A}| \leq \binom{n-1}{k-1}$, which is the classical Erdős–Ko–Rado Theorem [2]. For various proofs of it let us refer to a paper of Ron and the first author [5]. One of them is due to Daykin [1], who observed that the Erdős–Ko–Rado Theorem can be deduced from the Kruskal–Katona Theorem ([8, 7]) on shadows. Hilton [6] observed that the same theorem can be used to investigate cross-intersecting families. To present his result we need to introduce the so-called *lexicographic order* $\prec$ on k-element sets.

Definition 9.1. $A, B \in \binom{[n]}{k}$, $A \prec B$ iff the minimal element of $A \setminus B$ is smaller than that of $B \setminus A$.

Definition 9.2. For $0 \le m \le \binom{n}{k}$ let $\mathcal{L}(m)$ denote the first m sets from $\binom{[n]}{k}$ in the lexicographic order.

Lemma 9.1 (Hilton, [6]). *If* $\mathcal{A}, \mathcal{B} \subset \binom{[n]}{k}$ *are cross-intersecting then so are* $\mathcal{L}(|\mathcal{A}|)$ *and* $\mathcal{L}(|\mathcal{B}|)$ *as well.*

By Hilton's lemma, in order to prove (9.1), it is sufficient to consider the case when the set families are initial segments in the lexicographic order. That is what we shall do in the next section.

9.2 Proof of Theorem 9.1

By symmetry, we suppose $|\mathcal{A}| \le |\mathcal{B}|$. First note that if $|\mathcal{A}| \le \binom{n-2}{k-2}$, then

$$|\mathcal{A}||\mathcal{B}| \le \binom{n-2}{k-2}\binom{n}{k} = \binom{n-1}{k-1}^2 \frac{n}{k} \cdot \frac{k-1}{n-1}.$$

As $n(k-1) = nk - n < nk - k = (n-1)k$, (9.1) holds in this case with strict inequality.

From now on we assume that $\binom{n-2}{k-2} \le |\mathcal{A}| \le |\mathcal{B}|$. By Lemma 9.1 we suppose that $\mathcal{A} = \mathcal{L}(|\mathcal{A}|)$, $\mathcal{B} = \mathcal{L}(|\mathcal{B}|)$; i.e., both families are initial segments in the lexicographic order.

Note that the first $\binom{n-2}{k-2}$ sets in the lexicographic order are all the k-sets that contain 1 and 2. Since $\mathcal{A}, \mathcal{B}$ are cross-intersecting, we infer that all their members must contain either 1 or 2. We shall use this fact to prove the following.

Proposition 9.1. *We have*

$$|\mathcal{A}| + |\mathcal{B}| \le 2\binom{n-1}{k-1}. \tag{9.2}$$

Note that (9.2) implies (9.1) by the inequality between arithmetic and geometric means. One can even deduce that (9.1) is strict unless $|\mathcal{A}| = |\mathcal{B}| = \binom{n-1}{k-1}$ holds.

Proof of Proposition 9.1. If $|\mathcal{B}| \le \binom{n-1}{k-1}$ then (9.2) is obvious. Therefore, we assume $|\mathcal{B}| > \binom{n-1}{k-1}$. Note that the first $\binom{n-1}{k-1}$ members of $\binom{[n]}{k}$ are all the k-sets containing 1. Since $\mathcal{A}, \mathcal{B}$ are cross-intersecting, $1 \in A$ holds for all $A \in \mathcal{A}$. Let $\mathcal{B}'$ be the family of the remaining sets in $\mathcal{B}$, i.e.,

$$\mathcal{B}' = \{B \in \mathcal{B} : 1 \notin B\}.$$

Let $\mathcal{C} = \{C \in \binom{[n]}{k} : 1 \in C, C \notin \mathcal{A}\}$. To prove (9.2) we need to show that

$$|\mathcal{C}| \geq |\mathcal{B}'| \quad \text{holds.}$$

Recall that *all* k-sets containing both 1 and 2 are in $\mathcal{A}$ and therefore all members of $\mathcal{B}$ contain 1 or 2. We infer that $B \cap \{1, 2\} = \{2\}$ for all $B \in \mathcal{B}'$ and $C \cap \{1, 2\} = \{1\}$ for all $C \in \mathcal{C}$.

Let us now consider a bipartite graph $\mathcal{G} = (X_1, X_2, E)$, where $X_i := \left\{D_i \in \binom{[n]}{k} : D_i \cap \{1, 2\} = \{i\}\right\}$ and two vertices D_1 and D_2 are connected by an edge if and only if $D_1 \cap D_2 = \emptyset$ holds.

Note that $\mathcal{G}$ is regular of degree $\binom{n-k-1}{k-1}$, $\mathcal{C} \subseteq X_1$, $\mathcal{B}' \subseteq X_2$ hold. Moreover, the cross-intersecting property implies that if D_1 and D_2 are connected for some $D_2 \in \mathcal{B}'$ then $D_1 \in \mathcal{C}$. In other words, the full neighborhood of $\mathcal{B}'$ in the regular bipartite graph $\mathcal{G}$ is contained in $\mathcal{C}$. This implies $|\mathcal{C}| \geq |\mathcal{B}'|$ and concludes the proof. □

9.3 Some Remarks

Matsumoto and Tokushige [9] proved the following extension of (9.1).

Theorem 9.2 ([9]). *If n, k, l are positive integers with $n \geq 2k, n \geq 2l$ and $\mathcal{A} \subset \binom{[n]}{k}, \mathcal{B} \subset \binom{[n]}{l}$ are cross-intersecting then*

$$|\mathcal{A}||\mathcal{B}| \leq \binom{n-1}{k-1}\binom{n-1}{l-1} \quad holds. \tag{9.3}$$

Their proof, as the proof of Pyber, involves some nontrivial, rather lengthy computation. It would be nice to have a shorter, more elegant argument. We succeeded in using some general results from a recent paper [4] to somewhat shorten the calculations leading to (9.3). However, it is still far from being elegant.

From our proof of (9.1) it follows that equality is achieved only if $|\mathcal{A}| = |\mathcal{B}| = \binom{n-1}{k-1}$. Using a result of Füredi and Griggs [5] concerning uniqueness in the Kruskal–Katona Theorem it follows that the only way to achieve equality in (9.1) is letting both $\mathcal{A}$ and $\mathcal{B}$ consist of all k-subsets containing a fixed element.

Acknowledgments

This research was supported in part by the Swiss National Science Foundation Grants 200021-137574 and 200020-14453 and by the grant N 15-01-03530 of the Russian Foundation for Basic Research. 1

References

1. D. E. Daykin. Erdős-Ko-Rado from Kruskal-Katona. *J. Combin. Theory Ser. A* **17** (1974), 254–255.

2. P. Erdős, C. Ko, and R. Rado. Intersection theorems for systems of finite sets. *Quart. J. Math.* **12** (1961) N1, 313–320.

3. P. Frankl and R. L. Graham. Old and new proofs of the Erdos-Ko-Rado theorem. *Sichuan Daxue Xuebao* **26** (1989), 112–122.

4. P. Frankl and A. Kupavskii. Erdős-Ko-Rado theorem for {0, ±1}-vectors. *JCTA*, forthcoming. arXiv:1510.03912

5. Z. Furedi and J. R. Griggs. Families of finite sets with minimum shadows. *Combinatorica* **6** (1986), 355–363.

6. A. J. W. Hilton. The Erdos-Ko-Rado theorem with valency conditions. (1976), unpublished manuscript.

7. G. Katona. A theorem of finite sets. In "Theory of Graphs, Proc. Coll. Tihany, 1966." Akad, Kiado, Budapest, 1968; *Classic Papers in Combinatorics* (1987), 381–401.

8. J. B. Kruskal. The number of simplices in a complex. *Math. Optim. Techn.* **251** (1963), 251–278.

9. M. Matsumoto and N. Tokushige. The exact bound in the Erdos-Ko-Rado theorem for cross-intersecting families. *J. Combin. Theory Ser. A* **52** (1989) N1, 90–97.

10. L. Pyber. A new generalization of the Erdős-Ko-Rado theorem. *J. Combin. Theory Ser. A* **43** (1986), 85–90.

10

The Haight–Ruzsa Method for Sets with More Differences than Multiple Sums

Melvyn B. Nathanson

Abstract

Let h be a positive integer and let $\varepsilon > 0$. The Haight–Ruzsa method produces a positive integer m^* and a subset A of the additive abelian group $\mathbf{Z}/m^*\mathbf{Z}$ such that the difference set is large in the sense that $A - A = \mathbf{Z}/m^*\mathbf{Z}$ and the h-fold sumset is small in the sense that $|hA| < \varepsilon m^*$. This note describes, and in a modest way extends, the Haight–Ruzsa argument, and constructs sets with more differences than multiple sums in other additive abelian groups.

10.1 Sets with More Sums than Differences

Let W be an additive abelian group. For every subset A of W, we define the *difference set*

$$A - A = \{a_1 - a_2 : a_1, a_2 \in A\}$$

and, for every positive integer h, the *h-fold sumset*

$$hA = \{a_1 + a_2 + \cdots + a_h : a_i \in A \text{ for all } i = 1, 2, \ldots, h\}.$$

In particular,

$$2A = \{a_1 + a_2 : a_1, a_2 \in A\}.$$

Because

$$a_1 + a_2 = a_2 + a_1$$

but

$$a_1 - a_2 = -(a_2 - a_1)$$

173

it is reasonable to expect that

$$|A + A| \leq |A - A|$$

for "most" but not necessarily all finite nonempty subsets A of W. A set with more sums than differences is called an MSTD set. For example, in the additive group $\mathbf{Z}$ of integers, the set

$$A = \{0, 2, 3, 4, 7, 11, 12, 14\}$$

is an MSTD set:

$$A + A = [0, 28] \setminus \{1, 20, 27\}$$

$$A - A = [-14, 14] \setminus \{\pm 6, \pm 13\}$$

and so

$$|A + A| = 26 > 25 = |A - A|.$$

Several families of MSTD sets of integers have been constructed, but there is no classification of such sets and many unsolved problems remain (cf. Hegarty [3]; Hegarty and Miller [2]; Iyer, Lazarov, Miller, and Zhang [4, 5]; Martin and O'Bryant [6]; Nathanson [7, 8]).

A dual problem is to construct sets with more differences than multiple sums, that is, finite sets A in a group W such that the difference set $|A - A|$ is large but the h-fold sumset hA is small. In 1973, Haight [1] proved that for all positive integers h and ℓ there exists a modulus m^* and a subset A of $\mathbf{Z}/m^*\mathbf{Z}$ such that

$$A - A = \mathbf{Z}/m^*\mathbf{Z}$$

but hA omits ℓ consecutive congruence classes. Recently, Ruzsa [9] refined and improved Haight's method, and proved the following: For every positive integer h and every $\varepsilon > 0$, there exists a modulus m^* and a subset A of $\mathbf{Z}/m^*\mathbf{Z}$ such that

$$A - A = \mathbf{Z}/m^*\mathbf{Z}$$

and

$$|hA| < \varepsilon m^*.$$

The purpose of this chapter is to describe, and in a modest way extend, the Haight–Ruzsa argument, and construct sets with more differences than multiple sums in other additive abelian groups.

10.2 Haight–Ruzsa Method

Let W be an abelian group, and let $f : W \to W$ be a function, not necessarily a homomorphism. We define the subset

$$A(W, f) = \{w + f(w) : w \in W\} \cup \{f(w) : w \in W\}.$$

For all $w \in W$, we have

$$w = (w + f(w)) - f(w),$$

and so $A(W, f)$ is a *subtractive basis* for W, that is, $A(W, f)$ satisfies the difference set identity

$$A - A = W.$$

For every positive integer h and $\varepsilon > 0$, the Haight–Ruzsa method constructs a finite abelian group W and a function $f : W \to W$ such that the h-fold sumset of $A(W, f)$ is small in the sense that

$$|hA(W, f)| < \varepsilon|W|. \tag{10.1}$$

Let $z \in W$. For every positive integer h, the element z is in the h-fold sumset $hA(W, f)$ if and only if there exist functions

$$\alpha, \beta : W \to \{0, 1, 2, \ldots, h\} \tag{10.2}$$

such that

$$h = \sum_{w \in W} (\alpha(w) + \beta(w)) \tag{10.3}$$

and

$$z = \sum_{w \in W} (\alpha(w)(w + f(w)) + \beta(w)f(w)). \tag{10.4}$$

A pair of functions (α, β) that satisfies conditions (10.2) and (10.3) is called an *admissible pair for* W, and the element $z \in hA(W, f)$ defined by (10.4) is called the group element represented by the admissible pair (α, β). An element $z \in hA(W, f)$ can be represented by many different admissible pairs.

An admissible pair (α, β) has *level* ℓ if

$$\ell = \sum_{\substack{w \in W \\ \alpha(w) + \beta(w) \geq 1}} 1. \tag{10.5}$$

Condition (10.3) implies that $\alpha(w) + \beta(w) \geq 1$ for some $w \in W$, and so

$$\ell \in \{1, 2, \ldots, h\}.$$

If z in hA is represented by the pair (α, β), then the level of the pair counts the number of $w \in W$ such that at least one of the group elements $w + f(w)$ and $f(w)$ appears in the representation (10.4) of z.

Let $L_\ell(W, f)$ be the set of all $z \in W$ such that z can be represented by an admissible pair of level at most ℓ. We have

$$L_1(W, f) \subseteq L_2(W, f) \subseteq \cdots \subseteq L_h(W, f) = hA(W, f).$$

The Haight–Ruzsa method inductively constructs a sequence of groups

$$W_1 \subseteq W_2 \subseteq \cdots \subseteq W_h = W$$

and functions $f_i : W_i \to W_i$ for $i = 1, \ldots, h$ such that

$$A(W_1, f_1) \subseteq A(W_2, f_2) \subseteq \cdots \subseteq A(W_h, f_h)$$

and the sumset $hA(W_h, f_h) = L_h(W_h, f_h)$ satisfies inequality (10.1).

10.3 Preliminary

Let h be a positive integer, and let $\mathcal{R}_h$ be the set of finite commutative rings with identity such that, if $R \in \mathcal{R}_h$, then r is a unit in R for all $r \in \{1, 2, \ldots, h\}$. For example, if p is a prime number and $p > h$, then $\mathbf{Z}/p\mathbf{Z} \in \mathcal{R}_h$. Let $R_0, R_1, \ldots, R_n \in \mathcal{R}_h$. The direct sum $\bigoplus_{i=0}^n R_i$ is a finite ring with identity $(1_{R_0}, 1_{R_1}, \ldots, 1_{R_n})$, and $(r_0, r_1, \ldots, r_n)$ is a unit in $\bigoplus_{i=0}^n R_i$ if $r_i \in \{1, \ldots, h\}$ for all $i = 0, 1, \ldots, n$. Thus, $\bigoplus_{i=0}^n R_i \in \mathcal{R}_h$. If M_i is a finite R_i-module for $i = 0, 1, \ldots, n$, then $\bigoplus_{i=0}^n M_i$ is a finite $\bigoplus_{i=0}^n R_i$-module. The group W_h will be constructed as the direct sum of a finite number of finite R_i-modules M_i, where $R_i \in \mathcal{R}_h$ for $i = 0, 1, \ldots, n$.

We use the following simple combinatorial inequality.

Lemma 10.1. *Let $M_0, M_1, \ldots, M_n$ be finite sets, let*

$$W = M_0 \times M_1 \times \cdots \times M_n,$$

and let

$$(x_0^*, x_1^*, \ldots, x_n^*) \in W.$$

If

$$S = \{(x_0, x_1, \ldots, x_n) \in W : x_j = x_j^* \text{ for some } j \in \{0, 1, \ldots, n\}\}$$

then

$$|S| \leq |W| \sum_{j=0}^n \frac{1}{|M_j|}.$$

If $\varepsilon > 0$ and $|M_j| > (n+1)/\varepsilon$ for all $j \in \{0, 1, \ldots, n\}$, then $|S| < \varepsilon|W|$.

Proof. The cardinality of W is

$$|W| = \prod_{i=0}^{n} |M_i|.$$

For $j \in \{0, 1, \ldots, n\}$, the number of elements $(x_0, x_1, \ldots, x_n) \in W$ with $x_j = x_j^*$ is

$$\prod_{\substack{i=0 \\ i \neq j}}^{n} |M_i| = \frac{|W|}{|M_j|}.$$

If

$$S = \left\{ (x_0, x_1, \ldots, x_n) \in W : x_j = x_j^* \text{ for some } j \in \{0, 1, \ldots, n\} \right\}$$

then

$$|S| \leq \sum_{j=0}^{n} \frac{|W|}{|M_j|} = |W| \sum_{j=0}^{n} \frac{1}{|M_j|}.$$

If $\varepsilon > 0$ and $|M_j| > (n+1)/\varepsilon$ for all $j \in \{0, 1, \ldots, n\}$, then $|S| < \varepsilon |W|$. This completes the proof. $\square$

10.4 Initial Step

Let h be a positive integer and let $\varepsilon > 0$. Choose numbers $\varepsilon_1, \ldots, \varepsilon_h$ such that

$$0 < \varepsilon_1 < \varepsilon_2 < \cdots < \varepsilon_h < \varepsilon.$$

For $i = 0, 1, 2, \ldots, h$, let $R_i \in \mathcal{R}_h$, let M_i be a finite R_i-module such that $|M_i| > (h+1)/\varepsilon_1$, and let

$$W_1 = \bigoplus_{i=0}^{h} M_i.$$

We write the element $w_1 \in W_1$ as an $(h+1)$-tuple

$$w_1 = (x_0, x_1, \ldots, x_h) \tag{10.6}$$

where $x_i \in M_i$ for $i = 0, 1, \ldots, h$. Recall that h is a unit in R_i. For $i = 0, 1, \ldots, h$, we define the function $g_i : W_1 \to M_i$ by

$$g_i(w_1) = -\frac{i}{h} x_i$$

and we define the function $f_1 : W_1 \to W_1$ by

$$f_1(w_1) = (g_0(x_0), g_1(x_1), \ldots, g_h(x_h))$$

$$= \left(0, -\frac{1}{h}x_1, \ldots, -\frac{i}{h}x_i, \ldots, -\frac{h-1}{h}x_{h-1}, -x_h \right).$$

Let

$$A(W_1, f_1) = \{w_1 + f_1(w_1) : w_1 \in W_1\} \cup \{f_1(w_1) : w_1 \in W_1\}.$$

The level 1 set $L_1(W_1, f_1)$ is the set of all $x \in hA(W_1, f_1)$ of the form

$$x = j(w_1 + f_1(w_1)) + (h - j)f_1(w_1) = jw_1 + hf_1(w_1)$$

for some

$$j \in \{0, 1, 2, \ldots, h\}. \tag{10.7}$$

We have

$$jw_1 + hf_1(w_1)$$

$$= j(x_0, x_1, \ldots, x_h) + h\left(0, -\frac{1}{h}x_1, \ldots, -\frac{i}{h}x_i, \ldots, -\frac{h-1}{h}x_{h-1}, -x_h \right)$$

$$= (jx_0, jx_1, \ldots, jx_i, \ldots, jx_h) - (0, x_1, \ldots, ix_i, \ldots, hx_h)$$

$$= (jx_0, (j-1)x_1, \ldots, (j-i)x_i, \ldots, (j-h)x_h).$$

It follows from (10.7) that at least one coordinate of this $(h+1)$-tuple is 0. Applying Lemma 10.1, we obtain

$$|L_1(W_1, f_1)| \leq |W_1| \sum_{i=0}^{h} \frac{1}{|M_i|} < \varepsilon_1 |W_1|.$$

10.5 Inductive Step

Let $1 \leq k \leq h - 1$, and assume that we have a ring $R_0 \in \mathcal{R}_h$, a finite R_0-module W_k, and a function $f_k : W_k \to W_k$ such that the sets

$$A(W_k, f_k) = \{w_k + f_k(w_k) : w_k \in W_k\} \cup \{f_k(w_k) : w_k \in W_k\}$$

and

$$L_k(W_k, f_k) = \{w_k \in hA_k : \text{level}(w_k) \leq k\}$$

satisfy

$$|L_k(W_k, f_k)| < \varepsilon_k |W_k|.$$

Because W_k is a finite set, the number of admissible pairs on W_k is finite. Let n be the number of admissible pairs of level exactly $k+1$ with respect to W_k. We denote these pairs by (α_i, β_i) for $i = 1, 2, \ldots, n$. It follows from (10.3) that

$$\alpha_i(w_k) + \beta_i(w_k) \in \{0, 1, \ldots, h\}$$

for all $w_k \in W_k$.

For $i = 1, 2, \ldots, n$, let $R_i \in \mathcal{R}_h$, and let M_i be a finite R_i-module such that

$$|M_i| > \frac{n}{\varepsilon_{k+1} - \varepsilon_k}.$$

The set

$$W_{k+1} = W_k \oplus \bigoplus_{i=1}^{n} M_i$$

is a finite module over the finite ring $R_0 \oplus \bigoplus_{i=1}^{n} R_i \in \mathcal{R}_h$. We denote the components of $w_{k+1} \in W_{k+1}$ as follows:

$$w_{k+1} = (w_k, x_1, \ldots, x_n)$$

where $w_k \in W_k$ and $x_i \in M_i$ for $i = 1, \ldots, n$. Define the projection $\pi_0 : W_{k+1} \to W_k$ by

$$\pi_0(w_{k+1}) = w_k.$$

For $i = 1, 2, \ldots, n$, we define the projection $\pi_i : W_{k+1} \to M_i$ by

$$\pi_i(w_{k+1}) = x_i$$

and we define the function $g_i : W_{k+1} \to M_i$ as follows: If $\pi_0(w_{k+1}) = w_k$ and

$$\alpha_i(w_k) + \beta_i(w_k) = 0,$$

then

$$g_i(w_{k+1}) = 0.$$

If

$$\alpha_i(w_k) + \beta_i(w_k) \in \{1, \ldots, h\}$$

then

$$g_i(w_{k+1}) = -\frac{\alpha_i(w_k)x_i}{\alpha_i(w_k) + \beta_i(w_k)}.$$

The function g_i is well defined because r is a unit in R_i for all $r \in \{1, \ldots, h\}$.

Define the function $f_{k+1} : W_{k+1} \to W_{k+1}$ by

$$f_{k+1}(w_{k+1}) = (f_k(w_k), g_1(w_{k+1}), \ldots, g_n(w_{k+1})) .$$

We have

$$w_{k+1} + f_{k+1}(w_{k+1}) = (w_k + f(w_k), x_1 + g_1(w_{k+1}), \ldots, x_n + g_n(w_{k+1})) .$$

Consider the set

$$A(W_{k+1}, f_{k+1})$$
$$= \{w_{k+1} + f_{k+1}(w_{k+1}) : w_{k+1} \in W_{k+1}\} \cup \{f_{k+1}(w_{k+1}) : w_{k+1} \in W_{k+1}\} .$$

Because

$$\pi_0 \left(f_{k+1}(w_{k+1}) \right) = f(w_k)$$

and

$$\pi_0 \left(w_{k+1} + f_{k+1}(w_{k+1}) \right) = w_k + f(w_k)$$

it follows that

$$\pi_0(A_{k+1}(W_{k+1}, f_{k+1})) = A_k(W_k, f_k).$$

The $(k+1)$-level set $L_{k+1}(W_{k+1}, f_{k+1})$ is the set of all elements $z_{k+1} \in hA_{k+1}$ that can be represented by an admissible pair (γ, δ) of level at most $k + 1$. We define the functions

$$\hat{\gamma} : W_k \to \mathbf{N}_0 \qquad \text{and} \qquad \hat{\delta} : W_k \to \mathbf{N}_0$$

as follows: For $w_k \in W_k$, let

$$\hat{\gamma}(w_k) = \sum_{\substack{(x_1, \ldots, x_n) \\ \in M_1 \times \cdots \times M_n}} \gamma(w_k, x_1, \ldots, x_n)$$

and

$$\hat{\delta}(w_k) = \sum_{\substack{(x_1, \ldots, x_n) \\ \in M_1 \times \cdots \times M_n}} \delta(w_k, x_1, \ldots, x_n).$$

We have

$$\hat{\gamma}(w_k) + \hat{\delta}(w_k) = \sum_{\substack{(x_1,\ldots,x_n) \\ \in M_1 \times \cdots \times M_n}} (\gamma(w_k, x_1, \ldots, x_n) + \delta(w_k, x_1, \ldots, x_n)) \quad (10.8)$$

and

$$\sum_{w_k \in W_k} (\hat{\gamma}(w_k) + \hat{\delta}(w_k))$$

$$= \sum_{w_k \in W_k} \sum_{\substack{(x_1,\ldots,x_n) \\ \in M_1 \times \cdots \times M_n}} (\gamma(w_k, x_1, \ldots, x_n) + \delta(w_k, x_1, \ldots, x_n))$$

$$= \sum_{w_{k+1} \in W_{k+1}} (\gamma(w_{k+1}) + \delta(w_{k+1}))$$

$$= h.$$

Thus, $(\hat{\gamma}, \hat{\delta})$ is an admissible pair of functions on W_k.

Because (γ, δ) is an admissible pair of functions on W_{k+1} of level at most $k+1$, it follows that $\gamma(w_{k+1}) + \delta(w_{k+1}) \geq 1$ for at most $k+1$ elements $w_{k+1} \in W_{k+1}$. Identity (10.8) implies that if $w_k \in W_k$ and $\hat{\gamma}(w_k) + \hat{\delta}(w_k) \geq 1$, then there exists $w_{k+1} \in W_{k+1}$ such that $\pi_0(w_{k+1}) = w_k$ and $\gamma(w_{k+1}) + \delta(w_{k+1}) \geq 1$. It follows that $\hat{\gamma}(w_k) + \hat{\delta}(w_k) \geq 1$ for at most $k+1$ elements $w_k \in W_k$, and so the pair $(\hat{\gamma}, \hat{\delta})$ has level at most $k+1$. Similarly, if the pair (γ, δ) has level at most k, then the pair $(\hat{\gamma}, \hat{\delta})$ has level at most k.

If $z_{k+1} \in L(W_{k+1}, f_{k+1})$, then z_{k+1} is represented by an admissible pair (γ, δ) of level at most $k+1$. We have

$$z_{k+1} = \sum_{w_{k+1} \in W_{k+1}} (\gamma(w_{k+1})(w_{k+1} + f_{k+1}(w_{k+1})) + \delta(w_{k+1})f_{k+1}(w_{k+1}))$$

$$= \sum_{w_{k+1} \in W_{k+1}} ((\gamma(w_{k+1})(w_k + f(w_k), x_1 + g_1(w_{k+1}), \ldots, x_n + g_n(w_{k+1}))$$

$$+ \delta(w_{k+1})(f(w_k), g_1(w_{k+1}), \ldots, g_n(w_{k+1}))).$$

For $i = 1, \ldots, n$, we have

$$\pi_i(z_{k+1}) = \sum_{w_{k+1} \in W_{k+1}} (\gamma(w_{k+1})(x_i + g_i(w_{k+1}) + \delta(w_{k+1})g_i(w_{k+1}))$$

$$= \sum_{w_{k+1} \in W_{k+1}} (\gamma(w_{k+1})x_i + (\gamma(w_{k+1}) + \delta(w_{k+1}))g_i(w_{k+1})) \in M_i.$$

For $i = 0$, we have

$$z_k = \pi_0(z_{k+1})$$

$$= \sum_{w_{k+1} \in W_{k+1}} (\gamma(w_{k+1})(w_k + f(w_k)) + \delta(w_{k+1})f(w_k))$$

$$= \sum_{w_k \in W_k} \sum_{\substack{(x_1,\ldots,x_n) \\ \in M_1 \times \cdots \times M_n}} (\gamma(w_k, x_1, \ldots, x_n)(w_k + f(w_k)) + \delta(w_k, x_1, \ldots, x_n)f(w_k))$$

$$= \sum_{w_k \in W_k} \hat{\gamma}(w_k)(w_k + f(w_k)) + \hat{\delta}(w_k)f(w_k)$$

and so z_k is an element of $hA(W_k, f_k)$ that is represented by the admissible pair $(\hat{\gamma}, \hat{\delta})$ of level at most $k + 1$. If the admissible pair $(\hat{\gamma}, \hat{\delta})$ has level at most k, then $z_k \in L_k(W_k, f_k)$.

Suppose that $(\hat{\gamma}, \hat{\delta})$ is an admissible pair in W_k of level exactly $k + 1$. It follows that the admissible pair (γ, δ) in W_{k+1} also has level exactly $k + 1$. Because $\{(\alpha_i, \beta_i) : i = 1, \ldots, n\}$ is the set of all admissible pairs on W_k of level $k + 1$, there is a unique integer $j \in \{1, 2, \ldots, n\}$ such that $(\hat{\gamma}, \hat{\delta}) = (\alpha_j, \beta_j)$. Let

$$S = \{w_k \in W_k : \alpha_j(w_k) + \beta_j(w_k) \geq 1\}.$$

There are exactly $k + 1$ elements $w_k \in W_k$ that appear in the representation of z_k associated with the admissible pair (α, β), and so $|S| = k + 1$. If $w_k \in S$, then

$$1 \leq \alpha_j(w_k) + \beta_j(w_k) = \hat{\gamma}(w_k) + \hat{\delta}(w_k)$$

$$= \sum_{\substack{(x_1,\ldots,x_n) \\ \in M_1 \times \cdots \times M_n}} (\gamma(w_k, x_1, \ldots, x_n) + \delta(w_k, x_1, \ldots, x_n)).$$

Because the admissible pair (γ, δ) in W_{k+1} has level $k + 1$, for each $w_k \in S$, there is a unique n-tuple $(y_1, \ldots, y_n) \in M_1 \times \cdots \times M_n$ such that

$$\alpha_j(w_k) + \beta_j(w_k) = \hat{\gamma}(w_k) + \hat{\delta}(w_k)$$

$$= \gamma(w_k, y_1, \ldots, y_n) + \delta(w_k, y_1, \ldots, y_n)$$

$$\geq 1$$

and

$$\gamma(w_k, x_1, \ldots, x_n) + \delta(w_k, x_1, \ldots, x_n) = 0$$

for all $(x_1, \ldots, x_n) \in M_1 \times \cdots \times M_n$ with $(x_1, \ldots, x_n) \neq (y_1, \ldots, y_n)$. Therefore,

$$
\begin{aligned}
\pi_j(z_{k+1}) &= \sum_{w_{k+1} \in W_{k+1}} \big(\gamma(w_{k+1})x_j + (\gamma(w_{k+1}) + \delta(w_{k+1}))g_j(w_{k+1})\big) \\
&= \sum_{w_k \in W_k} \sum_{\substack{(x_1, \ldots, x_n) \in \\ M_1 \times \cdots \times M_n}} \gamma(w_k, x_1, \ldots, x_n)x_j \\
&\quad + \sum_{w_k \in W_k} \sum_{\substack{(x_1, \ldots, x_n) \in \\ M_1 \times \cdots \times M_n}} (\gamma(w_k, x_1, \ldots, x_n) + \delta(w_k, x_1, \ldots, x_n)) \, g_j(w_{k+1}) \\
&= \sum_{w_k \in W_k} \big(\gamma(w_k, y_1, \ldots, y_n)x_j + (\gamma(w_k, y_1, \ldots, y_n) \\
&\quad + \delta(w_k, y_1, \ldots, y_n))g_j(w_{k+1})\big) \\
&= \sum_{w_k \in W_k} \big(\alpha_j(w_k)x_j + (\alpha_j(w_k) + \beta_j(w_k))g_j(w_{k+1})\big) \\
&= \sum_{w_k \in W_k} \left(\alpha_j(w_k)x_j + (\alpha_j(w_k) + \beta_j(w_k))\left(-\frac{\alpha_j(w_k)x_j}{\alpha_j(w_k) + \beta_j(w_k)}\right)\right) \\
&= 0.
\end{aligned}
$$

To summarize, if $z_{k+1} \in L_{k+1}(W_{k+1}, f_{k+1})$, then $\pi_0(z_{k+1}) \in L_k(W_k, f_k)$ or $\pi_j(z_{k+1}) = 0$ for some $j \in \{1, \ldots, n\}$. Because $|L_k(W_k, f_k)| < \varepsilon_k |W_k|$, the number of elements $z_{k+1} \in W_{k+1}$ with $\pi_0(z_{k+1}) \in L_k(W_k, f_k)$ is at most

$$
\varepsilon_k |W_k| |M_1| \cdots |M_n| = \varepsilon_k |W_{k+1}|.
$$

For $j \in \{1, \ldots, n\}$, the number of elements in W_{k+1} with $\pi_j(z_{k+1}) = 0$ is

$$
|W_k| \prod_{\substack{i=1 \\ i \neq j}}^{n} |M_i| = \frac{|W_{k+1}|}{|M_j|} < \left(\frac{\varepsilon_{k+1} - \varepsilon_k}{n}\right) |W_{k+1}|
$$

and so

$$
|L_{k+1}(W_{k+1}, f_{k+1})| < \varepsilon_k |W_{k+1}| + \sum_{i=1}^{n} \left(\frac{\varepsilon_{k+1} - \varepsilon_k}{n}\right) |W_{k+1}| = \varepsilon_{k+1} |W_{k+1}|.
$$

This completes the induction. With $k = h$, we obtain

$$
hA(W_h, f_h) = L_h(W_h, f_h) < \varepsilon_h |W_h| < \varepsilon |W_h|.
$$

This completes the proof.

10.6 Examples

Example 10.1. Let h be a positive integer and let $\varepsilon > 0$. Let $\mathcal{M}_h$ be the set of positive integers m such that every prime divisor of m is greater than h. If $m \in \mathcal{M}_h$, then every integer $r \in \{1, 2, \ldots, h\}$ is a unit in the ring $\mathbf{Z}/m\mathbf{Z}$, and so $\mathbf{Z}/m\mathbf{Z} \in \mathcal{R}_h$ and $\mathbf{Z}/m\mathbf{Z}$ is a finite $\mathbf{Z}/m\mathbf{Z}$-module. In the Haight–Ruzsa construction, if we choose modules $M_i = \mathbf{Z}/m_i\mathbf{Z}$ with pairwise relatively prime moduli $m_i \in \mathcal{M}_h$, then we obtain a finite abelian group

$$W_h = \bigoplus_{i \in I} \mathbf{Z}/m_i\mathbf{Z} \cong \mathbf{Z}/m^*\mathbf{Z}$$

where

$$m^* = \prod_{i \in I} m_i$$

and a function $f_h : W_h \to W_h$ such that the set

$$A = A(W_h, f_h) = \{w_h + f_h(w_h) : w_h \in W_h\} \cup \{f_h(w_h) : w_h \in W_h\}$$

satisfies $A - A = W_h$ and $|hA| < \varepsilon|W_h|$. This is Ruzsa's result, with m_i prime for all i and m^* squarefree.

Example 10.2. Let h be a positive integer and let $\varepsilon > 0$. Let $\mathbf{F}_{q_i}$ be the finite field with $q_i = p_i^{k_i}$ and $p_i > h$. In the Haight–Ruzsa construction, we can choose modules M_i that are finite-dimensional vector spaces over the field $\mathbf{F}_{q_i}$ of sufficiently large dimension. We obtain a finite abelian group W_h that is a direct sum of vector spaces. If each of these vector spaces is a vector space over the same field $\mathbf{F}_q$, then W_h is a vector space over $\mathbf{F}_q$.

Example 10.3. Let h be a positive integer and let $\varepsilon > 0$. Let $\mathbf{F}_q$ be the finite field with $q = p^k$ and $p > h$, and let $\mathbf{F}_q[t]$ be the vector space of polynomials with coefficients in $\mathbf{F}_q$. Let $d_0 = 0 < d_1 < d_2 < \cdots$ be a sufficiently rapidly increasing sequence of integers. We choose finite subspaces of $\mathbf{F}_q[t]$ of the form

$$\left\{ \sum_{j=d_{i-1}}^{d_i - 1} c_j t^j : c_j \in \mathbf{F}_q \right\}.$$

The Haight–Ruzsa method produces a subspace W_h of the vector space $\mathbf{F}_q[t]$ consisting of all polynomials of degree less than d_h.

10.7 Open Problems

1. Let h be a positive integer and $\varepsilon > 0$. Does there exist a prime p and a set A in $\mathbf{Z}/p\mathbf{Z}$ such that $A - A = \mathbf{Z}/p\mathbf{Z}$ and $|hA| < \varepsilon p$? Do such sets exist for all sufficiently large primes p?

2. Let h be a positive integer, let $\varepsilon > 0$, and let $c > 0$. Does there exist an additive abelian group W with $|W| > c$ and a subset A of W such that $A - A = W$ and the set

$$G = \{2a_1 + a_2 + \cdots + a_h : a_i \in A \text{ for all } i = 1, 2, \ldots, h\}$$

satisfies

$$|G| < \varepsilon |W|?$$

3. Let $F(x_1, \ldots, x_h) = \sum_{i=1}^{h} r_i x_i$ be a linear form with coefficients in a ring R, and let W be an R-module. For every subset A of W, we define

$$F(A) = \left\{ \sum_{i=1}^{h} r_i a_i : a_i \in A \text{ for all } i = 1, \ldots, h \right\}.$$

It is an open problem to determine the pairs of linear forms (F, G) such that, for every $\varepsilon > 0$ and $c > 0$, there exists a finite R-module W with $|W| > c$, $F(A) = W$, and $|G(A)| < \varepsilon |W|$. The Haight–Ruzsa produces modules for which this condition is satisfied for the pair of linear forms (F, G), where

$$F(x_1, x_2) = x_1 - x_2$$

and

$$G(x_1, \ldots, x_h) = x_1 + \cdots + x_h.$$

4. If F and G are polynomials with coefficients in a commutative ring R, and if $\varepsilon > 0$ and $c > 0$, does there exist a finite R-algebra W with $|W| > c$ and a subset A of W such that $F(A) = W$ and $|G(A)| < \varepsilon |W|$.

Acknowledgments

Supported in part by a grant from the PSC-CUNY Research Award Program.

References

1. J. A. Haight. Difference covers which have small k-sums for any k. *Mathematika* **20** (1973), 109–118.
2. P. Hegarty and S. J. Miller. When almost all sets are difference dominated. *Random Struct. Algorith.* **35**, no. 1 (2009), 118–136.

3. P. V. Hegarty. Some explicit constructions of sets with more sums than differences. *Acta Arith.* **130** (2007), 61–77.
4. G. Iyer, O. Lazarev, S. J. Miller, and L. Zhang. Generalized more sums than differences sets. *J. Number Theory* **132**, no. 5 (2012), 1054–1073.
5. G. Iyer, O. Lazarev, S. J. Miller, and L. Zhang. *Finding and counting MSTD sets*. Combinatorial and additive number theory—CANT 2011 and 2012. Springer Proc. Math. Stat., Vol. 101, Springer, New York, 2014, pp. 79–98.
6. G. Martin and K. O'Bryant. *Many sets have more sums than differences*. Additive combinatorics. CRM Proc. Lecture Notes, Vol. 43, Amer. Math. Soc., Providence, RI, 2007, pp. 287–305.
7. M. B. Nathanson. *Problems in additive number theory. I.* Additive combinatorics. CRM Proc. Lecture Notes, Vol. 43. Amer. Math. Soc., Providence, RI, 2007, pp. 263–270.
8. M. B. Nathanson. Sets with more sums than differences. *Integers* **7** (2007), A5, 24.
9. I. Z. Ruzsa. *More differences than multiple sums*. arXiv: 1601.04146, 2016.

11

Dimension and Cut Vertices: An Application of Ramsey Theory

William T. Trotter, Bartosz Walczak, and Ruidong Wang

Abstract

Motivated by quite recent research involving the relationship between the dimension of a poset and graph-theoretic properties of its cover graph, we show that for every $d \geqslant 1$, if P is a poset and the dimension of a subposet B of P is at most d whenever the cover graph of B is a block of the cover graph of P, then the dimension of P is at most $d + 2$. We also construct examples that show that this inequality is best possible. We consider the proof of the upper bound to be fairly elegant and relatively compact. However, we know of no simple proof for the lower bound, and our argument requires a powerful tool known as the Product Ramsey Theorem. As a consequence, our constructions involve posets of enormous size.

11.1 Introduction

We assume that the reader is familiar with basic notation and terminology for partially ordered sets (here we use the short term *posets*), including chains and antichains, minimal and maximal elements, linear extensions, order diagrams, and cover graphs. Extensive background information on the combinatorics of posets can be found in [17, 18].

We will also assume that the reader is familiar with basic concepts of graph theory, including the following terms: connected and disconnected graphs, components, cut vertices, and k-connected graphs for an integer $k \geqslant 2$. Recall that when G is a connected graph, a connected induced subgraph H of G is called a *block* of G when H is 2-connected and there is no subgraph H' of G which contains H as a proper subgraph and is also 2-connected.

Here are the analogous concepts for posets. A poset P is said to be *connected* if its cover graph is connected. A subposet B of P is said to be *convex* if $y \in B$

187

whenever $x, z \in B$ and $x < y < z$ in P. Note that when B is a convex subposet of P, the cover graph of B is an induced subgraph of the cover graph of P. A convex subposet B of P is called a *component* of P when the cover graph of B is a component of the cover graph of P. A convex subposet B of P is called a *block* of P when the cover graph of B is a block in the cover graph of P.

Motivated by questions raised in recent papers exploring connections between the dimension of a poset P and graph-theoretic properties of the cover graph of P, our main theorem will be the following result.

Theorem 11.1. *For every $d \geqslant 1$, if P is a poset and every block in P has dimension at most d, then the dimension of P is at most $d + 2$. Furthermore, this inequality is best possible.*

The remainder of this chapter is organized as follows. In Section 11.2, we present a brief discussion of background material that serves to motivate this line of research and puts our theorem in historical perspective. Section 11.3 includes a compact summary of essential material from dimension theory. Section 11.4 contains the proof of the upper bound in our main theorem, and in Section 11.5, we give a construction that shows that our upper bound is best possible. This construction uses the Product Ramsey Theorem and produces posets of enormous size. We close in Section 11.6 with some brief remarks about challenges that remain.

11.2 Background Motivation

A family $\mathcal{F} = \{L_1, L_2, \ldots, L_d\}$ of linear extensions of a poset P is called a *realizer* of P when $x \leqslant y$ in P if and only if $x \leqslant y$ in L_i for each $i = 1, 2, \ldots, d$. The *dimension* of P, denoted by $\dim(P)$, is the least positive integer d for which P has a realizer of size d. For simplifying the details of arguments to follow, we consider families of linear extensions with repetition allowed. So if $\dim(P) = d'$, then P has a realizer of size d for every $d \geqslant d'$. For an integer $d \geqslant 2$, a poset P is said to be *d-irreducible* if $\dim(P) = d$ and $\dim(B) < d$ for every proper subposet B of P.

As is well known, the dimension of a poset P is just the maximum of the dimension of the components of P *except* when P is the disjoint sum of two or more chains. In the latter case, $\dim(P) = 2$ while all components of P have dimension 1. Accordingly, when $d \geqslant 3$, a poset P with $\dim(P) = d$ has a component Q with $\dim(Q) = d$.

It is easy to see that if the chromatic number of a connected graph G is r and $r \geqslant 2$, then there is a block H of G so that the chromatic number of H is

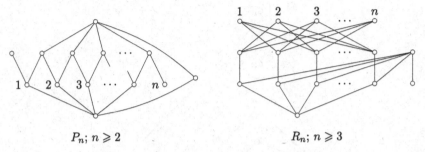

$P_n;\ n \geqslant 2$ $R_n;\ n \geqslant 3$

Fig. 11.1. Irreducible posets with cut vertices.

also r. The analogous statement for posets is not true. We show in Fig. 11.1 representatives of two infinite families of posets. When $n \geqslant 2$, the poset P_n shown on the left side is 3-irreducible (see [19] or [11] for the full list of 3-irreducible posets). For each $m \geqslant 3$, the poset Q_m shown on the right is $(m + 1)$-irreducible. This second example is part of an exercise given on p. 20 in [17]. Together, these examples show that for every $d \geqslant 2$, there are posets of dimension $d + 1$ every block of which has dimension at most d.

The following theorem is proved in [20].

Theorem 11.2. *If P is a poset and the cover graph of P is a tree, then* $\dim(P) \leqslant 3$.

In Fig. 11.2, we show two posets whose cover graphs are trees. These examples appear in [20], and we leave it as an exercise to verify that each of them has dimension 3. Accordingly, the inequality in Theorem 11.2 is best possible. In the language of this chapter, we note that when the cover graph of P is a tree and $|P| \geqslant 2$, then every block of P is a two-element chain and has dimension 1. Accordingly, in the case $d = 1$, our main theorem reduces to a result that has been known for nearly 40 years. However, we emphasize that the proof we give in Section 11.4 of the upper bound in Theorem 11.1 is not inductive and works for all $d \geqslant 1$ simultaneously. For this reason, it provides a new proof of Theorem 11.2 as a special case.

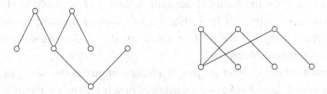

Fig. 11.2. 3-Dimensional posets whose cover graphs are trees.

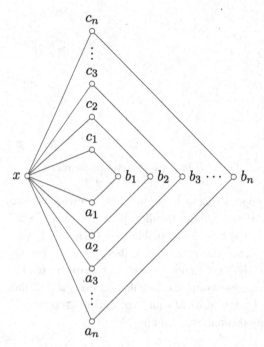

Fig. 11.3. A 4-dimensional poset (when $n \geqslant 17$) with outerplanar cover graph.

A second paper in which trees and cut vertices are discussed is [16], but the results of this chapter are considerably stronger. Here is a more recent result [5], and only recently has the connection with blocks and cut vertices become clear.

Theorem 11.3. *If P is a poset and the cover graph of P is outerplanar, then* $\dim(P) \leqslant 4$. *Furthermore, this inequality is best possible.*

In [5], the construction shown in Fig. 11.3 is given, and it is shown that when $n \geqslant 17$, the resulting poset has dimension 4. Note that the cover graph of this poset is outerplanar. As a consequence, the inequality in Theorem 11.3 is best possible. Moreover, every block of P_n is a 4-element subposet having dimension 2 (these subposets are called "diamonds"). We may then conclude that when $d = 2$, the inequality in our main theorem is best possible. However, the construction we present in Section 11.5 to show that our upper bound is best possible will again handle all values of d with $d \geqslant 2$ at the same time, so it will not use this result either.

The results of this chapter are part of a more comprehensive series of papers exploring connections between dimension of posets and graph-theoretic properties of their cover graphs. Recent related papers include [1, 8, 9, 10, 12, 13,

21, 22]. However, many of these modern research themes have their roots in results, such as Theorem 11.2, obtained in the 1970s or even earlier.

Here is one such example and again, Theorem 11.2 was the starting point. The result is given in [8].

Theorem 11.4. *For any positive integers t and h, there is a least positive d =* $d(t, h)$ *so that if P is a poset of height h and the cover graph of P has tree-width t, then* $\dim(P) \leqslant d$.

As discussed in greater detail in [8], the function $d(t, h)$ must go to infinity with h when $t \geqslant 3$, and in view of Theorem 11.2, it is bounded for all h when $t = 1$. These observations left open the question as to whether $d(2, h)$ is bounded or goes to infinity with h. It is now known that $d(2, h)$ is bounded [1, 9]. However, in attacking this problem, the fact that one can restrict their attention to posets with 2-connected cover graphs was a useful detail.

Also, the role of cut vertices in cover graphs surfaced in [22], where the following result, which is considerably stronger than Theorem 11.4, is proved.

Theorem 11.5. *For any positive integers t and h, there is a least positive integer d =* $d(n, h)$ *so that if P is a poset of height at most h and the cover graph of P does not contain the complete graph* K_n *as a minor, then* $\dim(P) \leqslant d$.

The proof given in [22] uses the machinery of structural graph theory. Subsequently, an alternative proof, using only elementary methods, was given in [12], and an extension to classes of graphs with bounded expansion was given in [10].

11.3 Dimension Theory Essentials

Let P be a poset with ground set X. Then, let $\mathrm{Inc}(P)$ denote the set of all ordered pairs $(x, y) \in X \times X$ where x is incomparable to y in P. The binary relation $\mathrm{Inc}(P)$ is of course symmetric, and it is empty when P is a total order—in this case, $\dim(P) = 1$.

A subset $R \subseteq \mathrm{Inc}(P)$ is *reversible* when there is a linear extension L of P so that $x > y$ in L for all $(x, y) \in R$. When $\mathrm{Inc}(P) \neq \emptyset$, the dimension of P is then the least positive integer d for which there is a covering

$$\mathrm{Inc}(P) = R_1 \cup R_2 \cup \cdots \cup R_d$$

such that R_j is reversible for each $j = 1, 2, \ldots, d$.

In the proof of the upper bound in our main theorem, we will apply these observations to show that a poset P has dimension at most $d + 2$ by first constructing a family $\mathcal{F} = \{L_1, L_2, \ldots, L_d\}$ of linear extensions of P and then setting $R = \{(x, y) \in P: x > y$ in L_j for all $j = 1, 2, \ldots, d\}$. If $R = \emptyset$, then $\dim(P) \leqslant d$, and when $R \neq \emptyset$, we will find a covering $R = R_{d+1} \cup R_{d+2}$, where both R_{d+1} and R_{d+2} are reversible. If L_{d+1} and L_{d+2} are linear extensions of P, so that $x > y$ in L_j whenever $(x, y) \in R_j$ for each $j = d+1, d+2$, then $\mathcal{R} = \{L_1, L_2, \ldots, L_d, L_{d+1}, L_{d+2}\}$ is a realizer of P, which shows $\dim(P) \leqslant d + 2$.

An indexed subset $\{(x_i, y_i): 1 \leqslant i \leqslant k\} \subseteq \mathrm{Inc}(P)$ is called an *alternating cycle* of length n when $x_i \leqslant y_{i+1}$ in P, for all $i = 1, 2, \ldots, k$ (here, subscripts are interpreted cyclically so that $x_n \leqslant y_1$ in P). In [20], the following elementary result is proved.

Lemma 11.1. *Let P be a poset and let $R \subseteq \mathrm{Inc}(P)$. Then R is reversible if and only if R does not contain an alternating cycle.*

The following construction was given in [3], where the concept of dimension was introduced. For an integer $d \geqslant 2$, let S_d be the following height 2 poset: S_d has d minimal elements $\{a_1, a_2, \ldots, a_d\}$ and d maximal elements $\{b_1, b_2, \ldots, b_d\}$; the partial ordering on S_d is defined by setting $a_i < b_j$ in S_d if and only if $i \neq j$. The poset S_d is called the *standard example* (of dimension d).

In dimension theory, standard examples play a role that in many ways parallels the role of complete graphs in the study of chromatic number, and we refer the reader to [2] for additional details on extremal problems for which results for graphs and results for posets have a similar flavor. In this chapter, we will need only the following basic information about standard examples, which was noted in [3].

Proposition 11.1. *For every $d \geqslant 2$, we have $\dim(S_d) = d$. In fact, if S_d is a standard example, and $a_i > b_i$ in a linear extension L of S_d then $a_j < b_j$ in L whenever $i \neq j$.*

11.4 Proof of the Upper Bound

Before launching into the main body of the proof, we pause to present an important proposition, which will be very useful in the argument to follow. When M is a linear extension of a poset P and $w \in P$, we will write $M = [A < w < B]$ when the elements of P can be labeled so that $M = [u_1 < u_2 < \cdots < u_m]$, $A = [u_1 < u_2 < \cdots < u_{k-1}]$, $w = u_k$, and $B = [u_{k+1} < u_{k+2} < \cdots < u_m]$. The generalization of this notation to an expression such as $M = [A < C < w < D < B]$ should be clear. Given this notation, the following is nearly self-evident, but it is stated for emphasis.

Proposition 11.2. *Let P be a poset, and let w be a cut vertex in P. Let P' and P'' be subposets of P such that P' ∩ P'' = {w} and the vertex w separates P' and P'' in the cover graph of P. If M' = [A < w < B] and M'' = [C < w < D] are linear extensions of P' and P'', respectively, then M = [A < C < w < D < B] is a linear extension of the subposet of P induced on P' ∪ P''. Furthermore, the restriction of M to P' is M' and the restriction of M to P'' is M''.*

The rule $M = [A < C < w < D < B]$ will be called the *merge rule*. When it is applied, we will consider $M' = [A < w < B]$ as a linear extension of an "old" subposet P' that shares a cut vertex w with a "new" subposet P'' for which $M'' = [C < w < D]$ is a linear extension. We apply the merge rule to form a linear extension $M = [A < C < w < D < B]$ of the union $P' \cup P''$ and note that M forces old points in $A \cup B$ to the outside while concentrating new points from $C \cup D$ close to w.

Now on to the proof. We fix a positive integer $d \geqslant 1$ and let P be a poset for which $\dim(B) \leqslant d$ for every block B of P. The remainder of the proof is directed toward proving that $\dim(P) \leqslant d + 2$. Let G be the cover graph of P. Since $d + 2 \geqslant 3$, we may assume that G is connected.

Let $\mathcal{B}$ be the family of blocks in P, and let $t = |\mathcal{B}|$. Then, let $\mathcal{B} = \{B_1, B_2, \ldots, B_t\}$ be any labeling of the blocks of P such that for every $i = 2, 3, \ldots, t$, one of the vertices of B_i belongs to some of the blocks $B_1, B_2, \ldots, B_{i-1}$. Such a vertex of B_i is unique and is a cut vertex of P—we call it the *root* of B_i and denote it by $\rho(B_i)$.

For every block $B_i \in \mathcal{B}$ and every element $u \in B_i$, we define the *tail* of u relative to B_i, denoted by $T(u, B_i)$, to be the subposet of P consisting of all elements $v \in \{u\} \cup B_{i+1} \cup B_{i+2} \cup \cdots \cup B_t$ for which every path from v to any vertex in B_i passes through u. Note that $T(u, B_i) = \{u\}$ if u is not a cut vertex. Also, if $u \in B_i$, $v \in B_{i'}$, and $(u, i) \neq (v, i')$, then either $T(u, B_i) \cap T(v, B_{i'}) = \emptyset$ or one of $T(u, B_i)$ and $T(v, B_{i'})$ is a proper subset of the other.

For every block $B_i \in \mathcal{B}$, using the fact that $\dim(B_i) \leqslant d$, we may choose a realizer $\mathcal{R}_i = \{L_j(B_i) : 1 \leqslant j \leqslant d\}$ of size d for B_i. For each $i = 1, 2, \ldots, t$, set $P_i = B_1 \cup B_2 \cup \cdots \cup B_i$. Note that when $2 \leqslant i \leqslant t$, we have $\rho(B_i) \in P_{i-1}$.

Fix an integer j with $1 \leqslant j \leqslant d$ and set $M_j(1) = L_j(B_1)$. Then, repeat the following for $i = 2, 3, \ldots, t$. Suppose that we have a linear extension $M_j(i - 1)$ of P_{i-1}. Let $w = \rho(B_i)$. Since $w \in P_{i-1}$, we can write $M_j(i - 1) = [A < w < B]$. If $L_j(B_i) = [C < w < D]$, we then use the merge rule to set $M_j(i) = [A < C < w < D < B]$. When the procedure halts, take $L_j = M_j(t)$. This construction is performed for all $j = 1, 2, \ldots, d$ to determine a family $\mathcal{F} = \{L_1, L_2, \ldots, L_d\}$ of linear extensions of P. The family $\mathcal{F}$ is a realizer for a poset P^* which is an extension of P. As outlined in the preceding section, we set $R = \{(x, y) \in \text{Inc}(P): x < y \text{ in } L_j \text{ for every } j = 1, 2, \ldots, d\}$. We will show that there is a covering $R = R_{d+1} \cup R_{d+2}$ of R by two reversible sets. This is enough to prove $\dim(P) \leqslant d + 2$.

Repeated application of Proposition 11.2 immediately yields the following.

Block Restriction Property. For each $j = 1, 2, \ldots, d$ and each block $B_i \in \mathcal{B}$, the restriction of L_j to B_i is $L_j(B_i)$.

When L is a linear order on a set X and $S \subseteq X$, we say S is an *interval* in L if $y \in S$ whenever $x, z \in S$ and $x < y < z$ in L. The next property follows easily from the observation that the merge rule concentrates new points close around the cut vertex w while pushing old points to the outside.

Interval Property for Tails. For every $j = 1, 2, \ldots, d$, and every pair (u, i) with $u \in B_i$, the tail $T(u, B_i)$ of u relative to B_i is an interval in L_j.

Let $(x, y) \in R$. Then, let i be the least positive integer for which every path from x to y in the cover graph of P contains at least two elements of the block B_i. We then define elements $u, v \in B_i$ by the following rules:

(1) u is the unique first common element of B_i with every path from x to y;
(2) v is the unique last common element of B_i with every path from x to y.

Note that $u \neq v$, $u = x$ when $x \in B_i$, and $v = y$ when $y \in B_i$.

Claim 11.1. *The following two statements hold:*

(1) $x \in T(u, B_i)$, $y \notin T(u, B_i)$, $y \in T(v, B_i)$, and $x \notin T(v, B_i)$;
(2) $u < v$ in P.

Proof. The first statement is an immediate consequence of the definition of tails. For the proof of the second statement, suppose to the contrary that $u \nless v$ in P. Since $u, v \in B_i$ and $u \neq v$, there is some j with $1 \leqslant j \leqslant d$ so that $u > v$ in $L_j(B_i)$. Therefore, $u > v$ in L_j. Since, $T(u, B_i)$ and $T(v, B_i)$ are disjoint intervals in L_j, we conclude that $x > y$ in L_j. This contradiction shows $u < v$ in P, as claimed. $\square$

Claim 11.2. *At least one of the following two statements holds:*

(1) *For all y' with $y' \geqslant x$ in P, we have $y' \in T(u, B_i)$ and $y' < y$ in P^*;*
(2) *For all x' with $x' \leqslant y$ in P, we have $x' \in T(v, B_i)$ and $x < x'$ in P^*.*

Proof. Suppose to the contrary that neither of the two statements holds. Since $T(u, B_i)$ is an interval in L_j, $x \in T(u, B_i)$, $y \notin T(u, B_i)$, and $x < y$ in L_j for each $j = 1, 2, \ldots, d$, there must exist some y' with $y' > x$ in P and $y' \notin T(u, B_i)$. Then a path in G from x to y' witnessing the inequality $x < y'$ in P must include the point u. In particular, this implies that $x \leqslant u$ in P. Similarly, we have $v \leqslant y$ in P. This and Claim 11.1 (2) yield $x \leqslant u < v \leqslant y$ in P, which is a contradiction.   □

Now, it is clear how to define the covering $R = R_{d+1} \cup R_{d+2}$. We assign (x, y) to R_{d+1} when the first statement in Claim 11.2 applies, and we assign it to R_{d+2} when the second statement applies. We show that R_{d+1} is reversible. The argument for R_{d+2} is analogous. Suppose to the contrary that R_{d+1} is not reversible. Then there is an integer $n \geqslant 2$ and an alternating cycle $\{(x_i, y_i): 1 \leqslant i \leqslant n\}$ contained in R_{d+1}. Then $x_i \leqslant y_{i+1}$ for each $i = 1, 2, \ldots, n$. However, since $(x_i, y_i) \in R_{d+1}$, we know that $y_{i+1} < y_i$ in P^* for every $i = 1, 2, \ldots, n$. This is impossible, because P^* is a partial order. The proof of the upper bound in Theorem 11.1 is now complete.

11.5 Proof That the Upper Bound Is Best Possible

As we noted previously, the examples shown in Fig. 11.2 show that our upper bound is best possible when $d = 1$. So in this section, we will fix an integer $d \geqslant 2$ and show that there is a poset P so that $\dim(P) = d + 2$ while $\dim(B) \leqslant d$ for every block B of P.

For a positive integer n, we let $\mathbf{n}$ denote the n-element chain $\{0 < 1 < \cdots < n - 1\}$. Also, we let $\mathbf{n}^d$ denote the Cartesian product of d copies of $\mathbf{n}$, that is, the elements of $\mathbf{n}^d$ are d-tuples of the form $u = (u_1, u_2, \ldots, u_d)$ where each coordinate u_i is an integer with $0 \leqslant u_i < n$. The partial order on $\mathbf{n}^d$ is defined by setting $u = (u_1, u_2, \ldots, u_d) \leqslant (v_1, v_2, \ldots, v_d) = v$ in $\mathbf{n}^d$ if and only if $u_i \leqslant v_i$ in $\mathbf{n}$ for all $i = 1, 2, \ldots, d$. As is well known, $\dim(\mathbf{n}^d) = d$ for all $n \geqslant 2$.

For each $n \geqslant 2$, we then construct a poset $P = P(n)$ as follows. We start with a base poset W which is a copy of $\mathbf{n}^d$. The base poset W will be a block in P, and W will also be the set of cut vertices in P. All other blocks in P will be "diamonds," that is, copies of the 2-dimensional poset on four points discussed in conjunction with Fig. 11.3. Namely, for each element $w \in W$, we attach a three-element chain $x_w < y_w < z_w$ so that $x_w < w < z_w$ while w is incomparable to y_w. In this way, the four-element subposet $\{w, x_w, y_w, z_w\}$ is a diamond.

We will now prove the following claim.

Claim 11.3. *If n is sufficiently large, then* $\dim(P) \geqslant d + 2$.

Proof. We must first gather some necessary tools from Ramsey theory. In particular, we need a special case of a result that has become known as the Product

Ramsey Theorem and appears in the classic text [7] as Theorem 5 on p. 113. However, we will use slightly different notation in discussing this result.

When $T_1, T_2, \ldots, T_d$ are k-element subsets of $X_1, X_2, \ldots, X_d$, respectively, we refer to the product $g = T_1 \times T_2 \times \cdots \times T_d$ as a $\mathbf{k}^d$-grid in $X_1 \times X_2 \times \cdots \times X_d$. Here is a formal statement of the version of the Product Ramsey Theorem we require for our proof.

Theorem 11.6. *For every 4-tuple (r, d, k, m) of positive integers with $m \geqslant k$, there is an integer $n_0 \geqslant k$ such that if $|X_i| \geqslant n_0$ for every $i = 1, 2, \ldots, d$, then whenever we have a coloring ϕ that assigns to each $\mathbf{k}^t$-grid g in $X_1 \times X_2 \times \cdots \times X_d$ a color $\phi(g)$ from a set R of r colors, then there is a color $\alpha \in R$ and there are m-element subsets $H_1, H_2, \ldots, H_d$ of $X_1, X_2, \ldots, X_d$, respectively, such that $\phi(g) = \alpha$ for every $\mathbf{k}^t$-grid g in $H_1 \times H_2 \times \cdots \times H_d$.*

We will apply this theorem with $k = 2$, and since k and d are now both fixed, we will just refer to a $\mathbf{k}^d$-grid as a grid. When $g = T_1 \times T_2 \times \cdots \times T_d$ is a grid, we consider the elements $w \in W$ with $w_j \in T_j$ for each $j = 1, 2, \ldots, d$. Clearly, there are 2^d such points. Counting the diamonds attached to these points, there are $4 \cdot 2^d$ points in P associated with the grid g. But we want to focus on $4d + 8$ of them.

First, we consider an antichain $A = \{a_1, a_2, \ldots, a_d\}$ defined as follows: for each $i, j = 1, 2, \ldots, d$, coordinate j of a_i is $\max(T_j)$ when $i = j$ and $\min(T_j)$ when $i \neq j$. Dually, the antichain $B = \{b_1, b_2, \ldots, b_d\}$ is defined as follows: for each $i, j = 1, 2, \ldots, d$, coordinate j of b_i is $\min(T_j)$ when $i = j$ and $\max(T_j)$ when $i \neq j$. We then note that when $1 \leqslant i, j \leqslant d$, we have $a_i < b_j$ in P if and only if $i \neq j$. As a consequence, the subposet of P determined by $A \cup B$ is the standard example S_d discussed previously. Note further that the points in the two antichains $\{x_{a_j} : 1 \leqslant j \leqslant d\}$ and $\{z_{b_j} : 1 \leqslant j \leqslant d\}$ also form a copy of S_d. Furthermore, if $x_{a_j} > z_{b_j}$ in some linear extension L of P, then $a_j > b_j$ in L.

Now, we consider two special points c and d in W associated with the grid g together with the points in the diamonds attached at c and d. First, we take c with $c_j = \min(T_j)$ for each $j = 1, 2, \ldots, d$, and then we take d with $d_j = \max(T_j)$ for each $j = 1, 2, \ldots, d$. We note that $x_c < c < d < z_d$ in P. However, we also note that both (x_c, y_d) and (y_c, z_d) are in $\mathrm{Inc}(P)$.

Now, suppose that $\dim(P) \leqslant d + 1$ and that $\mathcal{R} = \{L_1, L_2, \ldots, L_{d+1}\}$ is a realizer of P. We will argue to a contradiction provided n is sufficiently large. To accomplish this, we define a coloring ϕ of the grids in $\mathbf{n}^d$ using $(d + 1)^{d+2}$ colors. Let g be a grid and consider the $2d + 4$ points discussed earlier. The color $\phi(g)$ will be a vector $(\alpha_1, \alpha_2, \ldots, \alpha_{d+2})$ of length $d + 2$ defined as follows:

(1) For $j = 1, 2, \ldots, d$, α_j is the least index α for which $x_{a_j} > z_{b_j}$ in L_α;
(2) α_{d+1} is the least index α for which $x_c > y_d$ in L_α;
(3) α_{d+2} is the least index α for which $y_c > z_d$ in L_α.

We apply Theorem 11.6 with $r = (d+1)^{d+2}$ and $m = 3$. It follows that there is some fixed color $(\beta_1, \beta_2, \ldots, \beta_{d+2})$ such that for each $j = 1, 2, \ldots, d$, there is a three-element subset $H_j = \{u_{1,j} < u_{2,j} < u_{3,j}\} \subseteq \{0, 1, \ldots, n-1\}$ so that $\phi(g) = (\beta_1, \beta_2, \ldots, \beta_{d+2})$ for all grids g in $H_1 \times H_2 \times \cdots \times H_d$.

First, we claim that $\beta_{d+1} \neq \beta_{d+2}$. To see this, suppose that $\beta = \beta_{d+1} = \beta_{d+2}$. Then, let c, c', d, d' be vectors with $c_j = u_{1,j}, d_j = c'_j = u_{2,j}$ and $d'_j = u_{3,j}$ for each $j = 1, 2, \ldots, d$. Then

$$c > x_c > y_d = y_{c'} > z_{d'} > d' > c \quad \text{in } L_\beta.$$

Clearly, this is impossible.

In view of our earlier remarks concerning standard examples, we may relabel the linear extensions in the realizer so that $\beta_j = j$ for each $j = 1, 2, \ldots, d$. Given the fact that $\beta_{d+1} \neq \beta_{d+2}$, (at least) one of β_{d+1} and β_{d+2} is in $\{1, 2, \ldots, d\}$. We complete the argument assuming that $1 \leqslant j = \beta_{d+1} \leqslant d$, noting that the other case is analogous.

Now, let c, d, e be points in W defined as follows. First, set $c_j = u_{1,j}, e_j = u_{2,j}$ and $d_j = u_{3,j}$. Then, for each $i = 1, 2, \ldots, d$ with $i \neq j$, set $c_i = u_{1,i}, d_i = u_{2,i}$ and $e_i = u_{3,i}$. Note that $c < d$ and $c < e$ in P. It follows then that

$$c > x_c > y_d > x_d > z_e > e > c \quad \text{in } L_j,$$

where the inequality $x_d > z_e$ follows from the fact that $\phi(g) = (\beta_1, \beta_2, \ldots, \beta_{d+2})$ for the grid $g = \{u_{2,1}, u_{3,1}\} \times \{u_{2,2}, u_{3,2}\} \times \cdots \times \{u_{2,d}, u_{3,d}\}$. The contradiction completes the proof that the upper bound in our main theorem is best possible. □

11.6 Closing Comments

There are two other instances where the Product Ramsey Theorem has been applied to combinatorial problems for posets, although, to be completely accurate, the first uses it only implicitly. The following two inequalities are proved in [15].

Theorem 11.7. *Let P be a poset that is not an antichain, and let w be the width of the subposet $P - \mathrm{Max}(P)$, where $\mathrm{Max}(P)$ is the set of maximal elements of P. Then $\dim(P) \leqslant w + 1$.*

Theorem 11.8. *Let A be an antichain in a poset P. If $P - A \neq \emptyset$ and w is the width of $P - A$, then $\dim(P) \leqslant 2w + 1$.*

Both results admit quite simple proofs, and the only real challenge is to show that they are best possible. An explicit construction is given in [15] for a family of posets showing that Theorem 11.7 is tight, but the construction for the second is far more complicated and deferred to a separate paper [14]. Readers who are familiar with the details of this construction will recognize that it is an implicit application of the Product Ramsey Theorem.

The second application appears in [4], where it is shown that there is a *finite* three-dimensional poset that cannot be represented as a family of spheres in Euclidean space—of any dimension—ordered by inclusion. While it may in fact be the case that such posets exist with only a few hundred points, the proof produces an example that is extraordinarily large. This results from the fact that a further strengthening of the Product Ramsey Theorem to a lexicographic version (see [6]) is required.

We consider it a major challenge to construct examples of modest size for each of these three problems to replace the enormous posets resulting from the application of the Product Ramsey Theorem.

Acknowledgments

Bartosz Walczak was partially supported by National Science Center of Poland grant 2011/03/N/ST6/03111.

References

1. Csaba Bíró, Mitchel T. Keller, and Stephen J. Young. Posets with cover graph of pathwidth two have bounded dimension. *Order* **33**, no. 2 (2016), 195–212.
2. Csaba Bíró, Peter Hamburger, Attila Pór, and William T. Trotter. Forcing posets with large dimension to contain large standard examples. *Graphs Combin.* **32**, no. 3 (2016), 861–880.
3. Ben Dushnik and Edwin W. Miller. Partially ordered sets. *Amer. J. Math.* **63**, no. 3 (1941), 600–610.
4. Stefan Felsner, Peter C. Fishburn, and William T. Trotter. Finite three dimensional partial orders which are not sphere orders. *Discrete Math.* **201**, no. 1–3 (1999), 101–132.
5. Stefan Felsner, William T. Trotter, and Veit Wiechert. The dimension of posets with planar cover graphs. *Graphs Combin.* **31**, no. 4 (2015), 927–939.
6. Peter C. Fishburn and Ronald L. Graham. Lexicographic Ramsey theory. *J. Combin. Theory Ser. A* **62**, no. 2 (1993), 280–298.
7. Ronald L. Graham, Bruce L. Rothschild, and Joel H. Spencer. *Ramsey Theory*, 2nd edn. John Wiley & Sons, New York, 1990.
8. Gwenaël Joret, Piotr Micek, Kevin G. Milans, William T. Trotter, Bartosz Walczak, and Ruidong Wang. Tree-width and dimension. *Combinatorica* **36**, no. 4 (2016), 431–450.

9. Gwenaël Joret, Piotr Micek, William T. Trotter, Ruidong Wang, and Veit Wiechert. On the dimension of posets with cover graphs of treewidth 2. *Order* **34**, no. 2 (2017), 185–234.
10. Gwenaël Joret, Piotr Micek, and Veit Wiechert. Sparsity and dimension. *Combinatorica*, in press, doi: 10.1007/s00493-017-3638-4.
11. David Kelly. The 3-irreducible partially ordered sets. *Canad. J. Math.* **29** (1977), 367–383.
12. Piotr Micek and Veit Wiechert. Topological minors of cover graphs and dimension. *J. Graph Theory* **86**, no. 3 (2017), 295–314.
13. Noah Streib and William T. Trotter. Dimension and height for posets with planar cover graphs. *Eur. J. Combin.* **35** (2014), 474–489.
14. William T. Trotter. Irreducible posets with large height exist. *J. Combin. Theory Ser. A* **17**, no. 3 (1974), 337–344.
15. William T. Trotter. Inequalities in dimension theory for posets. *Proc. Amer. Math. Soc.* **47**, no. 2 (1975), 311–316.
16. William T. Trotter. Combinatorial problems in dimension theory for partially ordered sets. In *Problèmes Combinatoires et Théorie des Graphes*, Vol. 260 of *Colloques Internationeaux C.N.R.S.*, pp. 403–406, Éditions du C.N.R.S., Paris, 1978.
17. William T. Trotter. *Combinatorics and Partially Ordered Sets: Dimension Theory*. Johns Hopkins University Press, Baltimore, 1992.
18. William T. Trotter. Partially ordered sets. In Ronald L. Graham, Martin Grötschel, and László Lovász (eds.), *Handbook of Combinatorics*, Vol. I, pp. 433–480, North-Holland, Amsterdam, 1995.
19. William T. Trotter and John I. Moore. Characterization problems for graphs, partially ordered sets, lattices, and families of sets. *Discrete Math.* **16**, no. 4 (1976), 361–381.
20. William T. Trotter and John I. Moore. The dimension of planar posets. *J. Combin. Theory Ser. B* **22**, no. 1 (1977), 54–67.
21. William T. Trotter and Ruidong Wang. Dimension and matchings in comparability and incomparability graphs. *Order* **33**, no. 1 (2016), 101–119.
22. Bartosz Walczak. Minors and dimension. *J. Combin. Theory Ser. B* **122** (2017), 668–689.

12

Recent Results on Partition Regularity of Infinite Matrices

Neil Hindman

Abstract

We survey results obtained in the last 10 years on image and kernel partition regularity of infinite matrices.

12.1 Introduction

We let $\mathbb{N}$ be the set of positive integers and $\omega = \mathbb{N} \cup \{0\}$. We shall treat $u \in \mathbb{N}$ as an ordinal, so that $u = \{0, 1, \ldots, u-1\}$. Also $\omega = \{0, 1, 2, \ldots\}$ is the first infinite ordinal. Thus, if $u, v \in \mathbb{N} \cup \{\omega\}$, and A is a $u \times v$ matrix, the rows and columns of A will be indexed by $u = \{i : i < u\}$ and $v = \{i : i < v\}$, respectively.

As is standard in Ramsey Theory, a *finite coloring* of a set X is a function whose domain is X and whose range is finite. Similarly, a κ-coloring has range with cardinality κ. Given a coloring f of X, a subset B of X is *monochromatic* if and only if f is constant on B.

Definition 12.1. *Let $u, v \in \mathbb{N} \cup \{\omega\}$, let A be a $u \times v$ matrix with rational entries and finitely many nonzero entries per row, let S be a nontrivial subsemigroup of $(\mathbb{Q}, +)$, and let G be the subgroup of $\mathbb{Q}$ generated by S.*

(1) The matrix A is kernel partition regular over S *if and only if whenever $S \setminus \{0\}$ is finitely colored, there exists $\vec{x} \in (S \setminus \{0\})^v$ such that $A\vec{x} = \vec{0}$ and the entries of $\vec{x}$ are monochromatic.*

(2) The matrix A is image partition regular over S *if and only if whenever $S \setminus \{0\}$ is finitely colored, there exists $\vec{x} \in (S \setminus \{0\})^v$ such that the entries of $A\vec{x}$ are monochromatic.*

(3) The matrix A is weakly image partition regular over *S if and only if whenever S \ {0} is finitely colored, there exists $\vec{x} \in G^v$ such that the entries of $A\vec{x}$ are monochromatic.*

The definition of kernel partition regularity given is the only one that makes sense. However, different choices can be made for image partition regularity and weak image partition regularity, and the reader will find a sampling of these different choices in the references to this paper (unfortunately including some of the references with Hindman as an author). The reader will also find some places where one or the other of these notions is referred to simply as "partition regular."

Space consideration prevents me from explaining why these notions are interesting. The reader who needs convincing on this point is referred to the introduction to [11].

In 2005 I presented a survey [10] about image and kernel partition regularity of finite and infinite matrices at the *Integers Conference* 2005 *in Celebration of the 70th Birthday of Ron Graham*. The situation with respect to finite matrices was largely settled at that time, and consequently, there have not been many new results dealing with partition regularity of finite matrices published since then. In particular, characterizations of both kernel and image partition regularity were known (and included in [10]). And the relations among kernel and image partition regularity over various subsemigroups of $(\mathbb{R}, +)$ were largely settled.

In the case of infinite matrices, nothing close to a characterization of kernel or image partition regularity was known then, and that remains true today. (However, the main result of Section 12.3 completely characterizes kernel partition regularity in terms of image partition regularity.) And there were (and are) many open problems related to relationships among the notions. Progress on these problems is the subject of the current chapter.

Throughout this chapter, when I write that something "was known" without saying when, the reader may assume I mean "was known when [10] was written."

In Section 12.2 we present four examples showing that different suggestions for necessary or sufficient conditions for image or kernel partition regularity do not work.

In Section 12.3 we present a result showing that, given a matrix A there is a matrix B such that, for each nontrivial subsemigroup S of $\mathbb{Q}$, B is image partition regular over S if and only if A is kernel partition regular over S. We also present examples showing that two attempts to go in the other direction do not work.

In Section 12.4 we present a result establishing that if R and S are subrings of $\mathbb{Q}$ with 1 as a member and $R \setminus S \neq \emptyset$, then there is a matrix that is kernel partition regular over R but not kernel partition regular over S.

As a consequence of the results presented in Sections 12.3 and 12.4 one has immediately that if R and S are subrings of $\mathbb{Q}$ with 1 as a member and $R \setminus S \neq \emptyset$, then there is a matrix that is image partition regular over R but not image partition regular over S. We begin Section 12.5 with this observation. Section 12.5 also includes new results showing that an elaborate pattern of implications among various versions of image partition regularity has no valid implications except those diagramed.

Among the earliest known infinite matrices that are image partition regular over $\mathbb{N}$ are the Milliken–Taylor matrices. Section 12.6 consists of some new results about these matrices.

The final section, Section 12.7, presents some results establishing that certain matrices are or are not image partition regular.

Throughout the chapter, we will assume that hypothesized matrices have finitely many nonzero entries in each row. We will follow the custom of denoting the entries of a matrix with a capital letter name by the lowercase letter corresponding to that name. Given a semigroup S we will abbreviate "kernel partition regular over S," "image partition regular over S," and "weakly image partition regular over S" by KPR/S, IPR/S, and WIPR/S respectively.

12.2 In Search of Necessary or Sufficient Conditions

Rado in [21] and [22] characterized kernel partition regularity of a finite matrix A with rational entries over $\mathbb{N}$, $\mathbb{Z}$, $\mathbb{Q}$, or $\mathbb{R}$ via the *columns property*. (That is, if S is any one of $\mathbb{N}$, $\mathbb{Z}$, $\mathbb{Q}$, or $\mathbb{R}$, then A is KPR/S if and only if S satisfies the columns property.)

Definition 12.2. *Let $u, v \in \mathbb{N}$ and let A be a $u \times v$ matrix with entries from $\mathbb{Q}$. Denote the columns of A by $\langle \vec{c}_i \rangle_{i=0}^{v-1}$. The matrix A satisfies the* columns *property if and only if there exist $m \in \{1, 2, \ldots, v\}$ and a partition $\langle I_t \rangle_{t=0}^{m-1}$ of $\{0, 1, \ldots, v-1\}$ such that*

(1) $\sum_{i \in I_0} \vec{c}_i = \vec{0}$ *and*
(2) for each $t \in \{1, 2, \ldots, m-1\}$, $\sum_{i \in I_t} \vec{c}_i$ is a linear combination with coefficients from $\mathbb{Q}$ of $\{\vec{c}_i : i \in \bigcup_{j=0}^{t-1} I_j\}$.

The columns property has an obvious extension to infinite matrices.

Definition 12.3. *Let A be a countably infinite matrix with entries from $\mathbb{Q}$ and columns indexed by a set J. Denote the columns of A by $\langle \vec{c}_i \rangle_{i \in J}$. The matrix A satisfies the* columns property *if and only if there exists a partition $\langle I_\sigma \rangle_{\sigma < \mu}$ of J, where $\mu \in \mathbb{N} \cup \{\omega\}$, such that*

(1) $\sum_{i \in I_0} \vec{c}_i = \vec{0}$ and
(2) for each $t \in \mu \setminus \{0\}$, $\sum_{i \in I_t} \vec{c}_i$ is a linear combination with coefficients from $\mathbb{Q}$ of $\{\vec{c}_i : i \in \bigcup_{j<t} I_j\}$.

Note that the sums make sense even if I_j is infinite, since each row has only finitely many nonzero entries. It has been known that there are infinite matrices with integer entries satisfying the columns property that are not KPR/$\mathbb{N}$, but all previously known examples of matrices KPR/$\mathbb{N}$ had satisfied the columns property.

In the following theorem, B is an $\omega \times (\omega + \omega)$ matrix, so the definition of KPR/$\mathbb{N}$ requires an obvious adjustment.

Theorem 12.1. *Let A be the $\omega \times \omega$ matrix such that, for $i, j \in \omega$,*

$$a_{i,j} = \begin{cases} 2 & \text{if } j = i, \\ 1 & \text{if } 2^i \leq j < 2^{i+1}, \\ 0 & \text{otherwise,} \end{cases}$$

and let $B = (A \quad -I)$, where I is the $\omega \times \omega$ identity matrix. Then B is KPR/$\mathbb{N}$ and B does not satisfy the columns property. In fact, no nonempty set of columns of B sum to $\vec{0}$.

Proof. [4, Theorem 2.1]. □

On the other hand, it is shown in [4, Theorem 2.2] that if A is a countably infinite matrix with integer entries and bounded row sums, then some nonempty set of columns of A do sum to $\vec{0}$.

A version of a problem posed in [12] asked whether, if an infinite matrix A is IPR/$\mathbb{N}$, $\langle d_n \rangle_{n=0}^\infty$ is a sequence in $\mathbb{N}$, and $\mathbb{N}$ is finitely colored, must there exist $\vec{x}$ such that the entries of $A\vec{x}$ are monochromatic and for each $n < \omega$, $x_n \equiv 0 \pmod{d_n}$. In [5, Proposition 5], Barber and Leader give an example showing that the answer is "no" with $d_n = 2^n$ for each n.

All known examples of matrices that were KPR/$\mathbb{N}$ had bounded entries in each column. (More precisely, given a matrix A, one can first multiply each row by a constant so that the smallest absolute value of a nonzero entry in that row is 1, and then ask whether there is a column with unbounded entries.)

Theorem 12.2. *The matrix*

$$A = \begin{pmatrix} 2 & 1 & -1 & 0 & 0 & 0 & 0 & 0 & 0 & 0 & 0 & 0 & 0 & \cdots \\ 4 & 0 & 0 & 1 & 1 & -1 & -1 & 0 & 0 & 0 & 0 & 0 & 0 & \cdots \\ 8 & 0 & 0 & 0 & 0 & 0 & 0 & 1 & 1 & 1 & -1 & -1 & -1 & \cdots \\ \vdots & \vdots & \vdots & \vdots & \vdots & \vdots & \vdots & \vdots & \vdots & \vdots & \vdots & \vdots & \vdots & \ddots \end{pmatrix}$$

is KPR/$\mathbb{N}$.

Proof. [2, Theorem 9]. □

12.3 Relations Between Image and Kernel Partition Regularity

The new result that is the most definitive is a complete characterization of kernel partition regularity in terms of image partition regularity.

Theorem 12.3. *Let $u, v \in \mathbb{N} \cup \{\omega\}$ and let A be a $u \times v$ matrix with rational entries. Then there is a $v \times v$ matrix B with rational entries such that for each nontrivial subsemigroup S of $\mathbb{Q}$,*

$$\{\vec{x} \in (S \setminus \{0\})^v : A\vec{x} = \vec{0}\} = \{B\vec{y} : \vec{y} \in (S \setminus \{0\})^v\} \cap (S \setminus \{0\})^v.$$

Further for each such S, A is KPR/S if and only if B is IPR/S.

Proof. [16, Theorem 2.4]. □

Notice that in Theorem 12.3, the relevant kernel members of A are exactly equal to the relevant image members of B.

There is a corresponding result starting with an image partition regular matrix, but it only applies when $S = \mathbb{Q}$. In fact, it is shown in [16] that if S is a nontrivial proper subsemigroup of $\mathbb{Q}^+ = \{x \in \mathbb{Q} : x > 0\}$ or a nontrivial proper subgroup of $\mathbb{Q}$, then there is a 3×2 matrix B which is IPR/S for which the conclusion of Theorem 12.4 fails.

Theorem 12.4. *Let $u, v \in \mathbb{N} \cup \{\omega\}$, let B be a $u \times v$ matrix with rational entries. Then there exist $J \subseteq u$ and a $J \times u$ matrix A such that*

$$\{\vec{y} \in (\mathbb{Q} \setminus \{0\})^u : A\vec{y} = \vec{0}\} = \{B\vec{x} : \vec{x} \in \mathbb{Q}^v\} \cap (\mathbb{Q} \setminus \{0\})^u.$$

Further, A is WIPR/$\mathbb{Q}$ if and only if B is KPR/$\mathbb{Q}$.

Proof. [16, Theorem 2.8]. □

Given a matrix A that is IPR/$\mathbb{N}$ and has linearly dependent rows, there is a naturally associated matrix $B(A)$ which is KPR/$\mathbb{N}$. (This matrix is constructed using the linear dependence among the rows of A. See [16, Theorem 2.6] for details.) It was known that there exists an infinite matrix A with rational entries for which $B(A)$ is KPR/$\mathbb{N}$ but A is not IPR/$\mathbb{N}$. But what happened if the entries of A were integers was not known. In [5] Barber and Leader showed that there are matrices A_1 and A_2 with integer entries such that A_1 is IPR/$\mathbb{N}$, A_2 is not IPR/$\mathbb{N}$, but $B(A_2) = B(A_1)$ and so $B(A_2)$ is KPR/$\mathbb{N}$.

12.4 Relations Among Kernel Partition Regularity for Different Subsemigroups of $\mathbb{R}$

It was known ([17, Theorem 2.6]) that there exist infinite matrices that are IPR/$\mathbb{Q}$ but not IPR/$\mathbb{N}$. But the corresponding question for kernel partition regularity remained open. The main motivation for [13] was the question of whether the matrix

$$A = \begin{pmatrix} 1 & 1 & -1 & 0 & 0 & 0 & 0 & \cdots \\ \frac{1}{2} & 0 & 0 & 1 & -1 & 0 & 0 & \cdots \\ \frac{1}{3} & 0 & 0 & 0 & 0 & 1 & -1 & \cdots \\ \vdots & \vdots & \vdots & \vdots & \vdots & \vdots & \vdots & \ddots \end{pmatrix}$$

was KPR/$\mathbb{Q}$. (It is trivial that A is not KPR/$\mathbb{N}$.) Unfortunately (from the point of view of trying to answer the question), it is a consequence of the main result of [13] that this matrix is not even KPR/$\mathbb{R}$.

A few years later, the question of whether every matrix which is KPR/$\mathbb{Q}$ must also be KPR/$\mathbb{N}$ was answered.

Theorem 12.5. *The matrix*

$$A = \begin{pmatrix} \frac{1}{2} & 1 & -1 & 0 & 0 & 0 & 0 & 0 & 0 & 0 & 0 & 0 & \cdots \\ \frac{1}{4} & 0 & 0 & 1 & 1 & -1 & -1 & 0 & 0 & 0 & 0 & 0 & \cdots \\ \frac{1}{8} & 0 & 0 & 0 & 0 & 0 & 0 & 1 & 1 & 1 & -1 & -1 & -1 & \cdots \\ \vdots & \vdots & \vdots & \vdots & \vdots & \vdots & \vdots & \vdots & \vdots & \vdots & \vdots & \vdots & \vdots & \ddots \end{pmatrix}$$

is KPR/$\mathbb{Q}$ but not KPR/$\mathbb{N}$. In fact, A is KPR/$\mathbb{D}$, where $\mathbb{D}$ is the set of dyadic rationals.

Proof. [2, Theorem 12]. $\qquad\qquad\qquad\qquad\qquad\qquad\qquad\qquad\qquad\qquad$ $\square$

Theorem 12.5 was significantly extended in [3].

Definition 12.4. *Let P be the set of primes and let $F \subseteq P$. Then*

$$\mathbb{G}_F = \{a/b : a \in \mathbb{Z}, \ b \in \mathbb{N} \text{ and all prime factors of } b \text{ are in } F\}.$$

It is not hard to see that $\{\mathbb{G}_F : F \subseteq P\}$ is exactly the set of subrings R of $\mathbb{Q}$ with $1 \in R$.

Theorem 12.6. *Let* $\langle d_n \rangle_{n=1}^{\infty}$ *be a sequence in* $\mathbb{Q}$ *and let*

$$A = \begin{pmatrix} d_1 & 1 & 1 & -1 & 0 & 0 & 0 & 0 & 0 & 0 & 0 & 0 & 0 & \cdots \\ d_2 & 0 & 0 & 0 & 1 & 1 & 1 & -1 & 0 & 0 & 0 & 0 & 0 & \cdots \\ d_3 & 0 & 0 & 0 & 0 & 0 & 0 & 0 & 1 & 1 & 1 & 1 & -1 & \cdots \\ \vdots & \vdots & \vdots & \vdots & \vdots & \vdots & \vdots & \vdots & \vdots & \vdots & \vdots & \vdots & \vdots & \ddots \end{pmatrix}.$$

(1) Let F *and* H *be subsets of* P *with* $H \setminus F \neq \emptyset$, *pick* $q \in H \setminus F$, *and for each* $n \in \mathbb{N}$, *let* $d_n = \frac{1}{q^n}$. *Then* A *is KPR/*$\mathbb{G}_H$ *and* $A\vec{x} = \vec{0}$ *has no solutions in* $\mathbb{G}_F$.

(2) Enumerate P *as* $\langle p_n \rangle_{n=1}^{\infty}$. *For each* $n \in \mathbb{N}$, *let* $d_n = \prod_{t=1}^{n} \frac{1}{p_t}$. *Then* A *is KPR/*$\mathbb{Q}$ *and* $A\vec{x} = \vec{0}$ *has no solutions in* $\mathbb{G}_F$ *for any proper subset* F *of* P.

Proof. [3, Theorems 4.3 and 4.4]. □

12.5 Relations Among Image Partition Regularity for Different Subsemigroups of $\mathbb{R}$

We begin this section with an immediate consequence of results established earlier.

Theorem 12.7. *(1) Let* F *and* H *be subsets of* P *with* $H \setminus F \neq \emptyset$. *There is a matrix* B *such that* B *is IPR/*$\mathbb{G}_H$ *and no image of* C *is contained in* $\mathbb{G}_F$.

(2) There is a matrix B *such that* B *is IPR/*$\mathbb{Q}$, *and for each proper subset* F *of* $\mathbb{Q}$, B *has no image contained in* $\mathbb{G}_F$.

Proof. Theorems 12.3 and 12.6. □

In [6] the following two notions of image partition regularity near zero were introduced.

Definition 12.5. *Let* S *be a subsemigroup of* $(\mathbb{R}, +)$ *with* 0 *in the closure of* S, *let* $u, v \in \mathbb{N} \cup \{\omega\}$, *and let* A *be a* $u \times v$ *matrix with entries from* $\mathbb{Q}$.

(1) The matrix A *is image partition regular over* S *near zero (abbreviated IPR/*S_0) *if and only if, whenever* $S \setminus \{0\}$ *is finitely colored and* $\delta > 0$, *there exists* $\vec{x} \in S^v$ *such that the entries of* $A\vec{x}$ *are monochromatic and lie in the interval* $(-\delta, \delta)$.

(2) If $v = \omega$, *then* A *is image partition regular over* S *near zero in the strong sense (abbreviated IPR/*S_{0s}) *if and only if, whenever* $S \setminus \{0\}$ *is finitely*

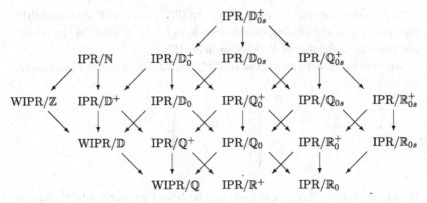

Fig. 12.1. Diagram of implications.

colored and $\delta > 0$, there exists $\vec{x} \in S^\omega$ such that $\lim\limits_{n \to \infty} x_n = 0$ and the entries of $A\vec{x}$ are monochromatic and lie in the interval $(-\delta, \delta)$.

It is trivial that all of the implications diagrammed in Fig. 12.1 hold. (In [6] IPR/S was defined so that, if S is a group, the notion is equivalent to what we have defined as WIPR/S.) Examples were presented in [6] showing that most of the missing implications were not valid in general. There were actually seventeen missing implications, but it was shown that if one had an example of a matrix that was IPR/$\mathbb{N}$ but not IPR/$\mathbb{R}_0$, then none of the seventeen missing implications was valid. (For example, it was not known whether every matrix which was IPR/$\mathbb{D}$ must be IPR/$\mathbb{Q}_0$. A matrix which is IPR/$\mathbb{N}$ but not IPR/$\mathbb{R}_0$ is IPR/$\mathbb{D}$ and is not IPR/$\mathbb{Q}_0$.) It was shown in [6] that the matrix

$$A = \begin{pmatrix}
1 & 0 & 0 & 0 & 0 & 0 & 0 & 0 & \dots \\
0 & 1 & 0 & 0 & 0 & 0 & 0 & 0 & \dots \\
2 & 1 & 0 & 0 & 0 & 0 & 0 & 0 & \dots \\
0 & 0 & 1 & 1 & 0 & 0 & 0 & 0 & \dots \\
4 & 0 & 1 & 0 & 0 & 0 & 0 & 0 & \dots \\
4 & 0 & 0 & 1 & 0 & 0 & 0 & 0 & \dots \\
0 & 0 & 0 & 0 & 1 & 1 & 1 & 1 & \dots \\
8 & 0 & 0 & 0 & 1 & 0 & 0 & 0 & \dots \\
8 & 0 & 0 & 0 & 0 & 1 & 0 & 0 & \dots \\
8 & 0 & 0 & 0 & 0 & 0 & 1 & 0 & \dots \\
8 & 0 & 0 & 0 & 0 & 0 & 0 & 1 & \dots \\
\vdots & \vdots & \vdots & \vdots & \vdots & \vdots & \vdots & \vdots & \ddots
\end{pmatrix}$$

is not IPR/$\mathbb{R}_0$, and the question was asked whether A is IPR/$\mathbb{N}$.

In [2] a different matrix was shown to be IPR/$\mathbb{N}$ but not IPR/$\mathbb{R}_0$, establishing that none of the missing implications in Fig. 12.1 is valid. In [1], Barber established that the matrix A shown earlier is IPR/$\mathbb{N}$.

It is easy to see that a matrix that is IPR/$\mathbb{Q}$ need not be WIPR/$\mathbb{N}$. For example, the matrix

$$\begin{pmatrix} 1 & 0 & 0 & \cdots \\ \frac{1}{2} & 1 & 0 & \cdots \\ \frac{1}{3} & 0 & 1 & \cdots \\ \vdots & \vdots & \vdots & \ddots \end{pmatrix}$$

from [17, Theorem 2.6] is easily seen to be IPR/$\mathbb{Q}$ but not WIPR/$\mathbb{N}$. But the question arises as to what happens if the entries of the matrix are assumed to be integers. Barber and Leader showed that the implication is still not valid.

Theorem 12.8. *There exists a sequence* $\langle c_n \rangle_{n=1}^{\infty}$ *in* $\mathbb{N}$ *such that the matrix*

$$\begin{pmatrix} 1 & 0 & 0 & 0 & 0 & 0 & 0 & \cdots \\ 0 & 2 & 0 & 0 & 0 & 0 & 0 & \cdots \\ c_1 & 2 & 0 & 0 & 0 & 0 & 0 & \cdots \\ 0 & 0 & 4 & 0 & 0 & 0 & 0 & \cdots \\ 0 & 0 & 0 & 4 & 0 & 0 & 0 & \cdots \\ c_2 & 0 & 4 & 4 & 0 & 0 & 0 & \cdots \\ 0 & 0 & 0 & 0 & 8 & 0 & 0 & \cdots \\ 0 & 0 & 0 & 0 & 0 & 8 & 0 & \cdots \\ 0 & 0 & 0 & 0 & 0 & 0 & 8 & \cdots \\ c_3 & 0 & 0 & 0 & 8 & 8 & 8 & \cdots \\ \vdots & \vdots & \vdots & \vdots & \vdots & \vdots & \vdots & \ddots \end{pmatrix}$$

is IPR/$\mathbb{Q}$ but not WIPR/$\mathbb{N}$.

Proof. [5, p. 296]. (The definition of image partition regularity they were using gave WIPR/$\mathbb{Q}$ rather than IPR/$\mathbb{Q}$, but it is easy to see that their proof establishes that the matrix is IPR/$\mathbb{Q}$.)   □

12.6 Milliken–Taylor Matrices

Among the earliest known infinite matrices that are IPR/$\mathbb{N}$ are the *Milliken–Taylor* matrices. They are so named because the fact that they are IPR/$\mathbb{N}$ follows easily from the Milliken–Taylor Theorem [19, Theorem 2.2], [23, Lemma 2.2].

Definition 12.6. *Let* $k \in \omega$ *and let* $\vec{a} = \langle a_0, a_1, \ldots, a_k \rangle$ *be a sequence in* $\mathbb{R}$ *such that* $\vec{a} \neq \vec{0}$. *The sequence* $\vec{a}$ *is compressed if and only if no* $a_i = 0$ *and for each* $i \in \{0, 1, \ldots, k-1\}$, $a_i \neq a_{i+1}$. *The sequence* $c(\vec{a}) = \langle c_0, c_1, \ldots, c_m \rangle$ *is the compressed sequence obtained from* $\vec{a}$ *by first deleting all occurrences of* 0 *and then deleting any entry which is equal to its successor. Then* $c(\vec{a})$ *is called the compressed form of* $\vec{a}$. *And* $\vec{a}$ *is said to be a compressed sequence if* $\vec{a} = c(\vec{a})$.

Definition 12.7. *Let* $k \in \omega$, *let* $\vec{a} = \langle a_0, a_1, \ldots, a_k \rangle$ *be a compressed sequence in* $\mathbb{R} \setminus \{0\}$, *and let* A *be an* $\omega \times \omega$ *matrix. Then* A *is an* $MT(\vec{a})$-matrix *if and only if the rows of* A *are all rows* $\vec{r} \in \mathbb{Z}^\omega$ *such that* $c(\vec{r}) = \vec{a}$. *The matrix* A *is a Milliken–Taylor matrix if and only if it is an* $MT(\vec{a})$-matrix *for some* $\vec{a}$.

Let $\vec{a}$ and $\vec{b}$ be compressed sequences in $\mathbb{Z} \setminus \{0\}$, let A be an $MT(\vec{a})$-matrix, and let B be an $MT(\vec{b})$-matrix. Two basic facts were known about A and B. First, A and B are IPR/$\mathbb{Z}$. Second, if $\vec{a}$ is not a rational multiple of $\vec{b}$, then there is a partition of $\mathbb{Z} \setminus \{0\}$ into two cells neither of which contains an image of both A and B.

In [7], De and Paul identified what is needed for a subfamily of a semigroup to contain images of all Milliken–Taylor matrices for compressed sequences in $\mathbb{N}$. (Restricting to positive entries in $\vec{a}$ was needed here since they were working with arbitrary semigroups, where $(-1)x$ may not mean anything.)

Theorem 12.9. *Let* $(S, +)$ *be an arbitrary (not necessarily commutative) semigroup, let* $\vec{a}$ *be a compressed sequence in* $\mathbb{N}$, *and let* $\mathcal{A} \subseteq \mathcal{P}(S)$ *such that*

(1) $(\forall A \in \mathcal{A})(\forall B \in \mathcal{A})(A \cap B \in \mathcal{A})$;
(2) $\mathcal{A} \neq \emptyset$ *and* $\emptyset \notin \mathcal{A}$;
(3) $(\forall A \in \mathcal{A})(\forall a \in A)(\exists B \in \mathcal{A})(a + B \subseteq A)$; *and*
(4) $(\forall A \in \mathcal{A})(\exists B \in \mathcal{A})(B + B \subseteq A)$.

Then whenever S *is finitely colored,* $A \in \mathcal{A}$, *and* M *is an* $MT(\vec{a})$ *matrix, there exists* $\vec{x} \in S^\omega$ *such that the entries of* $M\vec{x}$ *are monochromatic and in* A.

Proof. [7, Theorem 2.6]. □

In [8], the same authors extended the image partition regularity of Milliken–Taylor matrices to allow the entries of $\vec{a}$ to come from $\mathbb{R} \setminus \{0\}$.

Theorem 12.10. *Let* $\vec{a}$ *be a compressed sequence in* $\mathbb{R} \setminus \{0\}$ *with* $a_0 > 0$ *and let* M *be an* $MT(\vec{a})$ *matrix. Then* M *is IPR/$\mathbb{R}_0^+$.*

Proof. [8, Theorem 3.6]. □

Quite recently, the ability to separate Milliken–Taylor matrices was extended to the group $\mathbb{Q}$, allowing the compressed sequence to come from $\mathbb{Q} \setminus \{0\}$. (The fact that one can allow the terms of $\vec{a}$ to come from $\mathbb{Q} \setminus \{0\}$ is essentially trivial. The substance is producing the ability to appropriately color $\mathbb{Q} \setminus \{0\}$.)

Theorem 12.11. *Let $\vec{a}$ and $\vec{b}$ be compressed sequences in $\mathbb{Q} \setminus \{0\}$ such that $\vec{b}$ is not a multiple of $\vec{a}$, let A be an $MT(\vec{a})$-matrix, and let B be an $MT(\vec{b})$ matrix. There exists a 2-coloring of $\mathbb{Q} \setminus \{0\}$ such that there do not exist $\vec{x}$ and $\vec{y}$ in $(\mathbb{Q} \setminus \{0\})^\omega$ with the entries of $A\vec{x}$ together with the entries of $B\vec{y}$ monochromatic.*

Proof. [18, Corollary 4.5].   □

The conclusion of Theorem 12.11 (minus the information about the number of colors) can be restated as saying that the matrix $\begin{pmatrix} A & \mathbf{O} \\ \mathbf{O} & B \end{pmatrix}$ is not IPR/$\mathbb{Q}$. (The matrix is $(\omega + \omega) \times (\omega + \omega)$ rather than $\omega \times \omega$ so Definition 12.1 requires an obvious adjustment.)

Consequently, we find the following theorem to be very surprising. It says that Milliken–Taylor matrices determined by compressed sequences with final term equal to 1 are almost compatible. In this theorem, $\bar{1}$ and $\bar{0}$ are the length ω column vectors with constant value 1 and 0 respectively. Also $\mathbf{F}$ is an $MT(\langle 1 \rangle)$-matrix; that is a *finite sums* matrix. (The matrix B is an $\omega \cdot (m + 2) \times \omega$ matrix.)

Theorem 12.12. *Let $m \in \omega$ and for each $i \in \{0, 1, \ldots, m\}$, let $k(i) \in \mathbb{N}$, let $\vec{a}_i = \langle a_{i,0}, a_{i,1}, \ldots, a_{i,k(i)} \rangle$ be a compressed sequence in $\mathbb{Z} \setminus \{0\}$ with $a_{i,k(i)} = 1$, and let M_i be an $MT(\vec{a}_i)$-matrix. Then*

$$B = \begin{pmatrix} \bar{1} & \bar{0} & \ldots & \bar{0} & M_0 \\ \bar{0} & \bar{1} & \ldots & \bar{0} & M_1 \\ \vdots & \vdots & \ddots & \vdots & \vdots \\ \bar{0} & \bar{0} & \ldots & \bar{1} & M_m \\ \bar{0} & \bar{0} & \ldots & \bar{0} & \mathbf{F} \end{pmatrix}$$

is IPR/$\mathbb{N}$.

Proof. [15, Corollary 6.4].   □

12.7 Some Special Matrices

The first special kind of matrix with which we are concerned in this section is intimately related to the Milliken–Taylor matrices.

Definition 12.8. *Let $k \in \omega$ and let $\vec{a} = \langle a_0, a_1, \ldots, a_{k-1} \rangle$ be a sequence in $\mathbb{Z} \setminus \{0\}$. Then $M(\vec{a})$ is an $\omega \times \omega$ matrix that has all rows with a single 1 as the only nonzero entry as well as all rows whose nonzero entries are $a_0, a_1, \ldots, a_{k-1}$ in order, each occurring only once.*

Note that we are not assuming that $\vec{a}$ is a compressed sequence. If $\vec{x} \in \mathbb{N}^\omega$, then the entries of $M(\vec{a})\vec{x}$ are the entries of $\vec{x}$ together with all sums of the form $\sum_{i=0}^{k-1} a_i x_{n(i)}$, where $n(0) < n(1) < \ldots < n(k-1)$. We would like to characterize those sequences $\vec{a}$ for which $M(\vec{a})$ is IPR/$\mathbb{N}$ (and such a characterization is obtained in [9] for finite versions of $M(\vec{a})$). For infinite matrices, we are able to obtain a characterization only in the case that each a_i is plus or minus a power of a fixed integer.

Theorem 12.13. *Let $k, b \in \mathbb{N}$ and let $\vec{a} = \langle a_0, a_1, \ldots, a_{k-1} \rangle$ be a sequence in $\mathbb{Z} \setminus \{0\}$ such that for each $i \in \{0, 1, \ldots, k-1\}$, there is some $c \in \omega$ such that $a_i = b^c$ or $a_i = -b^c$. Then $M(\vec{a})$ is IPR/$\mathbb{N}$ if and only if one of*

(1). $a_0 + a_1 + \ldots + a_{k-1} = 0$ and $a_{k-1} = 1$;

(2) $a_0 + a_1 + \ldots + a_{k-1} = 1$; or

(3) $a_0 = a_1 = \ldots = a_{k-1} = 1$.

Proof. [9, Theorem 2.2]. $\qquad\qquad\qquad\qquad\qquad\qquad\qquad\qquad\qquad\qquad\square$

For the rest of this section we will be dealing with matrices that have constant row sums. Any such matrix will automatically be image partition regular via a constant sequence. (It is known – see [10, Theorem 4.8] – that a finite matrix A is IPR/$\mathbb{N}$ if and only if whenever $\mathbb{N}$ is finitely colored, there is some injective $\vec{x}$ such that the entries of $A\vec{x}$ are monochromatic.)

Definition 12.9. *Let $(S, +)$ be a semigroup, let κ be an infinite cardinal, and let A be a $\kappa \times \kappa$ matrix. Then A is injectively IPR/S if and only if, whenever S is finitely colored, there exists an injective $\vec{x} \in S^\kappa$ such that the entries of $A\vec{x}$ are monochromatic.*

Definition 12.10. *For $k \in \mathbb{N} \setminus \{1\}$ let R_k be an $\omega \times \omega$ matrix with entries from ω consisting of all rows, the sum of whose entries equals k.*

If $\vec{x} \in \mathbb{R}^\omega$, then the entries of $R_k\vec{x}$ are all sums of the form $\sum_{i=0}^{k-1} x_{n(i)}$ where $n(0) \leq n(1) \leq \ldots \leq n(k-1)$.

It is an old question of Owings [20] whether, whenever $\mathbb{N}$ is 2-colored, there must exist injective $\vec{x} \in \mathbb{N}^\omega$ with the entries of $F_2\vec{x}$ monochromatic. And it was known (and not hard to see) that for any $k \in \mathbb{N} \setminus \{1\}$, R_k is not injectively IPR/$\mathbb{N}$. It was not known whether R_k is injectively IPR/$\mathbb{Q}$ or injectively IPR/$\mathbb{R}$. (The latter is still not known without some special set theoretic assumptions.)

Since the entries of R_k are nonnegative integers, there is an obvious extension of the definition of injective image partition regularity to any semigroup $(S, +)$. We denote the nth infinite cardinal by ω_n, where as usual, a cardinal is the first ordinal of a given size, and we denote the cardinal of $\mathbb{R}$ by $\mathfrak{c}$.

Theorem 12.14. *Let $n \in \omega$, let $k \in \mathbb{N} \setminus \{1\}$, and let G be the direct sum of ω_n copies of $\mathbb{Q}$. Then R_k is not injectively IPR/G. In particular, if $\mathfrak{c} < \omega_{\omega+1}$, then R_k is not injectively IPR/$\mathbb{R}$.*

Proof. [14, Theorem 2.8]. □

Definition 12.11. *For $k \in \mathbb{N} \setminus \{1\}$ and an infinite cardinal κ, F_k^κ is a $\kappa \times \kappa$ matrix with entries from $\{0, 1\}$ consisting of all rows with exactly k occurrences of 1.*

Of course for any $k \in \mathbb{N} \setminus \{1\}$, F_k^ω is injectively IPR/$\mathbb{N}$ and therefore IPR/$\mathbb{R}$. Also, if $\kappa > \omega$, it is impossible for a $\kappa \times \kappa$ matrix to be injectively IPR/$\mathbb{Q}$. We do not know whether for some or all $k \in \mathbb{N} \setminus \{1\}$, $F_k^{\omega_1}$ is injectively IPR/$\mathbb{R}$, but the following result shows that $F_k^\mathfrak{c}$ is not. The coloring involved does not even depend on k.

Theorem 12.15. *There is a 2-coloring of $\mathbb{R}$ such that given any $k \in \mathbb{N} \setminus \{1\}$, there is no injective $\vec{x} \in \mathbb{R}^\mathfrak{c}$ with the entries of $(F_k^\mathfrak{c})\vec{x}$ monochromatic.*

Proof. [14, Theorem 3.2]. □

Acknowledgments

The author acknowledges support received from the National Science Foundation (USA) via Grants DMS-1160566 and DMS-1460023.

References

1. B. Barber. Partition regularity of a system of De and Hindman *Integers* **14** (2014), #A31.
2. B. Barber, N. Hindman, and I. Leader. Partition regularity in the rationals, *J. Combin. Theory Ser. A* **120** (2013), 1590–1599.
3. B. Barber, N. Hindman, I. Leader, and D. Strauss. Distinguishing subgroups of the rationals by their Ramsey properties. *J. Combin. Theory Ser. A* **129** (2014), 93–104.
4. B. Barber, N. Hindman, I. Leader, and D. Strauss. Partition regularity without the columns property. *Proc. Amer. Math. Soc.* **143** (2015), 3387–3399.
5. B. Barber and I. Leader. Partition regularity with congruence conditions. *J. Comb.* **4** (2013), 293–297.

6. D. De and N. Hindman. Image partition regularity near zero. *Discrete Math.* **309** (2009), 3219–3232.
7. D. De and R. Paul. Universally image partition regularity. *Electronic J. Combin.* **15** (2008), Research Paper 141.
8. D. De and R. Paul. Image partition regularity of matrices near 0 with real entries. *New York J. Math.* **17** (2011), 149–161.
9. D. Gunderson, N. Hindman, and H. Lefmann. Some partition theorems for infinite and finite matrices. *Integers* **14** (2014), #A12.
10. N. Hindman. Partition regularity of matrices. *Integers* **7(2)** (2007), #A18.
11. N. Hindman and I. Leader. Image partition regularity of matrices, *Combin. Prob. Comp.* **2** (1993), 437–463.
12. N. Hindman, I. Leader, and D. Strauss. Open problems in partition regularity. *Combin. Prob. Comp.* **12** (2003), 571–583.
13. N. Hindman, I. Leader, and D. Strauss. Forbidden distances in the rationals and the reals. *J. London Math. Soc.* **73** (2006), 273–286.
14. N. Hindman, I. Leader, and D. Strauss. Pairwise sums in colourings of the reals. *Abh. Math. Sem. Univ. Hamburg* **87** (2017), 275–287.
15. N. Hindman, I. Leader, and D. Strauss. Extensions of infinite partition regular systems. *Electron. J. Combin.* **22(2)** (2015), #P2.29.
16. N. Hindman, I. Leader, and D. Strauss. Duality for image and kernel partition regularity of infinite matrices. *J. Combin.* **8** (2017), 653–672.
17. N. Hindman and D. Strauss. Image partition regularity over the integers, rationals and reals. *New York J. Math.* **11** (2005), 519–538.
18. N. Hindman and D. Strauss. Separating Milliken-Taylor systems in $\mathbb{Q}$. *J. Combin.* **5** (2014), 305–333.
19. K. Milliken. Ramsey's Theorem with sums or unions. *J. Combin. Theory Ser. A* **18** (1975), 276–290.
20. J. Owings. Problem *E*2494. *Amer. Math. Monthly* **81** (1974), 902.
21. R. Rado. Studien zur Kombinatorik. *Math. Zeit.* **36** (1933), 242–280.
22. R. Rado. Note on combinatorial analysis. *Proc. London Math. Soc.* **48** (1943), 122–160.
23. A. Taylor. A canonical partition relation for finite subsets of ω. *J. Combin. Theory Ser. A* **21** (1976), 137–146.

13

Some Remarks on $\pi_{\wedge}$

Christian Reiher, Vojtěch Rödl, and Mathias Schacht

Abstract

We investigate extremal problems for hypergraphs satisfying the following density condition. A 3-uniform hypergraph $H = (V, E)$ is $(d, \eta, \wedge)$-*dense* if for any two subsets of pairs $P, Q \subseteq V \times V$ the number of pairs $((x, y), (x, z)) \in P \times Q$ with $\{x, y, z\} \in E$ is at least $d\,|\mathcal{K}_{\wedge}(P, Q)| - \eta\,|V|^3$, where $\mathcal{K}_{\wedge}(P, Q)$ denotes the set of pairs in $P \times Q$ of the form $((x, y), (x, z))$. For a given 3-uniform hypergraph F we are interested in the infimum $d \geq 0$ such that for sufficiently small η every sufficiently large $(d, \eta, \wedge)$-dense hypergraph H contains a copy of F and this infimum will be denoted by $\pi_{\wedge}(F)$. We present a few results for the case when $F = K_k^{(3)}$ is a complete 3-uniform hypergraph on k vertices. It will be shown that $\pi_{\wedge}(K_{2r}^{(3)}) \leq \frac{r-2}{r-1}$, which is sharp for $r = 2, 3, 4$, where the lower bound for $r = 4$ is based on a result of Chung and Graham [Edge-colored complete graphs with precisely colored subgraphs, *Combinatorica* **3**, 3–4 (1983), 315–324].

13.1 Introduction

13.1.1 Extremal Problems for Uniformly Dense Hypergraphs

We study extremal problems for 3-uniform hypergraphs and if not stated otherwise, by a hypergraph we always mean a 3-uniform hypergraph. Recall that for a given 3-uniform hypergraph F the *extremal number* $\text{ex}(n, F)$ denotes the maximal number of hyperedges a hypergraph $H = (V, E)$ on n vertices can have without containing a copy of F. Since the sequence $\text{ex}(n, F)/\binom{n}{3}$ is decreasing, the *Turán density*

$$\pi(F) = \lim_{n \to \infty} \frac{\text{ex}(n, F)}{\binom{n}{3}}$$

214

is well defined. The study of these extremal parameters was already initiated by Turán [12] more than 70 years ago, but despite a great deal of effort still only very few results are known and several variations were considered. In particular, Erdős and Sós (see, e.g., [2, 3]) suggested a variant, where one restricts to F-free hypergraphs H, that are *uniformly dense* on large subsets of the vertices. For reals $d \in [0, 1]$ and $\eta > 0$ we say a 3-uniform hypergraph $H = (V, E)$ is $(d, \eta, \therefore)$-*dense*, if all subsets $X, Y, Z \subseteq V$ induce at least $d|X||Y||Z| - \eta|V|^3$ triples $(x, y, z) \in X \times Y \times Z$ such that $\{x, y, z\}$ is a hyperedge of H. Restricting to $\therefore$-dense hypergraphs, the appropriate Turán density $\pi_{\therefore}(F)$ for a given hypergraph F can be defined as

$$\pi_{\therefore}(F) = \sup\{d \in [0, 1]: \text{for every } \eta > 0 \text{ and } n \in \mathbb{N} \text{ there exists a 3-uniform,}$$
$$F\text{-free}, (d, \eta, \therefore)\text{-dense hypergraph } H \text{ with } |V(H)| \geq n\}.$$

For $F = K_4^{(3)-}$, i.e., the 3-uniform hypergraph with three hyperedges on four vertices, it was shown by Glebov, Král, and Volec [5] that $\pi_{\therefore}(K_4^{(3)-}) = 1/4$ (see also [8]). However, when $F = K_4^{(3)}$ is the clique on four vertices, the interesting conjecture asking if

$$\pi_{\therefore}(K_4^{(3)}) = \frac{1}{2}$$

remains still open.

In [9] we considered the following stronger density notion. We say a 3-uniform hypergraph $H = (V, E)$ is $(d, \eta, \vdots)$-*dense*, if all subsets $X \subseteq V$ and sets of pairs $P \subseteq V \times V$ induce at least $d|X||P| - \eta|V|^3$ pairs $(x, (y, z)) \in X \times P$ such that $\{x, y, z\}$ is a hyperedge of H. Similarly as earlier, for this concept one defines the Turán density $\pi_{\vdots}(F)$ for a given hypergraph F by

$$\pi_{\vdots}(F) = \sup\{d \in [0, 1]: \text{for every } \eta > 0 \text{ and } n \in \mathbb{N} \text{ there exists a 3-uniform,}$$
$$F\text{-free}, (d, \eta, \vdots)\text{-dense hypergraph } H \text{ with } |V(H)| \geq n\}$$

and the main result in [9] asserts $\pi_{\vdots}(K_4^{(3)}) = 1/2$. Strictly speaking, in [8, 9] *quasirandom* hypergraphs H were considered, i.e., a matching upper bound on the number of hyperedges induced on $X \times Y \times Z$ (resp. $X \times P$) was formally required. However, in the proofs only the density condition (i.e., the lower bound on the number of induced hyperedges) is utilised. Moreover, for extremal problems involving embeddings of fixed hypergraphs it seems natural to restrict only to a lower bound on the density and here this path will be followed.

13.1.2 Main Result

Here we investigate the following density condition, which further strengthens the notion of $\cdot\!\!\cdot$-dense hypergraphs and is in some sense the strongest nontrivial density condition for extremal problems in 3-uniform hypergraphs (see, e.g., [9] for a more detailed discussion of the different conditions).

Definition 13.1. *A 3-uniform hypergraph $H = (V, E)$ on $n = |V|$ vertices is $(d, \eta, \boldsymbol{\wedge})$-dense if for any two subsets of pairs $P, Q \subseteq V \times V$ the number $e_{\boldsymbol{\wedge}}(P, Q)$ of pairs of pairs $((x, y), (x, z)) \in P \times Q$ with $\{x, y, z\} \in E$ satisfies*

$$e_{\boldsymbol{\wedge}}(P, Q) \geq d \, |\mathcal{K}_{\boldsymbol{\wedge}}(P, Q)| - \eta \, n^3, \qquad (13.1)$$

where $\mathcal{K}_{\boldsymbol{\wedge}}(P, Q)$ denotes the set of pairs in $P \times Q$ of the form $((x, y), (x, z))$. The class of $(d, \eta, \boldsymbol{\wedge})$-dense 3-uniform hypergraphs will be denoted by $\mathcal{D}(d, \eta, \boldsymbol{\wedge})$.

The corresponding Turán density for a given 3-uniform hypergraph F is then defined by

$$\pi_{\boldsymbol{\wedge}}(F) = \sup\{d \in [0, 1]\colon \text{for every } \eta > 0 \text{ and } n \in \mathbb{N} \text{ there exists a 3-uniform,}$$
$$F\text{-free hypergraph } H \in \mathcal{D}(d, \eta, \boldsymbol{\wedge}) \text{ with } |V(H)| \geq n\}.$$

Our main result establishes an upper bound on $\pi_{\boldsymbol{\wedge}}(F)$, when F is a complete 3-uniform hypergraph.

Theorem 13.1. *For every integer $r \geq 2$ and $\varepsilon > 0$ there exists an $\eta > 0$ and an integer n_0 such that every 3-uniform $(\frac{r-2}{r-1} + \varepsilon, \eta, \boldsymbol{\wedge})$-dense hypergraph H with at least n_0 vertices contains a copy of the complete 3-uniform hypergraph on 2^r vertices $K_{2^r}^{(3)}$, i.e., we have*

$$\pi_{\boldsymbol{\wedge}}(K_{2^r}^{(3)}) \leq \frac{r-2}{r-1}.$$

In particular, for $r = 2$ the theorem yields $\pi_{\boldsymbol{\wedge}}(K_4^{(3)}) = 0$ and in the text that follows we discuss lower bound constructions, which show that Theorem 13.1 is sharp for $r = 3$ and 4 as well. For a summary of our results for small cliques see (13.3). Moreover, a simple general construction given in the next section yields $\lim_{k \to \infty} \pi_{\boldsymbol{\wedge}}(K_k^{(3)}) = 1$.

The upper bound in Theorem 13.1 shows that the Turán densities for cliques for $\boldsymbol{\wedge}$-dense hypergraphs grow much slower than those for $\cdot\!\!\cdot$-dense hypergraphs. In fact, combined with a lower bound construction from [9] we have

$$\pi_{\boldsymbol{\wedge}}(K_{2^r}^{(3)}) \leq \frac{r-2}{r-1} \leq \pi_{\cdot\!\cdot}(K_{r+1}^{(3)}).$$

We also remark that the proof of Theorem 13.1 extends to k-colorable hypergraphs. Recall that a hypergraph F is *k-colorable*, if there exists a partition $V(F) = V_1 \cup \ldots \cup V_k$ such that no hyperedge of F is contained in some V_i for $i \in [k]$. For example, splitting the vertex set of $K_{2^r}^{(3)}$ into 2^{r-1} sets of size two shows that $K_{2^r}^{(3)}$ is 2^{r-1}-colorable and the proof of Theorem 13.1 presented here yields the same upper bound on $\pi_\wedge(F)$ for any 2^{r-1}-colorable hypergraph, i.e., if F is 2^{r-1}-colorable for some $r \geq 2$ then

$$\pi_\wedge(F) \leq \frac{r-2}{r-1}$$

(see also Remark 13.2).

13.1.3 Lower Bound Constructions

In this section we consider constructions yielding lower bounds for $\pi_\wedge(K_k^{(3)})$ where $k = 5$, 6, and 11. All constructions given here will be probabilistic, which will ensure the required $\wedge$-denseness (see Proposition 13.1). For the exclusion of the cliques of given order we shall utilize Ramsey-type arguments.

The following terminology will be useful. For a given finite set C of colors, by a *color pattern* we mean a multiset containing three (counting with repetition) elements from C and a set $\mathscr{P}$ of such patterns is called a *palette* over C. For example,

$$\mathscr{P} = \left\{ \{1, 1, 2\}, \{1, 1, 3\}, \{2, 2, 3\}, \{2, 2, 1\}, \{3, 3, 1\}, \{3, 3, 2\} \right\} \quad (13.2)$$

is a palette over $C = \{1, 2, 3\}$, which consists of all patterns using exactly two colors.

For $d \in [0, 1]$ we say a palette $\mathscr{P}$ over C is $(d, \wedge)$-*dense*, if any pair of (not necessarily distinct) colors of C appears in at least $d\,|C|$ patterns of $\mathscr{P}$. For example, it is easy to check that the palette given in (13.2) is $(2/3, \wedge)$-dense.

Next we describe the connection between $(d, \wedge)$-dense palettes and $(d, \eta, \wedge)$-dense hypergraphs. For a vertex set V and a coloring $\varphi \colon V^{(2)} \to C$ of the (unordered) pairs of V let $H_\varphi^{\mathscr{P}} = (V, E)$ be the 3-uniform hypergraph defined by

$$E = \left\{ \{x, y, z\} \in V^{(3)} \colon \{\varphi(x, y), \varphi(x, z), \varphi(y, z)\} \in \mathscr{P} \right\},$$

where $\{\varphi(x, y), \varphi(x, z), \varphi(y, z)\}$ is regarded as a multiset. Considering random colorings φ and a $(d, \wedge)$-dense palette $\mathscr{P}$ results for any given $\eta > 0$ with high probability for a sufficiently large set V in a $(d, \eta, \wedge)$-dense hypergraph $H_\varphi^{\mathscr{P}} = (V, E)$.

Proposition 13.1. *Suppose $\mathscr{P}$ is a $(d, \wedge)$-dense palette over some finite set of colors C. For every $\eta > 0$ and all sufficiently large sets V there exists a coloring $\varphi \colon V^{(2)} \to C$ of the unordered pairs of V such that the 3-uniform hypergraph $H_\varphi^{\mathscr{P}} = (V, E)$ is $(d, \eta, \wedge)$-dense.*

Moreover, if any coloring of the edges of the complete graph K_k with colors from C yields a triangle with a pattern not from $\mathscr{P}$, then $H_\varphi^{\mathscr{P}}$ contains no copy of $K_k^{(3)}$ and, consequently, $\pi_\wedge(K_k^{(3)}) \geq d$.

Before we prove the proposition we deduce some lower bounds for $\pi_\wedge(K_5^{(3)})$, $\pi_\wedge(K_6^{(3)})$, and $\pi_\wedge(K_{11}^{(3)})$ from it.

- We first show $\pi_\wedge(K_5^{(3)}) \geq 1/3$. For that we consider the $(1/3, \wedge)$-dense palette

$$\mathscr{P} = \big\{\{1, 1, 2\},\ \{2, 2, 3\},\ \{3, 3, 1\}\big\}$$

over $\{1, 2, 3\}$ and we have to show that the edges of the complete graph K_5 cannot be colored in such a way that all triangles get a pattern from $\mathscr{P}$. We consider three cases. If there is a vertex of K_5 incident with edges of each color, those three neighbors would have to span a triangle using all three colors, which is not a pattern in $\mathscr{P}$. Similarly, if there is a vertex incident with three edges of the same color, then the triangle in that neighborhood is monochromatic, which is also not contained in $\mathscr{P}$. Hence, every vertex has degree 2 or 0 in the three monochromatic subgraphs of K_5 given by the edge coloring. Therefore, the coloring induces a decomposition of K_5 into monochromatic cycles. But since monochromatic triangles are not allowed, this decomposition must consist of two cycles of length 5. However, this leads to two triangles with the patterns $\{a, a, b\}$ and $\{b, b, a\}$ for some $a \neq b$, but only one of these patterns is in the palette.

- Next we verify $\pi_\wedge(K_6^{(3)}) \geq 1/2$. Indeed, since every two-coloring of the edges of K_6 yields a monochromatic triangle, the assertion follows from the $(1/2, \wedge)$-dense palette $\{\{1, 1, 2\},\ \{2, 2, 1\}\}$ over $\{1, 2\}$. Combining this lower bound with the obvious monotonicity and with Theorem 13.1 for $r = 3$ leads to

$$\frac{1}{2} \leq \pi_\wedge(K_6^{(3)}) \leq \pi_\wedge(K_7^{(3)}) \leq \pi_\wedge(K_8^{(3)}) \leq \frac{1}{2}.$$

We also remark that this construction can be generalized for multi-color Ramsey numbers. For every integer ℓ let $k = R(3; \ell)$ be the ℓ-color Ramsey number for the triangle. Then the palette consisting of all but the monochromatic patterns over $[\ell] = \{1, \ldots, \ell\}$ yields $\pi_\wedge(K_k^{(3)}) \geq (\ell - 1)/\ell$, which shows that $\pi_\wedge(K_k^{(3)}) \to 1$ as k tends to infinity.

- The last construction establishes $\pi_\wedge(K_{11}^{(3)}) \geq 2/3$. For this we appeal to the palette given in (13.2), which is $(2/3, \wedge)$-dense over $\{1, 2, 3\}$. It follows from the result of Chung and Graham from [1] that every three-coloring of the edges of K_{11} yields either a monochromatic triangle or a rainbow triangle. Since these patterns are not in the palette in (13.2), it follows that $\pi_\wedge(K_{11}^{(3)}) \geq 2/3$. Together with the case $r = 4$ of Theorem 13.1 this yields

$$\frac{2}{3} \leq \pi_\wedge(K_{11}^{(3)}) \leq \cdots \leq \pi_\wedge(K_{16}^{(3)}) \leq \frac{2}{3}.$$

Summarizing the foregoing discussion for cliques of small size we established

$$\pi_\wedge(K_4^{(3)}) = 0$$
$$\tfrac{1}{3} \leq \pi_\wedge(K_5^{(3)})$$
$$\pi_\wedge(K_6^{(3)}) = \pi_\wedge(K_7^{(3)}) = \pi_\wedge(K_8^{(3)}) = \tfrac{1}{2} \leq \pi_\wedge(K_9^{(3)}) \leq \pi_\wedge(K_{10}^{(3)}) \quad (13.3)$$
$$\pi_\wedge(K_{11}^{(3)}) = \cdots = \pi_\wedge(K_{16}^{(3)}) = \tfrac{2}{3}$$

which leaves gaps for the values of $\pi_\wedge(K_k^{(3)})$ for $k = 5, 9, 10$ and it would be very interesting to close these. We conclude this introduction with the short proof of Proposition 13.1.

Proof of Proposition 13.1. For a given $(d, \wedge)$-dense palette $\mathscr{P}$ over some finite set of colors $\mathcal{C}$ and $\eta > 0$ we consider a random coloring $\varphi\colon V^{(2)} \to \mathcal{C}$ of the unordered pairs of some sufficiently large set V, where each pair is colored independently and uniformly with one of the colors from $\mathcal{C}$. We shall show that with probability tending to 1 as $|V| \to \infty$ the hypergraph $H_\varphi^{\mathscr{P}}$ is $(d, \eta, \wedge)$-dense.

We begin with the following observation. For any two subsets $Y, Z \subseteq V$ and any selection $\mathcal{C}' \subseteq \mathcal{C}$ of at least $d\,|\mathcal{C}|$ colors we expect at least $d(|Y||Z| - |Y \cap Z|)$ pairs $(y, z) \in Y \times Z$ with $y \neq z$ such that $\varphi(y, z) \in \mathcal{C}'$. Chernoff's inequality in the form

$$\mathbb{P}(X < \mathbb{E}X - t) \leq \exp(-\tfrac{t^2}{2\mathbb{E}X})$$

applied with $t = \eta|V|^2/|\mathcal{C}|^2 - |Y \cap Z|$ shows that at least

$$d|Y||Z| - \eta|V|^2/|\mathcal{C}|^2 \qquad (13.4)$$

such pairs will be present with probability at least $1 - \exp(-\frac{\eta^2|V|^2}{3|\mathcal{C}|^4})$. Consequently, applying the union bound over all choices of $Y, Z \subseteq V$ and $\mathcal{C}' \subseteq \mathcal{C}$ we infer that with probability tending to 1 as $|V| \to \infty$ that the bound in (13.4) holds for all these choices and for the rest of the proof we assume that φ satisfies this property.

Let $P, Q \subseteq V \times V$. For a color $c \in \mathcal{C}$ and a vertex $x \in V$ set

$$N_P^c(x) = \left\{ y \in V : (x, y) \in P \text{ and } \varphi(x, y) = c \right\}$$

and, similarly, define $N_Q^c(x)$. Clearly, we have

$$|\mathcal{K}_\wedge(P, Q)| = \sum_{x \in V} \sum_{c, c' \in \mathcal{C}} |N_P^c(x)||N_Q^{c'}(x)|.$$

Since the palette $\mathscr{P}$ is $(d, \wedge)$-dense, for any (not necessarily distinct) colors $c, c' \in \mathcal{C}$ the set $\mathcal{C}' = \{c'' \in \mathcal{C} : \{c, c', c''\} \in \mathscr{P}\}$ has size at least $d|\mathcal{C}|$. Hence, using the lower bound given in (13.4) for $Y = N_P^c(x)$ and $Z = N_Q^{c'}(x)$ yields that there are at least

$$d|N_P^c(x)||N_Q^{c'}(x)| - \eta|V|^2/|\mathcal{C}|^2.$$

triples (x, y, z) with $(x, y) \in P$, $\varphi(x, y) = c$, $(x, z) \in Q$, $\varphi(x, z) = c'$, and $\{x, y, z\} \in E(H_\varphi^{\mathscr{P}})$. Summing this estimate over all vertices $x \in V$ and colors $c, c' \in \mathcal{C}$ leads to

$$
\begin{aligned}
e_\wedge(P, Q) &\geq \sum_{x \in V} \sum_{c, c' \in \mathcal{C}} \left(d|N_P^c(x)||N_Q^{c'}(x)| - \eta|V|^2/|\mathcal{C}|^2 \right) \\
&= d \sum_{x \in V} \sum_{c, c' \in \mathcal{C}} |N_P^c(x)||N_Q^{c'}(x)| - \eta|V|^3 \\
&= d|\mathcal{K}_\wedge(P, Q)| - \eta|V|^3,
\end{aligned}
$$

which shows that $H_\varphi^{\mathscr{P}}$ is $(d, \eta, \wedge)$-dense.

The moreover-part follows directly from the assumed Ramsey-type property of K_k and the definitions of $H_\varphi^{\mathscr{P}}$ and $\pi_\wedge(\cdot)$. $\qquad\square$

13.2 Hypergraph Regularity Method

A key tool in the proof of Theorem 13.1 is the regularity lemma for 3-uniform hypergraphs. We follow the approach from [10, 11] combined with the results from [6] and [7] and in the text that follows we introduce the necessary notation.

For two disjoint sets X and Y we denote by $K(X, Y)$ the complete bipartite graph with that vertex partition. We say a bipartite graph $P = (X \cup Y, E)$ is (δ_2, d_2)-*regular* if for all subsets $X' \subseteq X$ and $Y' \subseteq Y$ we have

$$\left| e(X', Y') - d_2|X'||Y'| \right| \leq \delta_2|X||Y|,$$

where $e(X', Y')$ denotes the number of edges of P with one vertex in X' and one vertex in Y'. Moreover, for $k \geq 2$ we say a k-partite graph $P = (X_1 \cup \ldots \cup X_k, E)$ is (δ_2, d_2)-regular, if all of its $\binom{k}{2}$ naturally induced bipartite subgraphs $P[X_i, X_j]$ are (δ_2, d_2)-regular. For a tripartite graph

$P = (X \cup Y \cup Z, E)$ we denote by $\mathcal{K}_3(P)$ the triples of vertices spanning a triangle in P, i.e.,

$$\mathcal{K}_3(P) = \{\{x, y, z\} \subseteq X \cup Y \cup Z : xy, xz, yz \in E\}.$$

If the tripartite graph P is (δ_2, d_2)-regular, then the so-called *triangle counting lemma* implies

$$|\mathcal{K}_3(P)| \leq d_2^3 |X||Y||Z| + 3\delta_2 |X||Y||Z|. \tag{13.5}$$

We say a 3-uniform hypergraph $H = (V, E_H)$ is regular with respect to a tripartite graph P if it matches approximately the same proportion of triangles for every subgraph $Q \subseteq P$. This we make precise in the following definition.

Definition 13.2. *A 3-uniform hypergraph $H = (V, E_H)$ is (δ_3, d_3)-regular with respect to a tripartite graph $P = (X \cup Y \cup Z, E_P)$ with $V \supseteq X \cup Y \cup Z$ if for every tripartite subgraph $Q \subseteq P$ we have*

$$\big| |E_H \cap \mathcal{K}_3(Q)| - d_3 |\mathcal{K}_3(Q)| \big| \leq \delta_3 |\mathcal{K}_3(P)|.$$

Moreover, we simply say H is δ_3-regular with respect to P, if it is (δ_3, d_3)-regular for some $d_3 \geq 0$. We also define the relative density *of H with respect to P by*

$$d(H|P) = \frac{|E_H \cap \mathcal{K}_3(P)|}{|\mathcal{K}_3(P)|},$$

where we use the convention $d(H|P) = 0$ if $\mathcal{K}_3(P) = \varnothing$. If H is not δ_3-regular with respect to P, then we simply refer to it as δ_3-irregular.

The regularity lemma for 3-uniform hypergraphs, introduced by Frankl and Rödl in [4], provides for every hypergraph H a partition of its vertex set and a partition of the edge sets of the complete bipartite graphs induced by the vertex partition such that for appropriate constants δ_3, δ_2, and d_2

(1) The bipartite graphs given by the partitions are (δ_2, d_2)-regular and
(2) H is δ_3-regular with respect to "most" tripartite graphs P given by the partition.

In many proofs based on the regularity method it is convenient to "clean" the regular partition provided by the regularity lemma. In particular, we shall disregard hyperedges of H that belong to $\mathcal{K}_3(P)$ when H is not δ_3-regular or when $d(H|P)$ is very small. These properties are rendered in the following somewhat standard corollary of the regularity lemma.

Theorem 13.2. *For every $d_3 > 0$, $\delta_3 > 0$ and $m \in \mathbb{N}$, and every function $\delta_2 : \mathbb{N} \to (0, 1]$, there exist integers T_0 and n_0 such that for every $n \geq n_0$ and every n-vertex 3-uniform hypergraph $H = (V, E)$ the following holds.*

*There exists a subhypergraph $\hat{H} = (\hat{V}, \hat{E}) \subseteq H$, an integer $\ell \leq T_0$, a vertex
partition $V_1 \uplus \ldots \uplus V_m = \hat{V}$, and for all $1 \leq i < j \leq m$ there exists a partition*

$$\mathcal{P}^{ij} = \{P_\alpha^{ij} = (V_i \uplus V_j, E_\alpha^{ij}): 1 \leq \alpha \leq \ell\}$$

of $K(V_i, V_j)$ satisfying the following properties:

(1) $|V_1| = \cdots = |V_m| \geq (1 - \delta_3)n/T_0$.
(2) For every $1 \leq i < j \leq m$ and $\alpha \in [\ell]$ the bipartite graph P_α^{ij} is $(\delta_2(\ell), 1/\ell)$-regular.
(3) $\hat{H}$ is δ_3-regular with respect to $P_{\alpha\beta\gamma}^{ijk}$ for all tripartite graphs (which will be later referred to as triads)

$$P_{\alpha\beta\gamma}^{ijk} = P_\alpha^{ij} \uplus P_\beta^{ik} \uplus P_\gamma^{jk} = (V_i \uplus V_j \uplus V_k, E_\alpha^{ij} \uplus E_\beta^{ik} \uplus E_\gamma^{jk}), \qquad (13.6)$$

with $1 \leq i < j < k \leq m$ and $\alpha, \beta, \gamma \in [\ell]$, where the density $d(\hat{H}|P_{\alpha\beta\gamma}^{ijk})$ is either 0 or at least d_3.
(4) For every $1 \leq i < j < k \leq m$ there are at most $\delta_3 \ell^3$ triples $(\alpha, \beta, \gamma) \in [\ell]^3$ such that $d(\hat{H}|P_{\alpha\beta\gamma}^{ijk}) < d(H|P_{\alpha\beta\gamma}^{ijk}) - d_3$.

The standard proof of Theorem 13.2 based on a refined version of the regularity lemma from [10, Theorem 2.3] can be found in [8, Corollary 3.3]. Actually the statement there differs from the one given here in the final clause, but the proof from [8] shows the present version as well. In fact, the new version of (4) is a consequence of clause (a) in the definition of the hypergraph R in [8, Proof of Corollary 3.3], as we remove more than $d_3|\mathcal{K}(\mathcal{P}_{\alpha\beta\gamma}^{ijk})|$ hyperedges from H to obtain $\hat{H}$ only when H is δ_3-irregular with respect to $\mathcal{P}_{\alpha\beta\gamma}^{ijk}$.

We shall use a so-called *counting/embedding lemma*, which allows us to embed hypergraphs of fixed isomorphism type into appropriate and sufficiently regular and dense triads of the partition provided by Theorem 13.2. The following statement is a direct consequence of [7, Corollary 2.3].

Theorem 13.3 (Embedding Lemma). *For every 3-uniform hypergraph $F = (V_F, E_F)$ with vertex set $V_F = [f]$ and every $d_3 > 0$ there exists $\delta_3 > 0$, and functions $\delta_2 \colon \mathbb{N} \to (0, 1]$ and $N \colon \mathbb{N} \to \mathbb{N}$ such that the following holds for every $\ell \in \mathbb{N}$.*

Suppose $P = (V_1 \uplus \ldots \uplus V_f, E_P)$ is a $(\delta_2(\ell), \frac{1}{\ell})$-regular, f-partite graph with vertex classes satisfying $|V_1| = \cdots = |V_f| \geq N(\ell)$ and suppose H is an f-partite, 3-uniform hypergraph such that for every edge $ijk \in E_F$ we have

(1) H is δ_3-regular with respect to to the tripartite subgraph $P[V_i \uplus V_j \uplus V_k]$ and
(2) $d(H|P[V_i \uplus V_j \uplus V_k]) \geq d_3$

then H contains a copy of F, where for every $i \in [f] = V_F$ the image of i is contained in V_i.

In an application of Theorem 13.3 the tripartite graphs $P[V_i \cup V_j \cup V_k]$ in (*1*) and (*2*) will be given by triads $P_{\alpha\beta\gamma}^{ijk}$ from the partition given by Theorem 13.2.

For the proof of Theorem 13.1 we consider a $\wedge$-dense hypergraph H of density $\frac{r-2}{r-1} + \varepsilon$. We will apply the regularity lemma in the form of Theorem 13.2 to H. The main part of the proof concerns the appropriate selection of dense and regular triads, that are ready for an application of the embedding lemma with $F = K_{2r}^{(3)}$. This will be the focus in Sections 13.3 and 13.4.

13.3 Reduced Hypergraphs

Like many other proofs based on the regularity method, the proof of Theorem 13.1 will factor naturally through an auxiliary statement speaking about certain "reduced hypergraphs" that we would like to describe next.

Consider any finite set of indices I, suppose that associated with any two distinct indices $i, j \in I$ we have a finite nonempty set of vertices $\mathcal{P}^{ij} = \mathcal{P}^{ji}$, and that for distinct pairs of indices the corresponding vertex classes are disjoint. Assume further that for any three distinct indices $i, j, k \in I$ we are given a tripartite 3-uniform hypergraph $\mathcal{A}^{ijk}$ with vertex classes $\mathcal{P}^{ij}$, $\mathcal{P}^{ik}$, and $\mathcal{P}^{jk}$. Under such circumstances we call the $\binom{|I|}{2}$-partite 3-uniform hypergraph $\mathcal{A}$ defined by

$$V(\mathcal{A}) = \bigcup_{\{i,j\} \in I^{(2)}} \mathcal{P}^{ij} \quad \text{and} \quad E(\mathcal{A}) = \bigcup_{\{i,j,k\} \in I^{(3)}} E(\mathcal{A}^{ijk})$$

a *reduced hypergraph*. We also refer to I as the *index set* of $\mathcal{A}$, to the sets $\mathcal{P}^{ij}$ as the *vertex classes* of $\mathcal{A}$, and to the hypergraphs $\mathcal{A}^{ijk}$ as the *constituents* of $\mathcal{A}$.

This concept of a reduced hypergraph might look a bit artificial at first, especially since only $\binom{|I|}{3}$ out of the $\binom{\binom{|I|}{2}}{3}$ naturally induced tripartite subhypergraphs are inhabited. However, as it turns out these reduced hypergraphs are well suited for analyzing the structure of the partition provided by Theorem 13.2 applied to a given hypergraph H.

Now, when H happens to be $(d, \eta, \wedge)$-dense, then the corresponding reduced hypergraph $\mathcal{A}$ inherits a property reflecting this. We are thus led to the notion of a reduced hypergraph $\mathcal{A}$ being $(d, \delta, \wedge)$-dense. Roughly speaking, this means that all constituents of $\mathcal{A}$ are required to satisfy a δ-approximate pair-degree condition with proportion d, which is rendered in the following definition.

Definition 13.3. *A reduced hypergraph $\mathcal{A}$ with index set I is $(d, \delta, \boldsymbol{\lambda})$-dense for some $d \in [0, 1]$ and $\delta > 0$, if for any three distinct $i, j, k \in I$ the following is true:*

There are at most $\delta |\mathcal{P}^{ij}| |\mathcal{P}^{ik}|$ pairs of vertices $(P^{ij}, P^{ik}) \in \mathcal{P}^{ij} \times \mathcal{P}^{ik}$ with the property that there are fewer than $d |\mathcal{P}^{jk}|$ vertices $P^{jk} \in \mathcal{P}^{jk}$ for which $\{P^{ij}, P^{ik}, P^{jk}\} \in E(\mathcal{A}^{ijk})$ holds.

For an integer $t \geq 3$ we say that a reduced hypergraph $\mathcal{A}$ *contains a clique of order t* if there are

- a set $J \subseteq I$ with $|J| = t$
- and for any two distinct indices $i, j \in J$ a vertex $P^{ij} \in \mathcal{P}^{ij}$

such that we have $\{P^{ij}, P^{ik}, P^{jk}\} \in E(\mathcal{A}^{ijk})$ for any three distinct $i, j, k \in J$.

Now the statement to which we may reduce Theorem 13.1 via the regularity method is the following.

Proposition 13.2. *Given an integer $r \geq 2$ and a real $\varepsilon > 0$, there exists a real $\delta > 0$ and an integer m such that every $\left(\frac{r-2}{r-1} + \varepsilon, \delta, \boldsymbol{\lambda}\right)$-dense reduced hypergraph whose index set has size at least m contains a clique of order 2^r.*

Proof of Theorem 13.1 assuming Proposition 13.2. Roughly speaking, this reduction consists of two parts. Given a $\left(\frac{r-2}{r-1} + \varepsilon, \eta, \boldsymbol{\lambda}\right)$-dense hypergraph H we will apply the regularity lemma in the form of Theorem 13.2 and obtain a reduced hypergraph $\mathcal{A}$ for $\hat{H} \subseteq H$. In the first part we then verify that for an appropriate choice of the involved constants the reduced hypergraph $\mathcal{A}$ is indeed $\left(\frac{r-2}{r-1} + \varepsilon/4, \delta, \boldsymbol{\lambda}\right)$-dense. This allows for an application of Proposition 13.2 yielding a clique of order 2^r in $\mathcal{A}$. In the second part it remains to check that this clique of order 2^r in $\mathcal{A}$ defines an appropriate collection of triads ready for an application of the embedding lemma (Theorem 13.3) yielding a copy of $K_{2^r}^{(3)}$ in $\hat{H} \subseteq H$. Below we give the details of this proof.

Given $r \geq 2$ and $\varepsilon > 0$ we fix auxiliary constants and functions to satisfy the hierarchy

$$\frac{1}{r}, \varepsilon \gg \delta \gg \frac{1}{m}, d_3 \gg \delta_3 \gg \frac{1}{\ell} \gg \delta_2(\ell), \frac{1}{N(\ell)} \gg \frac{1}{T_0} \gg \eta, \tag{13.7}$$

where δ and m are given by Proposition 13.2 applied with r and $\varepsilon/4$; and δ_3, and the functions $\delta_2(\cdot)$, and $N(\cdot)$ are given by Theorem 13.3 applied for $F = K_{2^r}^{(3)}$ and d_3; and T_0 is given by Theorem 13.2.

For a $\left(\frac{r-2}{r-1} + \varepsilon, \eta, \boldsymbol{\lambda}\right)$-dense hypergraph $H = (V, E)$ on sufficiently many vertices, we apply the regularity lemma in the form of Theorem 13.2 and obtain a subhypergraph

$$\hat{H} = (\hat{V}, \hat{E}) \subseteq H,$$

some integer $\ell \leq T_0$, a vertex partition $V_1 \cup \ldots \cup V_m = \hat{V}$, and bipartite graphs P_α^{ij} for all i and j with $1 \leq i < j \leq m$ and every $\alpha \in [\ell]$ satisfying properties (1)–(4) of Theorem 13.2. For the index set $I = [m]$ we consider the naturally given reduced hypergraph $\mathcal{A}$ for the regular partition of $\hat{H}$ with vertex classes

$$\mathcal{P}^{ij} = \{P_\alpha^{ij} : \alpha \in [\ell]\}$$

for all distinct $i, j \in I$ and with constituents $\mathcal{A}^{ijk}$ for distinct $i, j, k \in I$ with

$$E(\mathcal{A}^{ijk}) = \{\{P_\alpha^{ij}, P_\beta^{ik}, P_\gamma^{jk}\} : (\alpha, \beta, \gamma) \in [\ell]^3 \text{ and } d(\hat{H}|P_{\alpha\beta\gamma}^{ijk}) \geq d_3\}.$$

Next we check that the reduced hypergraph $\mathcal{A}$ is $\left(\frac{r-2}{r-1} + \frac{\varepsilon}{4}, \delta, \Lambda\right)$-dense. Given distinct indices $i, j, k \in I$ and $P^{ij} \in \mathcal{P}^{ij}$ and $P^{ik} \in \mathcal{P}^{ik}$ it follows from the so-called graph counting lemma for graphs that for the $(\delta_2(\ell), 1/\ell)$-regular bipartite graphs P^{ij} and P^{ik} we have

$$\left| |\mathcal{K}_\Lambda(P^{ij}, P^{ik})| - \frac{1}{\ell^2} |V_i||V_j||V_k| \right| \leq 2\delta_2(\ell)|V_i||V_j||V_k|. \tag{13.8}$$

Consequently, the $(\frac{r-2}{r-1} + \varepsilon, \eta, \Lambda)$-denseness of H implies that the number $e_\Lambda^H(P^{ij}, P^{ik})$ of hyperedges in H matching P_2's from $\mathcal{K}_\Lambda(P^{ij}, P^{ik})$ satisfies

$$e_\Lambda^H(P^{ij}, P^{ik}) \geq \left(\frac{r-2}{r-1} + \varepsilon\right) \frac{1}{\ell^2} |V_i||V_j||V_k| - 2\delta_2(\ell)|V_i||V_j||V_k| - \eta n^3$$

$$\geq \left(\frac{r-2}{r-1} + \frac{\varepsilon}{2}\right) \frac{1}{\ell^2} |V_i||V_j||V_k|.$$

Owing to (13.5) we have $|\mathcal{K}_3(P^{ij} \cup P^{ik} \cup P^{jk})| \leq |V_i||V_j||V_k|/\ell^3 + 3\delta_2(\ell)|V_i||V_j||V_k|$ for every $P^{jk} \in \mathcal{P}^{jk}$ and combined with the upper bound in (13.8) and $\delta_2(\ell) \ll 1/\ell \ll \varepsilon$ we obtain

$$\left| \{P^{jk} \in \mathcal{P}^{jk} : d(H|P^{ij} \cup P^{ik} \cup P^{jk}) \geq d_3\} \right| \geq \left(\frac{r-2}{r-1} + \frac{\varepsilon}{3} - d_3\right) \ell \tag{13.9}$$

for any given pair $(P^{ij}, P^{ik}) \in \mathcal{P}^{ij} \times \mathcal{P}^{ik}$.

Property (4) of Theorem 13.2 implies that for all but up to at most $\sqrt{\delta_3}\ell^2$ pairs $(P^{ij}, P^{ik}) \in \mathcal{P}^{ij} \times \mathcal{P}^{ik}$ there are at most $\sqrt{\delta_3}\ell$ graphs $P^{jk} \in \mathcal{P}^{jk}$ such that

$$d(\hat{H}|P^{ij} \cup P^{ik} \cup P^{jk}) = 0, \text{ while } d(H|P^{ij} \cup P^{ik} \cup P^{jk}) \geq d_3.$$

Consequently, from (13.9) it follows that for all but at most $\sqrt{\delta_3}\ell^2$ pairs $(P^{ij}, P^{ik}) \in \mathcal{P}^{ij} \times \mathcal{P}^{ik}$ there are at least

$$\left(\frac{r-2}{r-1} + \frac{\varepsilon}{3} - d_3 - \sqrt{\delta_3}\right) \ell \geq \left(\frac{r-2}{r-1} + \frac{\varepsilon}{4}\right) |\mathcal{P}^{jk}|$$

graphs $P^{jk} \in \mathcal{P}^{jk}$ such that $\{P^{ij}, P^{ik}, P^{jk}\} \in E(\mathcal{A})$. In other words, since $\sqrt{\delta_3} \leq \delta$ the reduced hypergraph $\mathcal{A}$ for $\hat{H}$ is $\left(\frac{r-2}{r-1} + \frac{\varepsilon}{4}, \delta, \boldsymbol{\lambda}\right)$-dense.

Proposition 13.2 then shows that $\mathcal{A}$ contains a clique of order 2^r, i.e., there exists $J \subseteq [m]$ of size 2^r and bipartite graphs $P^{ij} \in \mathcal{P}^{ij}$ for any distinct $i, j \in J$ such that $\{P^{ij}, P^{ik}, P^{jk}\}$ is a hyperdge of $\mathcal{A}$ for all distinct $i, j, k \in J$. By the definition of $\mathcal{A}$ this shows that $d(\hat{H}|P^{ij} \cup P^{ik} \cup P^{jk}) \geq d_3$ and, hence, we may apply the embedding lemma (Theorem 13.3) to $\hat{H}[\bigcup_{j \in J} V_j]$ and $P = \bigcup_{\{i,j\} \in J^{(2)}} P^{ij}$ to obtain the desired clique $K_{2^r}^{(3)}$ in $\hat{H} \subseteq H$. $\qquad \square$

It is left to verify Proposition 13.2, which will be the content of the next section.

13.4 Embedding Cliques in the Reduced Hypergraph

In this section we shall provide a proof of Proposition 13.2. This will involve several inductions, which will require to prove a more general and somewhat technical statement (see Proposition 13.3). Instead of proving it directly it appears preferable to state and prove an even more general Proposition 13.4. We will show that

$$\text{Proposition } 13.4 \Longrightarrow \text{Proposition } 13.3 \Longrightarrow \text{Proposition } 13.2,$$

and thus the proof of Theorem 13.1 will be complete with the proof of Proposition 13.4.

To facilitate the wording of these generalizations, we introduce some further concepts. We will frequently deal with finite sequences of the form $a = (a_1, \ldots, a_k)$, where k is a nonnegative integer. If $k = 0$, then a is the *empty sequence* denoted by $\varnothing$. Generally, k is called the *length* of a and we express this by writing $k = |a|$. For an integer $\ell \in [0, k]$ the *restriction* $a|\ell$ is defined to be the *initial segment* $(a_1, \ldots, a_\ell)$ of a and for $\ell \in [k]$ we denote the ℓth element of a by $a(\ell)$, i.e., $a(\ell) = a_\ell$. A *direct continuation* of a is a finite sequence b obtainable from a by appending an arbitrary further term to it, so that b satisfies $|b| = k + 1$ and $b|k = a$. The *concatenation* of two finite sequences $a = (a_1, \ldots, a_k)$ and $b = (b_1, \ldots, b_\ell)$ is defined to be $(a_1, \ldots, a_k, b_1, \ldots, b_\ell)$ and denoted by $a \circ b$. Moreover, the longest common initial segment of a and b is denoted by $a \wedge b$. The following notion of regular trees will be useful.

Definition 13.4 ([k, M]-system). *For integers $k, M \geq 1$ an M-ary tree of height k is a set T of finite sequences whose length is at most k, such that*

- $\varnothing \in T$, *sometimes called the* root *of T, and*
- *every $a \in T$ with $|a| < k$ has precisely M direct continuations in T.*

The set of elements σ that extend some $a = (a_1, \ldots, a_\ell) \in T$ to a direct continuation $a \circ (\sigma) \in T$ are the successors *of a denoted by*

$$\mathscr{S}_T(a) = \{\sigma : (a_1, \ldots, a_\ell, \sigma) \in T\}.$$

The leaves *of T are its elements of length k and the set of these leaves is denoted by $[T]$. We say a set S is a $[k, M]$-system if $S = [T]$ holds for some M-ary tree T of height k.*

Giving two examples, we would like to mention that any set consisting of M elements can be viewed as a $[1, M]$-system, while the boolean cube $\{0, 1\}^k$ is a $[k, 2]$-system. Moreover, we notice that we have $|S| = M^k$ for any $[k, M]$-system S.

In the iterated Ramsey-type arguments that we use in the proofs of this section we will move from $[k, M]$-systems to $[k, m]$-systems for some $m \ll M$. However, although we can preserve the tree structure, we have no control about the subtree of the original $[k, M]$-system that will be kept after a Ramsey argument. In fact, this is the reason why we prefer to work with trees and $[k, M]$-systems, instead of sets S of the form $S = \mathcal{M}^k$ for some M-element set $\mathcal{M}$.

For the proof of Proposition 13.2 we are given a $\left(\frac{r-2}{r-1} + \varepsilon, \delta, \Lambda\right)$-dense reduced hypergraph $\mathcal{A}$ with index set I and we may assume that $\delta \ll |I|^{-1} \ll \varepsilon$. We then need to obtain some $J \subseteq I$ with $|J| = 2^r$ that spans a clique in $\mathcal{A}$. For that we will view I as an $[r, M]$-system for some large integer M and use Ramsey-type arguments for shrinking I down to an appropriate $[r, 2]$-system J, such that the required vertices and edges of $\mathcal{A}$ exist (see Definition 13.5 and Fact 13.1). We begin with the following observation, which follows by a simple averaging argument, and that will be utilized in the proof of Proposition 13.4.

Lemma 13.1. *Given two integers $k, M \geq 1$, let S be a $[k, M]$-system and let X be a subset of S satisfying $|X| \geq \varepsilon M^k$ for some $\varepsilon > 0$. Then for some integer $m \geq \varepsilon M/k$ there exists a $[k, m]$-system $S' \subseteq X$.*

Proof. We argue by induction on k. The base case $k = 1$ is clear because X is automatically going to be a $[1, |X|]$-system. Now suppose that $k \geq 2$ and that the lemma holds for $k - 1$ in place of k. Let S and X be as earlier and denote the underlying tree of S by T. Clearly the set $A = \{a \in T : |a| = 1\}$ has size M and for every $a \in A$ the set $S_a = \{b : a \circ b \in S\}$ is a $[k - 1, M]$-system. Consider for every $a \in A$ the set

$$X_a = \{b \in S_a : a \circ b \in X\}$$

and let

$$A' = \left\{ a \in A : |X_a| \geq \tfrac{(k-1)\varepsilon}{k} \cdot M^{k-1} \right\}.$$

Due to

$$\varepsilon M^k \leq |X| = \sum_{a \in A} |X_a| \leq \tfrac{(k-1)\varepsilon}{k} \cdot M^k + |A'| \cdot M^{k-1},$$

we have $|A'| \geq \varepsilon M/k$ and we can select a subset $A'' \subseteq A'$ with $|A''| = \lceil \varepsilon M/k \rceil =: m$. Moreover, for every $a \in A''$ we may apply the induction hypothesis to $X_a \subseteq S_a$ with $\varepsilon' = (k-1)\varepsilon/k$, thus obtaining a $[k-1, m]$-system $S'_a \subseteq X_a$. Consequently,

$$S' = \{a \circ b : a \in A'' \text{ and } b \in S'_a\}$$

is a $[k, m]$-system contained in X. $\qquad\qquad\qquad\qquad\qquad\qquad\square$

Next we introduce the somewhat technical notion of a *fortress* (see Definition 13.5), in a reduced hypergraph $\mathcal{A}$. The additional structural requirements for $[k, M]$-systems to support a fortress serve two purposes: first a $[r, 2]$-system that supports a fortress will give rise to a clique of order 2^r in $\mathcal{A}$ and second $[r, M]$-systems for $M \geq 2$ that support fortresses will be "rich enough" for the intended inductive arguments.

Consider an M-ary tree T of height k and the associated $[k, M]$-system $[T]$. For every sequence $c = (c_1, \ldots, c_\ell) \in T$ we set

$$Q(c) = \left\{ (d_1, \ldots, d_\ell) : d_i \in \mathscr{S}_T(c|(i-1)) \smallsetminus \{c_i\} \text{ for every } i \in [\ell] \right\}.$$

Since T is an M-ary tree, there are $M - 1$ successors of $c|(i-1)$ different from c_i for every $i \in [\ell]$ and thus we have $|Q(c)| = (M-1)^{|c|}$ for each $c \in T$. Moreover, it follows from the definition that for the empty sequence $\varnothing$ we have $Q(\varnothing) = \{\varnothing\}$. We also remark that $Q(c)$ is not necessarily a subset of T. For example, if $(a, \alpha_1), (a, \alpha_2), (a, \alpha_3), (b, \beta_1), \ldots, (c, \gamma_3)$ are the leaves of a ternary tree of height 2, then $Q((b, \beta_2))$ consists of $(a, \beta_1), (a, \beta_3), (c, \beta_1)$, and (c, β_3). In the next definition we will make use of the fact, that if $d \in Q(c)$, then $d|s \in Q(c|s)$ for any $s = 0, \ldots, |c|$.

Definition 13.5 (fortress). *Let T be an M-ary tree of height k and let $\mathcal{A}$ be a reduced hypergraph whose index set I contains the $[k, M]$-system $S = [T]$. We say S supports a fortress in $\mathcal{A}$ if for every $a, b \in S$ with $a \neq b$ and for every $d \in Q(a \wedge b)$ there exists some vertex $P_d^{ab} \in \mathcal{P}^{ab} \subseteq V(\mathcal{A})^1$ such that*

---

[1] To be consistent with the notation in Section 13.2 we should maybe write something like $P_{\alpha_{ab}(d)}^{ab}$ where $1 \leq \alpha_{ab}(d) \leq |\mathcal{P}^{ab}|$. However, for a simpler notation we will suppress such functions $\alpha_{ab} : Q(a \wedge b) \to [|\mathcal{P}^{ab}|]$ and simply write P_d^{ab}.

(F) for all distinct $a, b, c \in S$ satisfying

$$s := |a \wedge b| = |a \wedge c| < |b \wedge c|, \tag{13.10}$$

and for every $d \in Q(b \wedge c)$ with $d(s+1) = a(s+1)$ we have

$$\{P^{ab}_{d|s}, P^{ac}_{d|s}, P^{bc}_d\} \in E(\mathcal{A}^{abc}).$$

We refer to the set of vertices $\mathcal{F} = \{P^{ab}_d : a, b \in S,\ a \neq b,\ and\ d \in Q(a \wedge b)\}$ as a fortress *and we say that $\mathcal{A}$ contains a $[k, M]$-fortress if some subset of I is a $[k, M]$-system supporting a fortress.*

Remark 13.1. Note that property *(F)* is void for any $[1, M]$-fortress, since for a $[1, M]$-system S we have $a \wedge b = \varnothing$ for any two distinct $a, b \in S$. Therefore, (13.10) will never hold for distinct $a, b, c \in S$ and we can select $P^{ab}_\varnothing \in \mathcal{P}^{ab}$ arbitrarily.

As mentioned earlier, we now show that for $r \geq 2$ a $[r, 2]$-fortress yields a clique of order 2^r in $\mathcal{A}$.

Fact 13.1. *For every integer $r \geq 2$ a reduced hypergraph $\mathcal{A}$ contains a $[r, 2]$-fortress if and only if it contains a clique of order 2^r.*

Proof. For a binary tree T of height r we have $|Q(c)| = 1$ for any $c \in T$. Consequently, an $[r, 2]$-fortress contained in $\mathcal{A}$ corresponds to a subset J of the index set of $\mathcal{A}$ of size $|J| = 2^r$ and a selection $\{P^{ab} \in \mathcal{P}^{ab} : a, b \in J$ and $a \neq b\}$ such that $\{P^{ab}, P^{ac}, P^{bc}\} \in E(\mathcal{A})$, whenever (13.10) holds. In fact, the condition $d(s+1) = a(s+1)$ for the unique $d \in Q(b \wedge c)$ follows for binary trees directly from (13.10). Moreover, any three distinct leaves of a binary tree can be labeled a, b, c in such a way that (13.10) holds and, consequently, an $[r, 2]$-fortress corresponds to a clique of order 2^r in $\mathcal{A}$.

On the other hand, if $\mathcal{A}$ with index set I contains a clique of order 2^r, then there exists a subset $J \subseteq I$ of size 2^r and vertices $P^{ij} \in \mathcal{P}^{ij}$ for all distinct i, $j \in J$ such that $\{P^{ij}, P^{ik}, P^{jk}\}$ is a hyperedge of $\mathcal{A}$ for all distinct $i, j, k \in J$. Relabeling all elements of J by binary sequences of length r gives rise to a binary tree of height r that carries a fortress in $\mathcal{A}$. $\quad\square$

Remark 13.2. It is not hard to show that a $[2, m]$-fortress in $\mathcal{A}$ gives rise to a situation in which the embedding lemma (Theorem 13.3) can be applied for any hypergraph F on at most m vertices with the property that one can color the vertices and the pairs of vertices of F with m colors such that every hyperedge $\{x, y, z\}$ of F contains exactly two vertices, say x and y, that are colored by the same color, say red, and the pair $\{x, y\}$ and the vertex z have the same color different from red.

Note that this includes for example all 2-colorable hypergraphs F and in view of Proposition 13.3 for $r = 2$ this can be used to show that $\pi_\Lambda(F) = 0$ holds for any 2-colorable 3-uniform hypergraph F. In fact, for general r one can show that $[r, m]$-fortresses allow the embedding of 2^{r-1}-partite 3-uniform hypergraphs F with m vertices and, as a result, one can deduce $\pi_\Lambda(F) = \frac{r-2}{r-1}$ for such F. We omit the details here.

Proposition 13.2 follows by Fact 13.1g from the case $m = 2$ of Proposition 13.3.

Proposition 13.3. *Suppose that integers $r, m \geq 2$ and a real $\varepsilon > 0$ are given. Then there are a real $\delta > 0$ and an integer M with the property that every $\left(\frac{r-2}{r-1} + \varepsilon, \delta, \Lambda\right)$-dense reduced hypergraph whose index set is an $[r, M]$-system contains an $[r, m]$-fortress.*

The proof of Proposition 13.3 in turn proceeds in r steps. The first idea of this kind one might come up with is to wish proving by induction on k that for $k \in [r]$ and $m \ll M \ll \delta^{-1}$ every $\left(\frac{r-2}{r-1} + \varepsilon, \delta, \Lambda\right)$-dense reduced hypergraph, whose index set I is a $[k, M]$-system, contains a $[k, m]$-fortress. However, for $k < r$ we have $\frac{k-2}{k-1} < \frac{r-2}{r-1}$ and Proposition 13.3 asserts, that a $[k, m]$-fortress already appear in $\left(\frac{k-2}{k-1} + \varepsilon, \delta, \Lambda\right)$-dense reduced hypergraphs. This seems to indicate that for $k < r$ we can insist on additional side conditions, which will be utilized in the inductive step and that become weaker for larger k.

The condition that turned out to work for us says roughly the following: Suppose that the index set I contains, besides the $[k, M]$-system X_0 under discussion, also some further sets $X_1, \ldots, X_{r-k}$ that we know to be "well-attached" to X_0 in the sense that we are given for every $x \in X_0$ and every $y \in \bigcup_{j \in [r-k]} X_j$ a vertex $P^{xy} \in \mathcal{P}^{xy}$ such that for all choices $x, x' \in X_0$ and $y \in \bigcup_{j \in [r-k]} X_j$ the vertices P^{xy} and $P^{x'y}$ have high pair-degree in $\mathcal{P}^{xx'}$ (see Definition 13.6). Note that this property would be given automatically for any choice of P^{xy}, if there would be no exceptional pairs in $\mathcal{A}$ in the sense of Definition 13.3.

This "well-attachedness" allows us to shrink the sets $X_1, \ldots, X_{r-k}$ down to linearly sized subsets $Y_1, \ldots, Y_{r-k}$ of themselves, such that later on one can find the vertices $P_d^{xx'}$ of the desired $[k, m]$-fortress in the neighborhood of P^{xy} and $P^{x'y}$ for all $y \in \bigcup_{j \in [r-k]} Y_j$. This additional property will be crucial for the inductive construction of the fortress in the proof of Proposition 13.4.

Definition 13.6. *Let X and Y be two disjoint subsets of the index set of a reduced hypergraph $\mathcal{A}$ and suppose that $d \in [0, 1]$. A d-admissible (X, Y)-selection is a collection*

$$\mathscr{C} = \{P^{xy} \in \mathcal{P}^{xy} : x \in X \text{ and } y \in Y\}$$

of vertices of $\mathcal{A}$ such that

- if $x, x' \in X$ are distinct and $y \in Y$, then the pair-degree of P^{xy} and $P^{x'y}$ in $\mathcal{P}^{xx'}$ is at least $d\,|\mathcal{P}^{xx'}|$.

We remark that this notion of an admissible selection is not symmetric in X and Y, i.e., a d-admissible (X, Y)-selection $\mathscr{C}$ is not necessarily also a d-admissible (Y, X)-selection.

Our next immediate objective is to formulate and prove the first step of the induction for $k = 1$ (see Lemma 13.2) of the upcoming Proposition 13.4. As it turns out the case $k = 1$ is a bit simpler than the case of general k, because (1) the arising constants can be calculated rather easily, (2) one of the assumptions and consequently of the variables turn out to be unnecessary when $k = 1$, and (3) the notion of a $[1, m]$-fortress is especially simple. However, for later purposes it is better to prove a probabilistic strengthening of the statement for $k = 1$. For all these reasons, we deal with this case separately.

Lemma 13.2. *Let integers $r, m \geq 2$ and a real $\varepsilon > 0$ be given. Assume further*

- *that $X_0, X_1, \ldots, X_{r-1}$ are disjoint subsets of the index set of a reduced hypergraph $\mathcal{A}$ with $|X_0| = m$,*
- *and that $\mathscr{C}_j = \{P^{xy}\colon x \in X_0 \text{ and } y \in X_j\}$ is an $\left(\frac{r-2}{r-1} + \varepsilon\right)$-admissible (X_0, X_j)-selection for every $j \in [r-1]$.*

For a collection of vertices

$$\mathscr{C} = \left\{P^{xx'} \in \mathcal{P}^{xx'} \colon x, x' \in X_0 \text{ and } x \neq x'\right\}$$

and $j \in [r-1]$ set

$$Y_j(\mathscr{C}) = \left\{y \in X_j \colon \{P^{xx'}, P^{xy}, P^{x'y}\} \in E(\mathcal{A}^{xx'y}) \text{ holds for all distinct } x, x' \in X_0\right\}.$$

Then for a selection $\mathscr{C}$ chosen uniformly at random from $\prod_{x \neq x' \in X_0} \mathcal{P}^{xx'}$ the events

$$|Y_j(\mathscr{C})| \geq \left(\tfrac{\varepsilon}{2}\right)^{\binom{m}{2}} |X_j|$$

hold simultaneously for all $j \in [r-1]$ with probability at least $\left(\tfrac{\varepsilon}{2}\right)^{\binom{m}{2}}$.

Proof. We commence by treating the case $m = 2$ and $X_0 = \{x, x'\}$, say. For each vertex $P^{xx'} \in \mathcal{P}^{xx'}$ and each $j \in [r-1]$ we set

$$Y_j(P^{xx'}) = \left\{y \in X_j \colon \{P^{xx'}, P^{xy}, P^{x'y}\} \in E(\mathcal{A}^{xx'y})\right\}.$$

Notice that for every fixed $j \in [r-1]$ we have

$$\sum_{P^{xx'} \in \mathcal{P}^{xx'}} |Y_j(P^{xx'})| \geq \left(\tfrac{r-2}{r-1} + \varepsilon\right) |\mathcal{P}^{xx'}|\,|X_j|$$

owing to our admissibility assumption, and thus the set

$$A_j = \{P^{xx'} \in \mathcal{P}^{xx'} : |Y_j(P^{xx'})| \geq \tfrac{\varepsilon}{2}|X_j|\}$$

satisfies $|A_j| \geq \left(\frac{r-2}{r-1} + \frac{\varepsilon}{2}\right) \cdot |\mathcal{P}^{xx'}|$. Consequently for their intersection $A = \bigcap_{j \in [r-1]} A_j$ we obtain $|A| \geq \frac{(r-1)\varepsilon}{2} \cdot |\mathcal{P}^{xx'}| \geq \frac{\varepsilon}{2}|\mathcal{P}^{xx'}|$. Hence, when $P^{xx'} \in \mathcal{P}^{xx'}$ gets chosen uniformly at random, the event $P^{xx'} \in A$ happens with probability at least $\frac{\varepsilon}{2}$. Thereby the case $m = 2$ of our lemma is proved.

To obtain the general case we iterate this argument $\binom{m}{2}$ many times. This means that we make a list $e_1, \ldots, e_{\binom{m}{2}}$ of the two-element subsets of X_0, say $e_i = \{x_i, x'_i\}$. Now imagine that rather than picking the vertices $\{P^{x_i x'_i} \in \mathcal{P}^{x_i x'_i} : i \in \binom{m}{2}\}$ simultaneously we would pick them one by one, each choice being uniformly at random and independent from all previous choices. For $h = 0, \ldots, \binom{m}{2}$ let $\mathscr{C}^h = \{P^{x_1 x'_1}, \ldots, P^{x_h x'_h}\}$ and set

$$Y_j(\mathscr{C}^h) = \{y \in X_j : \{P^{x_i x'_i}, P^{x_i y}, P^{x'_i y}\} \in E(\mathcal{A}^{x_i x'_i y}) \text{ holds for all } i \in [h]\}.$$

Thereby we get for every $j \in [r-1]$ a sequence of sets

$$X_j = Y_j(\mathscr{C}^0) \supseteq Y_j(\mathscr{C}^1) \supseteq \ldots \supseteq Y_j(\mathscr{C}^{\binom{m}{2}}) = Y_j.$$

By the case $m = 2$ of our lemma, for every $h \in \left[\binom{m}{2}\right]$ the event

$$\mathcal{E}^h = \left\{ \text{for every } j \in [r-1] \text{ we have } |Y_j(\mathscr{C}^h)| \geq \tfrac{\varepsilon}{2}|Y_j(\mathscr{C}^{h-1})| \right\}$$

has the property that

$$\mathbb{P}\left(\mathcal{E}^h \mid \mathscr{C}^{h-1}\right) \geq \tfrac{\varepsilon}{2}$$

holds for every fixed choice of $P^{x_1, x'_1}, \ldots, P^{x_{h-1}, x'_{h-1}}$. It follows that the event $\mathcal{E}$ that all the events $\mathcal{E}^1, \ldots, \mathcal{E}^{\binom{m}{2}}$ happen has at least the probability $(\frac{\varepsilon}{2})^{\binom{m}{2}}$. Moreover, since $\mathcal{E}$ implies for every $j \in [r-1]$

$$|Y_j| \geq (\tfrac{\varepsilon}{2})^{\binom{m}{2}}|X_j|,$$

we are thereby done. $\square$

The next and final Proposition tells what we can achieve in the kth step of the proof of Proposition 13.3. In particular, for the special case $k = r$ the second items in the assumption and in the conclusion of Proposition 13.4 are void and the statement coincides with Proposition 13.3.

Proposition 13.4. *Given any integers $r, m \geq 2$, some $k \in [r]$, and a real $\varepsilon > 0$, there exists an integer M and reals $\delta, \eta > 0$ such that the following holds:*

Suppose

- *that $X_0, X_1, \ldots, X_{r-k}$ are disjoint subsets of the index set of some $\left(\frac{r-2}{r-1} + \varepsilon, \delta, \Lambda\right)$-dense reduced hypergraph $\mathcal{A}$, where X_0 is a $[k, M]$-system,*
- *and that $\mathcal{C}_j = \{P^{xy} : x \in X_0 \text{ and } y \in X_j\}$ is an $\left(\frac{r-2}{r-1} + \varepsilon\right)$-admissible (X_0, X_j)-selection for $j \in [r - k]$.*

Then there are

- *a $[k, m]$-subsystem $Z_0 \subseteq X_0$ carrying a $[k, m]$-fortress*

$$\mathcal{F} = \left\{P_d^{zz'} : z, z' \in Z_0, z \neq z', \text{ and } d \in Q(z \wedge z')\right\}$$

- *and sets $Y_j \subseteq X_j$ with $|Y_j| \geq \eta \, |X_j|$ for $j \in [r - k]$*

such that for every distinct $z, z' \in Z_0$, $d \in Q(z \wedge z')$, and $j \in [r - k]$ we have

$$\{P_d^{zz'}, P^{zy}, P^{z'y}\} \in E(\mathcal{A}^{zz'y}) \quad \text{for every } y \in Y_j, \tag{13.11}$$

where P^{zy} and $P^{z'y}$ are given by $\mathcal{C}_j$.

Proof. We consider r and ε to be fixed and proceed by induction on k.

To deal with the base case $k = 1$ let $m \geq 2$ be given and set $M = m$. Moreover, set $\eta = (\frac{\varepsilon}{2})^{\binom{m}{2}}$. In this case the density assumption of $\mathcal{A}$ will not be utilized and, hence, $\delta > 0$ can be chosen arbitrarily. Then we are given disjoint subsets $X_0, X_1, \ldots, X_{r-1}$ of the index set of $\mathcal{A}$, where X_0 forms a $[1, M]$-system, and $\left(\frac{r-2}{r-1} + \varepsilon\right)$-admissible (X_0, X_j)-selections $\mathcal{C}_j$ for $j \in [r - 1]$.

Set $Z_0 = X_0$ and by Lemma 13.2 applied with r, m, ε, $X_0, X_1, \ldots, X_{r-1}$, and $\mathcal{C}_1, \ldots, \mathcal{C}_{r-1}$ there exists a collection $\mathcal{C}$ of vertices $P_\varnothing^{zz'} \in \mathcal{P}^{zz'}$ for distinct $z, z' \in X_0$ and subsets $Y_j \subseteq X_j$ with $|Y_j| \geq \eta \, |X_j|$ for $j \in [r - 1]$ such that for every $j \in [r - 1]$ we have

$$\{P_\varnothing^{zz'}, P^{zy}, P^{z'y}\} \in E(\mathcal{A}^{zz'y})$$

for all distinct $z, z' \in X_0$, $y \in Y_j$, and $P^{zy}, P^{z'y} \in \mathcal{C}_j$.

Owing to $k = 1$, the collection

$$\mathcal{F} = \left\{P_\varnothing^{zz'} : z, z' \in X_0 \text{ and } z \neq z'\right\}$$

is a $[1, m]$-fortress on X_0 for trivial reasons (see Remark 13.1). As stated before, $\mathcal{F}$ has the required property and this establishes the induction start.

For the induction step we suppose that $2 \leq k \leq r$, and that the proposition is valid for $k - 1$ in place of k. Whenever we apply this case of Proposition 13.4 to some integer $m \geq 2$ it returns an integer $M(m)$ and two positive reals called $\delta(m)$ and $\eta(m)$.

From now on, we fix an integer $m \geq 2$ for which we would like to complete the induction step. We divide the argument into five parts.

Part I. Choice of the constants. To begin with, we define a decreasing sequence of integers $M_0, \ldots, M_{m(m-1)}$ by backwards induction as follows:

$$M_{m(m-1)} = m \quad \text{and} \quad M_{h-1} = M(M_h) + \left\lceil \frac{(k-1)M_h}{\eta(M_h)} \right\rceil \quad \text{for } h \in [m(m-1)].$$
(13.12)

We set $M = M_0$ and define

$$\eta_0 = (\tfrac{\varepsilon}{2})^{m^2 M^{2(k-1)}} \quad \text{and} \quad \eta_h = \eta_{h-1} \cdot \eta(M_h) \quad \text{for } h \in [m(m-1)].$$
(13.13)

Finally, let

$$\delta = \min \left(\{ m^{-2} M^{-3(k-1)} \eta_0 \} \cup \{ \delta(M_h) \colon h \in [m(m-1)] \} \right) \text{ and } \eta = \eta_{m(m-1)}.$$
(13.14)

We shall prove the proposition for this choice of the constants M, δ, and η.

Part II. The first level of the $[k, m]$-tree underlying Z_0. Now let $\mathcal{A}$, $X_0, \ldots, X_{r-k}$ as well as the $(\frac{r-2}{r-1} + \varepsilon)$-admissible (X_0, X_j)-selections $\mathscr{C}_j = \{ P^{xy} \in \mathcal{P}^{xy} \colon x \in X_0 \text{ and } y \in X_j \}$ for $j \in [r-k]$ be as described in the statement of the proposition. We denote the underlying tree of the $[k, M]$-system X_0 by T. Let

$$A \subseteq \{ a \in T \colon |a| = 1 \} \text{ of size } |A| = m$$

be an arbitrary subset. We intend to construct the $[k, m]$-tree underlying Z_0 in such a way that the direct continuations of its root is A.

For every $a \in A$ we set

$$X_0^a = \{ z \colon a \circ z \in X_0 \}$$

and note that these sets are $[k - 1, M]$-systems. In Part IV we shall apply the induction hypothesis (several times) to X_0^a for every $a \in A$, which then will lead to the desired $[k, m]$-system Z_0.

Strictly speaking, X_0^a is not a subset of the index set I of $\mathcal{A}$, which would be required for the application of the induction hypothesis. However, in such situations we can simply identify X_0^a with $\{ \tilde{x} \in X_0 \colon \tilde{x}|1 = a \} \subseteq I$ to circumvent this technicality and in the text that follows we will suppress this identification.

Part III. The selection of some vertices $P_\varnothing^{aoz, a'oz'}$. Our next objective is to select the elements $P_\varnothing^{aoz, a'oz'}$, i.e., those labeled with $\varnothing$, of the desired fortress $\mathcal{F}$. In view of the conclusion of Proposition 13.4 we shall choose

(1) For any distinct $a, a' \in A$, $z \in X_0^a$, and $z' \in X_0^{a'}$ a vertex $P_\varnothing^{aoz,a'oz'} \in \mathcal{P}^{aoz,a'oz'}$ such that for all distinct $a, a' \in A$ the set

$$\left\{ P_\varnothing^{aoz,a'oz'} : z \in X_0^a \text{ and } z' \in X_0^{a'} \right\}$$

is a $\left(\frac{r-2}{r-1} + \varepsilon \right)$-admissible $(X_0^a, X_0^{a'})$-selection.
(2) Subsets $Y_j^0 \subseteq X_j$ with $|Y_j^0| \geq \eta_0 |X_j|$ for $j \in [r-k]$

such that for all distinct $a, a' \in A$, $z \in X_0^a$, $z' \in X_0^{a'}$ and $j \in [r-k]$ we have

$$\left\{ P_\varnothing^{aoz,a'oz'}, P^{aoz,y}, P^{a'oz',y} \right\} \in E(\mathcal{A}^{aoz,a'oz',y}) \quad \text{for every } y \in Y_j^0, \qquad (13.15)$$

where $P^{aoz,y}$ and $P^{a'oz',y}$ are given by $\mathscr{C}_j$.

We choose $P_\varnothing^{aoz,a'oz'} \in \mathcal{P}^{aoz,a'oz'}$ independently and uniformly at random and in the text that follows we show that property (1) holds with probability bigger than $1 - \eta_0$ and property (2) is satisfied with probability at least η_0.

Dealing with property (1) first, we consider the set

$$K = \left\{ (\{a \circ z, a \circ z'\}, b \circ y) : a, b \in A, \ a \neq b, \ z, z' \in X_0^a, \ z \neq z', \text{ and } y \in X_0^b \right\}$$

consisting of those combinations of indices for which property (1) has to be checked. Since $|X_0^a| = M^{k-1}$ for all $a \in A$ and $|A| = m$, we have

$$|K| = m(m-1) \binom{M^{k-1}}{2} M^{k-1} < m^2 M^{3(k-1)}. \qquad (13.16)$$

Moreover, in view of the $\left(\frac{r-2}{r-1} + \varepsilon, \delta, \Lambda \right)$-denseness of $\mathcal{A}$ for each

$$(\{a \circ z, a \circ z'\}, b \circ y) \in K$$

the "bad event" $\mathcal{E}_{abzz'y}$ that the pair-degree of $P_\varnothing^{aoz,boy}$ and $P_\varnothing^{aoz',boy}$ in $\mathcal{P}^{aoz,aoz'}$ is smaller than $\left(\frac{r-2}{r-1} + \varepsilon \right) |\mathcal{P}^{aoz,aoz'}|$ has probability at most δ. So the union bound together with (13.16) and (13.14) yields

$$\mathbb{P}\left(\mathcal{E}_{abzz'y} \text{ occurs for some } (\{a \circ z, a \circ z'\}, b \circ y) \in K \right) \leq |K|\delta < \eta_0, \qquad (13.17)$$

which shows that with probability greater than $1 - \eta_0$ the random selection satisfies property (1).

Next we turn to property (2). For this it is convenient to introduce the set

$$L = \left\{ \{a \circ z, a' \circ z'\} : a, a' \in A, \ a \neq a', \ z \in X_0^a, \text{ and } z' \in X_0^{a'} \right\}.$$

For $j \in [r-k]$ we consider the random subsets

$$Y_j^0 = \left\{ y \in X_j : \{P_\varnothing^{aoz,a'oz'}, P^{aoz,y}, P^{a'oz',y}\} \right.$$
$$\left. \times \in E(\mathcal{A}^{zz'y}) \text{ holds for all } \{a \circ z, a' \circ z'\} \in L \right\},$$

where the randomness is induced by the random choice of $P_\varnothing^{aoz,a'oz'} \in \mathcal{P}^{aoz,a'oz'}$ in the foregoing.

Now we apply Lemma 13.2 with $r - k + 1$, mM^{k-1} and $\bigcup_{a \in A} X_0^a$ in place of r, m, and X_0 and with X_j and $\mathscr{C}_j' = \{P^{a \circ x, y} \in \mathscr{C}_j : a \in A, \ x \in X_0^a, \ \text{and} \ y \in X_j\}$ for $j \in [r - k]$. It follows from our choice of η_0 in (13.13) that the event

$$\mathcal{E} = \{|Y_j^0| \geq \eta_0 |X_j| \text{ for every } j \in [r - k]\}$$

holds with probability at least η_0. In fact, Lemma 13.2 is more general, but at this point we only need it in this simpler form.

Now $\mathbb{P}(\mathcal{E}) \geq \eta_0$ combined with (13.17) implies that there are the vertices $P_{\varnothing}^{a \circ z, a' \circ z'} \in \mathcal{P}^{a \circ z, a' \circ z'}$ satisfying both properties (1) and (2) promised earlier.

Part IV. Inductive construction of subfortresses. Let $e_i = (a_i, b_i)$ for $i \in [m(m-1)]$ enumerate all ordered pairs of distinct elements from A. We will show that for every nonnegative integer $h \leq m(m-1)$ the following statement is true:

$(*)_h$ There are

- for each $a \in A$ a sequence of sets

$$Z^{a,h} \subseteq Z^{a,h-1} \subseteq \ldots \subseteq Z^{a,0} = X_0^a$$

with $Z^{a,i}$ being a $[k-1, M_i]$-system for every nonnegative $i \leq h$,
- and for each $j \in [r - k]$ a subset $Y_j^h \subseteq Y_j^0$ with $|Y_j^h| \geq \eta_h |X_j|$,

such that for each $i \in [h]$ the $[k-1, M_h]$-system $Z^{a_i,h}$ carries a $[k-1, M_h]$-fortress

$$\mathcal{F}^{i,h} = \{P_{b_i \circ d}^{a_i \circ z, \, a_i \circ z'} : z, z' \in Z^{a_i,h}, z \neq z', \text{ and } d \in Q(z \wedge z')\}$$

such that for every distinct $z, z' \in Z^{a_i,h}$ and $d \in Q(z \wedge z')$ we have

$$\{P_{b_i \circ d}^{a_i \circ z, \, a_i \circ z'}, P_{\varnothing}^{a_i \circ z, \, b_i \circ w}, P_{\varnothing}^{a_i \circ z', \, b_i \circ w}\} \in E(\mathcal{A}^{a_i \circ z, \, a_i \circ z', \, b_i \circ w}) \quad \text{for every } w \in Z^{b_i, h} \tag{13.18}$$

and for every $j \in [r - k]$ we have

$$\{P_{b_i \circ d}^{a_i \circ z, \, a_i \circ z'}, P^{a_i \circ z, \, y}, P^{a_i \circ z', \, y}\} \in E(\mathcal{A}^{a_i \circ z, \, a_i \circ z', \, y}) \quad \text{for every } y \in Y_j^h. \tag{13.19}$$

To show this we argue by induction on h. In the base case $h = 0$ we have to take $Z^{a,0} = X_0^a$ for all $a \in A$ and the sets Y_j^0 obtained in Part III. The assertion about the existence of fortresses holds vacuously.

Now suppose that some $h \in [m(m-1)]$ is such that $(*)_{h-1}$ holds with $Z^{a,h-1} \subseteq X_0^a$ for $a \in A$, with $Y_j^{h-1} \subseteq Y_j^0$ for $j \in [r-k]$, and with the

$[k-1, M_i]$-fortresses $\mathcal{F}^{i,h-1}$ for $i \in [h-1]$. Now we apply the outer induction hypothesis from the proof of Proposition 13.4 with M_h in place of m

- to the $[k-1, M_{h-1}]$-system $Z^{a_h,h-1}$ (in place of X_0) and the $r - (k-1)$ further subsets $Z^{b_h,h-1}, Y_1^{h-1}, \ldots, Y_{r-k}^{h-1}$ of I (in place of $X_1, \ldots, X_{r-(k-1)}$),
- to the $\left(\frac{r-2}{r-1} + \varepsilon\right)$-admissible $(Z^{a_h,h-1}, Z^{b_h,h-1})$-selection

$$\left\{ P_\varnothing^{a_h \circ z, b_h \circ z'} : z \in Z^{a_h,h-1} \text{ and } z' \in Z^{b_h,h-1} \right\}$$

obtained in Part III, and for $j \in [r-k]$ to the $\left(\frac{r-2}{r-1} + \varepsilon\right)$-admissible $(Z^{a_h,h-1}, Y_j^{h-1})$-selections

$$\left\{ P^{a_h \circ z, y} : z \in Z^{a_h,h-1} \text{ and } y \in Y_j^{h-1} \right\} \subseteq \mathscr{C}_j$$

provided by the assumption.

Since $M_{h-1} \geq M(M_h)$ holds by (13.12), this yields in particular

- a $[k-1, M_h]$-system $Z^{a_h,h} \subseteq Z^{a_h,h-1}$ carrying a $[k-1, M_h]$-fortress

$$\mathcal{F}^{h,h} = \left\{ P_{b_h \circ d}^{a_h \circ z, a_h \circ z'} : z, z' \in Z^{a_h,h}, z \neq z', \text{ and } d \in Q(z \wedge z') \right\},$$

- a subset $W \subseteq Z^{b_h,h-1}$ and subsets $Y_j^h \subseteq Y_j^{h-1}$ for $j \in [r-k]$ with

$$|W| \geq \eta(M_h) |M_{h-1}|^{k-1} \quad \text{and} \quad |Y_j^h| \geq \eta(M_h) |Y_j^{h-1}|, \tag{13.20}$$

such that for every distinct $z, z' \in Z^{a_h,h}$ and $d \in Q(z \wedge z')$ we have

$$\left\{ P_{b_h \circ d}^{a_h \circ z, a_h \circ z'}, P_\varnothing^{a_h \circ z, b_h \circ w}, P_\varnothing^{a_h \circ z', b_h \circ w} \right\} \in E\left(\mathcal{A}^{a_h \circ z, a_h \circ z', b_h \circ w} \right) \quad \text{for every } w \in W \tag{13.21}$$

and for every $j \in [r-k]$ we have

$$\left\{ P_{b_h \circ d}^{a_h \circ z, a_h \circ z'}, P^{a_h \circ z, y}, P^{a_h \circ z', y} \right\} \in E\left(\mathcal{A}^{a_h \circ z, a_h \circ z', y} \right) \quad \text{for every } y \in Y_j^h. \tag{13.22}$$

Now we are ready to define the remaining entities verifying $(*)_h$. Recall that we have already obtained the $[k-1, M_h]$-system $Z^{a_h,h}$, the sets Y_j^h for $j \in [r-k]$, and the fortress $\mathcal{F}^{h,h}$. Note that (13.12) yields $\eta(M_h)M_{h-1}/(k-1) \geq M_h$.

Thus Lemma 13.1 applied with $k-1$ and M_{h-1} to the $[k-1, M_{h-1}]$-system $Z^{b_h,h-1}$ and $W \subseteq Z^{b_h,h-1}$ which has size $\eta(M_h)|M_{h-1}|^{k-1}$ (see (13.20)) tells us that there exists a $[k-1, M_h]$-system $Z^{b_h,h} \subseteq W$.

Finally, for definiteness (somewhat wastefully) for any $c \in A \setminus \{a_h, b_h\}$ we let $Z^{c,h}$ be an arbitrary $[k-1, M_h]$-subsystem of $Z^{c,h-1}$ and for $i \in [h-1]$ we let $\mathcal{F}^{i,h}$ denote the "restriction" of $\mathcal{F}^{i,h-1}$ to $Z^{a_i,h}$.

It remains to check that we have indeed met all conditions mentioned in $(*)_h$. That $Z^{a,h} \subseteq Z^{a,h-1}$ holds for every $a \in A$ follows from our construction. Due to the choice of Y_j^h, the description of Y_j^{h-1} in $(*)_{h-1}$, and (13.13) we have

$$|Y_j^h| \overset{(13.20)}{\geq} \eta(M_h) |Y_j^{h-1}| \geq \eta(M_h)\eta_{h-1} |X_j| = \eta_h |X_j|$$

for every $j \in [r-k]$. Statements (13.18) and (13.19) hold for $i \neq h$ by $(*)_{h-1}$ and for $i = h$ by (13.21) and (13.22) respectively. This concludes the proof of $(*)_h$.

Part V. Conclusion of the argument. We will show that the statement $(*)_{m(m-1)}$ from Part IV yields the conclusion of the proposition. For that set $Z^a = Z^{a,m(m-1)}$ for $a \in A$, and $Y_j = Y_j^{m(m-1)}$ for $j \in [r-k]$. It follows from $M_{m(m-1)} = m$ that Z^a is a $[k-1, m]$-system for each $a \in A$ and consequently the set

$$Z_0 = \{a \circ z : a \in A \text{ and } z \in Z^a\}$$

is a $[k, m]$-system. Moreover, a quick thought reveals that the collection of vertices

$$\mathcal{F} = \{P_\varnothing^{aoz,a'oz'} : a, a' \in A, a \neq a', z \in Z^a \text{ and } z' \in Z^{a'}\}$$
$$\cup \{P_{bod}^{aoz,aoz'} : a, b \in A, \ a \neq b, \ z, z' \in Z^a, \ z \neq z', \text{ and } d \in Q(z \wedge z')\}$$

has the correct index structure for being a fortress on Z_0. To see that $\mathcal{F}$ actually is a fortress we need to verify the axiom (F); if $s = 0$ it follows from (13.18) and for $s > 0$ it follows from all the $\mathcal{F}^{i,m(m-1)}$ with $i \in [m(m-1)]$ being fortresses.

We contend that the system Z_0, the fortress $\mathcal{F}$, and the sets Y_j with $j \in [r-k]$ are as desired. The latter are large enough because of $\eta = \eta_{m(m-1)}$ and $(*)_{m(m-1)}$. Finally (13.11) was obtained in (13.15) for $z|1 \neq z'|1$ and otherwise in (13.19). This completes the induction step and thus the proof of Proposition 13.4. □

Acknowledgments

Vojtěch Rödl was supported by NSF grants DMS 1301698 and 1102086. Mathias Schacht was supported through the Heisenberg-Programme of the DFG.

References

1. F. R. K. Chung and R. L. Graham. Edge-colored complete graphs with precisely colored subgraphs. *Combinatorica* **3**, no. 3–4, (1983), 315–324.
2. P. Erdős. Problems and results on graphs and hypergraphs: Similarities and differences. *Mathematics of Ramsey theory*. Algorithms Combin., Vol. 5, Springer, Berlin, 1990, pp. 12–28.
3. P. Erdős and V. T. Sós. On Ramsey-Turán type theorems for hypergraphs. *Combinatorica* **2**, no. 3 (1982), 289–295.
4. P. Frankl and V. Rödl. Extremal problems on set systems. *Random Struct. Algorithms* **20** (2002), no. 2, 131–164.
5. R. Glebov, D. Král, and J. Volec. A problem of Erdős and Sós on 3-graphs. *Israel J. Math.* **211**, no. 1 (2016), 349–366.
6. W. T. Gowers. Quasirandomness, counting and regularity for 3-uniform hypergraphs. *Combin. Probab. Comput.* **15**, no. 1–2 (2006), 143–184.
7. B. Nagle, A. Poerschke, V. Rödl, and M. Schacht. Hypergraph regularity and quasirandomness. *Proceedings of the Twentieth Annual ACM-SIAM Symposium on Discrete Algorithms.* SIAM, Philadelphia, PA, 2009, pp. 227–235.
8. Chr. Reiher, V. Rödl, and M. Schacht. On a Turán problem in weakly quasirandom 3-uniform hypergraphs. Available at arXiv:1602.02290. Submitted.
9. Chr. Reiher, V. Rödl, and M. Schacht. Embedding tetrahedra into quasirandom hypergraphs. *J. Combin. Theory Ser. B* **121** (2016), 229–247.
10. V. Rödl and M. Schacht. Regular partitions of hypergraphs: Regularity lemmas. *Combin. Probab. Comput.* **16**, no. 6 (2007), 833–885.
11. V. Rödl and M. Schacht. Regular partitions of hypergraphs: counting lemmas. *Combin. Probab. Comput.* **16**, no. 6 (2007), 887–901.
12. P. Turán. Eine Extremalaufgabe aus der Graphentheorie. *Mat. Fiz. Lapok* **48** (1941), 436–452 (Hungarian, with German summary).

14

Ramsey Classes with Closure Operations (Selected Combinatorial Applications)

Jan Hubička and Jaroslav Nešetřil

Abstract

We state the Ramsey property of classes of ordered structures with closures and given local properties. This generalizes many old and new results: the Nešetřil–Rödl Theorem, the authors Ramsey lift of bowtie-free graphs as well as the Ramsey Theorem for Finite Models (i.e., structures with both functions and relations), thus providing the ultimate generalization of Structural Ramsey Theorem. We give here a more concise reformulation of the recent paper "All those Ramsey classes (Ramsey classes with closures and forbidden homomorphisms)," and the main purpose of this chapter is to show several applications. In particular, we prove the Ramsey property of ordered sets with equivalences on the power set, Ramsey theorem for Steiner systems, Ramsey theorem for resolvable designs and a partial Ramsey type results for H-factorizable graphs. All of these results are natural and easy to state, yet proofs involve most of the theory developed.

14.1 Introduction

Extending classical early Ramsey-type results [8, 7, 21, 9, 1, 17], the structural Ramsey theory originated at the beginning of the 1970s; see [18] for references of the early history of the subject.

Let us start with the key definition of this chapter. Let $\mathcal{K}$ be a class of structures endowed with embeddings. For objects $\mathbf{A}, \mathbf{B} \in \mathcal{K}$ denote by $\binom{\mathbf{B}}{\mathbf{A}}$ the set of all subobjects $\widetilde{\mathbf{A}}$ of $\mathbf{B}$ that are isomorphic to $\mathbf{A}$. (By a subobject we mean that the inclusion is an embedding.) Using this notation the central definition of this chapter gets the following form: A class $\mathcal{K}$ is a *Ramsey class* if for every two objects $\mathbf{A}$ and $\mathbf{B}$ and for every positive integer k there exists object $\mathbf{C} \in \mathcal{K}$ such that the following holds: For every partition $\binom{\mathbf{B}}{\mathbf{A}}$ in k classes there exists

$\widetilde{\mathbf{B}} \in \binom{\mathbf{C}}{\mathbf{B}}$ such that $\binom{\widetilde{\mathbf{B}}}{\mathbf{A}}$ belongs to one class of the partition. It is usual to shorten the last part of the definition as $\mathbf{C} \longrightarrow (\mathbf{B})_k^{\mathbf{A}}$.

In [8, 7, 21, 18, 9, 1, 17] are given many examples of Ramsey classes and the recent interest of people in topological dynamics, model theory [14, 24, 29] and, of course, combinatorists led to a hunt for more and more Ramsey classes. Particularly the following theorem has been recently proved in [12] (see also a strengthening in [6]).

Theorem 14.1 (Ramsey Theorem for Finite Models). *For every language L involving both relations and functions the class of all linearly ordered L structures is a Ramsey class.*

In Section 14.2 we formulate this result more precisely as Theorem 14.2 after introducing all relevant notions. But already at this place let us add that Theorem 14.1 nicely complements results for relational structures (Abramson-Harrington [1], Nešetřil–Rödl [23]), who proved this result for finite relational systems. It is exactly the presence of functions (and function symbols) that makes Theorem 14.1 interesting and that presents some challenging problems. In turn this aspect has some interesting applications. This we want to demonstrate here by several combinatorial examples: equivalences on the power set (thus solving a problem of [13]), Steiner systems (proving [3] in a different way), Ramsey theorem for resolvable designs, degenerate graphs and more generally $\mathcal{H}$-decomposable graphs. As we shall demonstrate all these structures form Ramsey classes. These all seemingly very special theorems can be uniformly treated by means of the main result of [12]. Although the nature of [12] is a blend of combinatorics and model theory we concentrate here on combinatorial side of the subject. This chapter is self-contained and in the next section we shall explain all the relevant notions.

14.2 Preliminaries

We work with the following structures involving relations and symmetric partial functions.

Let $L = L_{\mathcal{R}} \cup L_{\mathcal{F}}$ be a language involving relational symbols $R \in L_{\mathcal{R}}$ and function symbols $F \in L_{\mathcal{F}}$ each having associated arities denoted by $a(R) > 0$ for relations and $d(F) > 0$, $r(F) > 0$ for functions. An *L-model*, or sometimes *L-structure*, is a structure $\mathbf{A}$ with *vertex set* A, functions $F_{\mathbf{A}} : \mathrm{Dom}(F_{\mathbf{A}}) \to \binom{A}{r(F)}$, $\mathrm{Dom}(F_{\mathbf{A}}) \subseteq A^{d(F)}$ for $F \in L_F$ and relations $R_{\mathbf{A}} \subseteq A^{a(R)}$ for $R \in L_R$. (Note that by $\binom{A}{r(F)}$ we denote, as it is usual in this context, the set of all $r(F)$-element subsets of A.) Set $\mathrm{Dom}(F_{\mathbf{A}})$ is called the *domain* of function F in $\mathbf{A}$.

Note also that we have chosen to have the range of the function symbols to be the set of subsets (not tuples). This is motivated by [6], where we deal with (Hrushovski) extension properties and we need a "symmetric" range. However, from the point of view of Ramsey theory this is not an important issue.

The language is usually fixed and understood from the context (and it is in most cases denoted by L). If set A is finite we call $\mathbf{A}$ a *finite L-model* or *L-structure*. We consider only structures with countably many vertices. If language L contains no function symbols, we call L a *relational language* and an L-structure is also called a *relational L-structure*. Every function symbol F such that $d(F) = 1$ is a *unary function*. Unary relation is of course just defining a subset of elements of A.

A *homomorphism* $f : \mathbf{A} \to \mathbf{B}$ is a mapping $f : A \to B$ satisfying for every $R \in L_R$ and for every $F \in L_F$ the following two statements:

(1) $(x_1, x_2, \ldots, x_{a(R)}) \in R_\mathbf{A} \implies (f(x_1), f(x_2), \ldots, f(x_{a(R)})) \in R_\mathbf{B}$, and
(2) $f(\mathrm{Dom}(F_\mathbf{A})) \subseteq \mathrm{Dom}(F_\mathbf{B})$ and $f(F_\mathbf{A}(x_1, x_2, \ldots, x_{d(F)})) = F_\mathbf{B}(f(x_1), f(x_2), \ldots, f(x_{d(F)}))$ for every $(x_1, x_2, \ldots, x_{d(F)}) \in \mathrm{Dom}(F_\mathbf{A})$.

For a subset $A' \subseteq A$ we denote by $f(A')$ the set $\{f(x); x \in A'\}$ and by $f(\mathbf{A})$ the homomorphic image of a structure.

If f is injective, then f is called a *monomorphism*. A monomorphism is called an *embedding* if for every $R \in L_R$ and $F \in L_F$ the following hold:

(1) $(x_1, x_2, \ldots, x_{a(R)}) \in R_\mathbf{A} \iff (f(x_1), f(x_2), \ldots, f(x_{a(R)})) \in R_\mathbf{B}$, and,
(2) $(x_1, x_2, \ldots, x_{d(F)}) \in \mathrm{Dom}(F_\mathbf{A}) \iff (f(x_1), \ldots, f(x_{d(F)})) \in \mathrm{Dom}(F_\mathbf{B})$.

If f is an embedding that is an inclusion then $\mathbf{A}$ is a *substructure* (or *subobject*) of $\mathbf{B}$. For an embedding $f : \mathbf{A} \to \mathbf{B}$ we say that $\mathbf{A}$ is *isomorphic* to $f(\mathbf{A})$ and $f(\mathbf{A})$ is also called a *copy* of $\mathbf{A}$ in $\mathbf{B}$. Thus $\binom{\mathbf{B}}{\mathbf{A}}$ is defined as the set of all copies of $\mathbf{A}$ in $\mathbf{B}$. Finally, $\mathrm{Str}(L)$ denotes the class of all finite L-models and all their embeddings.

For L containing a binary relation $R^\leq$ we denote by $\overrightarrow{\mathrm{Str}}(L)$ the class of all finite models (or L-structures) $\mathbf{A} \in \mathrm{Str}(L)$ where the set A is linearly ordered by the relation $R^\leq$. $\overrightarrow{\mathrm{Str}}(L)$ is of course considered with all monotone (i.e., order preserving) embeddings.

The Ramsey theorem of finite models can now be stated as follows ([6]).

Theorem 14.2. *For every language L involving both relations and functions and containing a binary relation $R^\leq$ the class of all ordered L-structures $\overrightarrow{\mathrm{Str}}(L)$ is a Ramsey class.*

Ramsey Theorem for Finite Models (i.e., for structures with both functions and relations) thus provides the ultimate generalization of structural Ramsey

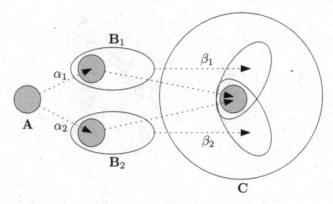

Fig. 14.1. An amalgamation of $\mathbf{B}_1$ and $\mathbf{B}_2$ over $\mathbf{A}$.

Theorem for relational structures [1, 21]. In this paper this theorem will be our starting point. We remark that a generalization of Nešetřil–Rödl Theorem for languages involving both functions and relations is also a main result of [28]. Our definition of functions, however, differs from the one used in [28].

Before stating more special (and stronger) Ramsey type statements we need to review some more model-theoretic notions (see, e.g., [10]).

Let $\mathbf{A}$, $\mathbf{B}_1$ and $\mathbf{B}_2$ be relational structures and α_1 an embedding of $\mathbf{A}$ into $\mathbf{B}_1$, α_2 an embedding of $\mathbf{A}$ into $\mathbf{B}_2$, then every structure $\mathbf{C}$ with embeddings $\beta_1 : \mathbf{B}_1 \to \mathbf{C}$ and $\beta_2 : \mathbf{B}_2 \to \mathbf{C}$ such that $\beta_1 \circ \alpha_1 = \beta_2 \circ \alpha_2$ is called an *amalgamation* of $\mathbf{B}_1$ and $\mathbf{B}_2$ over $\mathbf{A}$ with respect to α_1 and α_2. See Fig. 14.1. We will call $\mathbf{C}$ simply an *amalgamation* of $\mathbf{B}_1$ and $\mathbf{B}_2$ over $\mathbf{A}$ (as in the most cases α_1 and α_2 can be chosen to be inclusion embeddings).

We say that an amalgamation is *strong* when $\beta_1(x_1) = \beta_2(x_2)$ only if $x_1 \in \alpha_1(A)$ and $x_2 \in \alpha_2(A)$. Less formally, a strong amalgamation glues together $\mathbf{B}_1$ and $\mathbf{B}_2$ with the overlap no greater than the copy of $\mathbf{A}$ itself. A strong amalgamation is *free* if there are no tuples in any relations of $\mathbf{C}$ and no tuples in $\mathrm{Dom}(F_{\mathbf{C}})$, $F \in L$, using vertices of both $\beta_1(B_1 \setminus \alpha_1(A))$ and $\beta_2(B_2 \setminus \alpha_2(A))$.

An *amalgamation class* is a class $\mathcal{K}$ of finite structures satisfying the following three conditions:

1. Hereditary property: For every $\mathbf{A} \in \mathcal{K}$ and a substructure $\mathbf{B}$ of $\mathbf{A}$ we have $\mathbf{B} \in \mathcal{K}$;
2. Joint embedding property: For every $\mathbf{A}, \mathbf{B} \in \mathcal{K}$ there exists $\mathbf{C} \in \mathcal{K}$ such that $\mathbf{C}$ contains both $\mathbf{A}$ and $\mathbf{B}$ as substructures;
3. Amalgamation property: For $\mathbf{A}, \mathbf{B}_1, \mathbf{B}_2 \in \mathcal{K}$ and α_1 embedding of $\mathbf{A}$ into $\mathbf{B}_1$, α_2 embedding of $\mathbf{A}$ into $\mathbf{B}_2$, there is $\mathbf{C} \in \mathcal{K}$ which is an amalgamation of $\mathbf{B}_1$ and $\mathbf{B}_2$ over $\mathbf{A}$ with respect to α_1 and α_2.

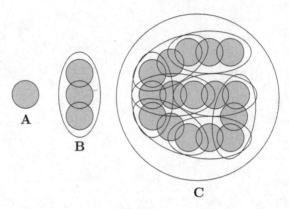

Fig. 14.2. Construction of a Ramsey object by multiamalgamation.

14.3 Previous Work: Multiamalgamation

We now refine amalgamation classes. Our aim is to describe even stronger sufficient criteria for Ramsey classes.

First, we develop a generalized notion of amalgamation that will serve as a useful tool for the construction of Ramsey objects. As schematically depicted in Fig. 14.2, Ramsey objects are a result of amalgamation of multiple copies of a given structure that are all performed at once. In a nontrivial class this leads to many problems. Instead of working with complicated amalgamation diagrams we split the amalgamation into two steps—the construction of (up to isomorphism unique) free amalgamation (which yields an incomplete or "partial" structure) followed then by a completion. Formally this is done as follows.

Definition 14.1. *An L-structure* **A** *is* irreducible *if* **A** *is not a free amalgam of two proper substructures of* **A**.

Remark. In the case of relational structures this definition allows a combinatorial description: **A** is irreducible if for every pair of distinct vertices u, v there is a tuple $\vec{t} \in R_{\mathbf{A}}$ (of some relation $R \in L_{\mathcal{R}}$) such that $\vec{t}$ contains both u and v. In the case of structures with function symbols this is more complicated as we have to consider closures defined in the text that follows.

Thus the irreducibility is meant with respect to the free amalgamation. The irreducible structures are our building blocks. Moreover, in structural Ramsey theory we are fortunate that most structures are (or may be interpreted as) irreducible. And in the most interesting case, the structures may be completed to irreducible structures. This will be introduced now by means of the following variant of the homomorphism notion.

Definition 14.2. *A homomorphism* $f : \mathbf{A} \to \mathbf{B}$ *is* homomorphism-embedding *if f restricted to any irreducible substructure of* $\mathbf{A}$ *is an embedding to* $\mathbf{B}$.

Whereas for (undirected) graphs the homomorphism and homomorphism-embedding coincide, for structures they differ. For example, any homomorphism-embedding of the Fano plane into a hypergraph is actually an embedding.

Definition 14.3. *Let* $\mathbf{C}$ *be a structure. An irreducible structure* $\mathbf{C}'$ *is a completion of* $\mathbf{C}$ *if there exists a homomorphism-embedding* $\mathbf{C} \to \mathbf{C}'$.

Let $\mathbf{B}$ *be an irreducible substructure of* $\mathbf{C}$*. We say that irreducible structure* $\mathbf{C}'$ *is a* completion of $\mathbf{C}$ *with respect to copies of* $\mathbf{B}$ *if there exists a function* $f : C \to C'$ *such that for every* $\widetilde{\mathbf{B}} \in \binom{\mathbf{C}}{\mathbf{B}}$ *the function f restricted to* $\widetilde{B}$ *is an embedding of* $\widetilde{\mathbf{B}}$ *to* $\mathbf{C}'$.

If $\mathbf{C}'$ *belongs to a given class* $\mathcal{K}$, *then* $\mathbf{C}'$ *is called a* $\mathcal{K}$-completion of $\mathbf{C}$ with respect to copies of $\mathbf{B}$.

Remark (on completion and holes). Completion may be seen as a generalized form of amalgamation. To see that let $\mathcal{K}$ be a class of irreducible structures. The amalgamation property of $\mathcal{K}$ can be equivalently formulated as follows: For $\mathbf{A}$, $\mathbf{B}_1, \mathbf{B}_2 \in \mathcal{K}$ and α_1 embedding of $\mathbf{A}$ into $\mathbf{B}_1$, α_2 embedding of $\mathbf{A}$ into $\mathbf{B}_2$, there is $\mathbf{C} \in \mathcal{K}$ which is a completion of the free amalgamation (which itself is not necessarily in $\mathcal{K}$) of $\mathbf{B}_1$ and $\mathbf{B}_2$ over $\mathbf{A}$ with respect to α_1 and α_2.

Free amalgamation may result in a reducible structure. The pairs of vertices where one vertex belong to $\mathbf{B}_1 \setminus \alpha_1(\mathbf{A})$ and the other to $\mathbf{B}_2 \setminus \alpha_2(\mathbf{A})$ are never both contained in a single tuple of any relation. Such pairs can be thought of as *holes* and a completion is then a process of filling in the holes to obtain irreducible structures (in a given class $\mathcal{K}$) while preserving all embeddings of irreducible structures.

Completion with respect to copies of $\mathbf{B}$ is the weakest notion of completion that preserves the Ramsey property for given structures $\mathbf{A}$ and $\mathbf{B}$. Note that in this case f does not need to be a homomorphism-embedding (and even homomorphism).

We now state all necessary conditions for our main result stated in this section:

Definition 14.4. *Let L be a language,* $\mathcal{R}$ *be a Ramsey class of finite irreducible L-structures. We say that a subclass* $\mathcal{K}$ *of* $\mathcal{R}$ *is an* $\mathcal{R}$-multiamalgamation class *if the following conditions are satisfied:*

1. Hereditary property: *For every* $\mathbf{A} \in \mathcal{K}$ *and a substructure* $\mathbf{B}$ *of* $\mathbf{A}$ *we have* $\mathbf{B} \in \mathcal{K}$.

2. Strong amalgamation property: *For* $\mathbf{A}, \mathbf{B}_1, \mathbf{B}_2 \in \mathcal{K}$ *and* α_1 *embedding of* $\mathbf{A}$ *into* $\mathbf{B}_1$, α_2 *embedding of* $\mathbf{A}$ *into* $\mathbf{B}_2$, *there is* $\mathbf{C} \in \mathcal{K}$ *which contains a strong amalgamation of* $\mathbf{B}_1$ *and* $\mathbf{B}_2$ *over* $\mathbf{A}$ *with respect to* α_1 *and* α_2 *as a substructure.*

3. Locally finite completion property: *Let* $\mathbf{B} \in \mathcal{K}$ *and* $\mathbf{C}_0 \in \mathcal{R}$. *Then there exists* $n = n(\mathbf{B}, \mathbf{C}_0)$ *such that if closed L-structure* $\mathbf{C}$ *satisfies the following:*

   (a) *There is a homomorphism-embedding from* $\mathbf{C}$ *to* $\mathbf{C}_0$, *(in other words,* $\mathbf{C}_0$ *is a completion of* $\mathbf{C}$*), and*

   (b) *Every substructure of* $\mathbf{C}$ *with at most n vertices has a* $\mathcal{K}$*-completion.*

   Then there exists $\mathbf{C}' \in \mathcal{K}$ *that is a completion of* $\mathbf{C}$ *with respect to copies of* $\mathbf{B}$.

We can now state the main result of [12] as:

Theorem 14.3 ([12]). *Every* $\mathcal{R}$*-multiamalgamation class* $\mathcal{K}$ *is Ramsey.*

The proof of this result is not easy and involves interplay of several key constructions of structural Ramsey theory, particularly Partite Lemma and Partite Construction (see [12] for details).

Paper [12] also contains several applications of this result. Here we add some new combinatorial ones in three areas mentioned in the introduction. All our results share a common pattern: We start with a carefully chosen special case of Theorem 14.2 and check that all defining restrictions fit into Theorem 14.3

We find it convenient (and intuitive) also to use the terminology of closed sets. We say that a structure $\mathbf{A}$ is *closed in* $\mathbf{B}$ if there is an embedding $\mathbf{A} \to \mathbf{B}$. So all our Ramsey theorems deal with closed sets.

Remark. For amalgamation classes of irreducible ordered structures our definition of closure is equivalent with the model-theoretic definition of the algebraic closure considered in the Fraïssé limit of the class (see, e.g., [10]). This follows from the fact that the closure relations are definable in the structure. Note that in [12] the closures are handled equivalently by means of "closure descriptions." The advantage of such a definition is that it involves the degree condition which is easier to control than the abstract closure one. Here we proceed differently as our starting point is Theorem 14.3.

14.4 Power Set Equivalences

In this section we consider objects that consist of sets on which there are partitions of elements, pairs or subsets of larger size.

The essential part of more complex Ramsey classes is handling structures with equivalences defined on vertices (and even tuples of vertices). Such an equivalence may be present latently (as, e.g., in S-metric spaces with jump numbers [26], antipodal metric spaces [2], or bowtie-free graphs [11, 4, 16]). It is an important fact that such equivalences may have unboundedly many equivalence classes and thus one cannot assign labels to them and use Theorem 14.3 directly. Here we show how to handle this.

We start our analysis with the following definition.

Definition 14.5. *A k-equivalence* **A** *consists of a linearly ordered set* $(A, \leq_\mathbf{A})$ *together with an equivalence $E_\mathbf{A}$ on the set* $\binom{A}{k}$. *Given two partition structures* **A** *and* **B** *a mapping $f : A \to B$ is said to be an* embedding *if f is a monotone injection and if for every two sets $M, M' \in \binom{A}{k}$ holds*

$$(M, M') \in E_\mathbf{A} \text{ if and only if } (f(M), f(M')) \in E_\mathbf{B}.$$

Note that 1-equivalence is just an equivalence on set A. Already 2-equivalences are interesting: they may be defined as edges vs. nonedges of a graph. We denote by $\mathcal{EQ}_k$ the class of all finite k-equivalences and all embeddings between them.

It is easy to see that $\mathcal{EQ}_k$ fails to be a Ramsey class. To see this it suffices to consider already equivalences defined by sets of vertices (i.e., equivalences on singletons only). This corresponds to colorings of disjoint copies of complete graphs. (In this particular case $(k = 1)$ one can easily save being Ramsey by a choice of special (admissible or convex) orderings.)

In fact for our purposes (i.e., application of Theorem 14.3) the equivalences with unboundedly many classes have to be interpreted so they can be viewed as structures over a finite language. This is a place where functions (and closure operations) may be used effectively.

This can be done as follows.

Definition 14.6. *A labeled k-equivalence* **A** *consists of a linearly ordered set* $(A, \leq_\mathbf{A})$ *together with a mapping $\ell_\mathbf{A} : \binom{A}{k} \to \mathbb{N}$ ("labeling" is thus symbolized by ℓ). We tacitly assume that the sets A and $\mathbb{N}$ are disjoint. By $\mathrm{Rg}(\ell_\mathbf{A})$ we denote the range of labeling $\ell_\mathbf{A}$.*

Given two labeled k-equivalences **A** *and* **B** *a mapping $f : A \to B$ is said to be an* embedding *if f is a monotone injection and if for every subset $M \in \binom{A}{k}$ holds $\iota(\ell_\mathbf{A}(M)) = \ell_\mathbf{B}(f(M))$, where $\iota : \mathrm{Rg}(\ell_\mathbf{A}) \to \mathrm{Rg}(\ell_\mathbf{B})$ (which is defined by the preceding formula) is monotone and injective.*

We denote by $\mathcal{LEQ}_k$ the class of all finite labeled k-equivalences and all embeddings between them.

Thus the equality of labels and their order is also preserved. Subsets with the same label represent an equivalence class. (In [12] labeled partition structures are called *pointed equivalences*.) Explicitly, the correspondence with equivalences is as follows:

For an equivalence E on set $\binom{A}{k}$ we assign to every equivalence class C of E a label $i_C \in \mathbb{N}$ and a mapping $\ell_E : \binom{A}{k} \to \mathbb{N}$ that maps every element of C to i_C (that is, $\ell_E(M) = i_C$ for every $M \in C$). What we obtain is a structure $\mathbf{A}(E)$ in the language $L = \{R, \leq, F\}$ consisting of a function symbol F (corresponding to labeling ℓ), binary relational symbol $\leq$ and unary relational symbol R (representing the labels). The class, denoted for the moment by $\mathcal{R} = \overrightarrow{\mathrm{Str}}(L)$, of all ordered L-models is a Ramsey class by virtue of Theorem 14.2. The class $\mathcal{LEQ}_k$ is described by local conditions in $\mathcal{R}$. More precisely $\mathbf{A} \in \mathcal{R}$ is in $\mathcal{LEQ}_k$ if and only if it satisfies:

1. $x \notin \vec{t} \in \mathrm{Dom}(F_{\mathbf{A}})$ whenever $(x) \in R$.
2. Every $\vec{t} \in Dom(F_{\mathbf{A}})$ consists of distinct elements of A.
3. $(y) \in R$ whenever $F_{\mathbf{A}}(\vec{t}) = \{y\}$;
4. If $F_{\mathbf{A}}(\vec{t})$ is defined and k-tuple $\vec{t_2}$ is created by reordering vertices of $\vec{t}$, then $F_{\mathbf{A}}(\vec{t}) = F_{\mathbf{A}}(\vec{t_2})$.

This then axiomatizes objects of $\mathcal{LEQ}_k$ within $\mathcal{R}$. Applying Theorem 14.3 we now have the following.

Theorem 14.4. *The class $\mathcal{LEQ}_k$ is a Ramsey class.*

Proof. We verify individual conditions of Definition 14.4 and show that $\mathcal{LEQ}_k$ is a $\mathcal{R}$-multiamalgamation class. $\mathcal{LEQ}_k$ is clearly hereditary and closed for amalgamation (which is free for F and where $\leq_{\mathbf{A}}$ is completed to a linear order). The only nontrivial condition is the completion property.

Given $\mathbf{A}, \mathbf{B} \in \mathcal{LEQ}_k$ and $\mathbf{C}_0 \in \mathcal{R}$ put $n(\mathbf{B}, \mathbf{C}_0) = 0$. For $\mathbf{C} \longrightarrow (\mathbf{B})_2^{\mathbf{A}}$ with a homomorphism-embedding to $\mathbf{C}_0$. Construct $\mathbf{C}'$ on vertex set C by completing $\leq_{\mathbf{C}}$ to a linear order $\leq_{\mathbf{C}'}$ (by an existence of homomorphism-embedding from $\mathbf{C}$ to ordered $\mathbf{C}_0$ this is always possible because $\leq_{\mathbf{C}}$ must be acyclic) and define $F_{\mathbf{C}'}(\vec{t}) = \{v\}$ if and only if $F_{\mathbf{C}}(\vec{t}) = \{v\}$ and all vertices of $\vec{t}$ as well as v are part of one copy of $\mathbf{B}$ in $\mathbf{C}$. It is easy to check $\mathbf{C}'$ is ordered L-model (and thus $\mathbf{C}' \in \mathcal{R}$), all three axioms of $\mathcal{LEQ}_k$ are satisfied (and thus $\mathbf{C}' \in \mathcal{LEQ}_k$), all copies of $\mathbf{B}$ in $\mathbf{C}$ are preserved and consequently the identity is $\mathcal{K}$-completion of $\mathbf{C}$ to $\mathbf{C}' \in \mathcal{LEQ}_k$ with respect to copies of $\mathbf{B}$. $\qquad\square$

Remark. Theorem 14.4 is a strong result. For example already for finitely many labels it is close to the Ramsey theorem for relational structures [21]:

The labeling function distinguishes between edges and nonedges. For relational structures with more relations we can always, for each arity, replace the overlap structure by disjoint types of edges thus forming an equivalence.

On the other hand, the labeling may be interpreted as a set of imaginaries which is in line with some standard model theoretic constructions, see [12].

In [13] it is proved that the automorphism group of the Fraïssé limit of the class $\mathcal{E}Q_k$ is amenable group (for every k). The proof is via (Hrushovski) extension property for partial automorphisms (EPPA). It is suggested in [13] that a related group does not contain an ω-categorical extremely amenable subgroup. This is not the case. This fact is a consequence of our Theorem 14.4 using Kechris, Pestov, Todorčević correspondence [14, 24]. (Another examples of ω-categorical groups not containing ω-categorical amenable or extremely amenable subgroup are given in [5].)

14.5 Steiner Systems

For fixed integers $k > t \geq 2$ a *partial (k, t)-Steiner system* is a k-uniform hypergraph $G = (V, E)$ that satisfies that every t-element subset of V is contained in at most one edge $e \in E$. Denote by $\mathcal{S}_{k,t}$ the class of all partial (k, t)-Steiner systems and, as usual, $\vec{\mathcal{S}}_{k,t}$ the class of linearly ordered partial (k, t)-Steiner systems. The class $\mathcal{S}_{k,t}$ is considered with embeddings and the class $\vec{\mathcal{S}}_{k,t}$ with monotone embeddings.

It is easy to see that neither $\mathcal{S}_{k,t}$ nor $\vec{\mathcal{S}}_{k,t}$ is a Ramsey class. For example, we can distinguish pairs of vertices by whether they belong to a hyperedge or not. However, this is basically the only obstacle, as we can show by an application of Theorem 14.3.

Consider the language $L = \{\leq, F\}$ containing one binary relational symbol $\leq$ and one function symbol F of domain arity t and range arity k. Let $\mathcal{R}$ be the class of all finite ordered L-structures. $\mathcal{R}$ is a Ramsey class by Theorem 14.2. It is perhaps surprising that within this class one can axiomatize $\vec{\mathcal{S}}_{k,t}$. Consider the following class $\mathcal{K}$ of structures $\mathbf{A} = (A, \leq_\mathbf{A}, F_\mathbf{A})$ where:

1. A is a linearly ordered set by $\leq_\mathbf{A}$.
2. $F_\mathbf{A}$ is a partial function from A^t to $\binom{A}{k}$.
3. Every $\vec{t} \in \mathrm{Dom}(F_\mathbf{A})$ has no repeated vertices.
4. For every $\vec{t} \in \mathrm{Dom}(F_\mathbf{A})$ it holds that every vertex of $\vec{t}$ is in $F_\mathbf{A}(\vec{t})$ and every r-tuple $\vec{t_2}$ of distinct vertices of $F_\mathbf{A}(\vec{t})$ is in $\mathrm{Dom}(F_\mathbf{A})$ and $F_\mathbf{A}(\vec{t}) = F_\mathbf{A}(\vec{t_2})$.

The embeddings and subobjects for structures in $\mathcal{K}$ are defined as in Section 14.2.

Let us formulate explicitly what is the meaning of embeddings in this case: According to the aforementioned definitions a monotone injection $f : A \to B$ is an embedding if and only if $F_\mathbf{B}((f(x_1), f(x_2), \ldots, f(x_t))) = f(F_\mathbf{A}(x_1, x_2, \ldots, x_t))$ whenever one of the sides of this equation makes sense. This then translates to the following notion of strong embedding: For partial linearly ordered (k, t)-Steiner systems $G = (V, E)$ and $G' = (V', E')$ a monotone injective mapping $h : V \to V'$ is a *strong embedding* if h is a monotone embedding (as in relational structures, i.e., in this case hypergraphs) of G into G' and with the additional property that no t-tuple of V belongs to a k-tuple of G' outside of G. Strong embeddings and strong subobjects correspond exactly to embeddings and (closed) substructures of the foregoing structures $\mathbf{A} = (A, \leq_\mathbf{A}, F_\mathbf{A})$. Denote (for a second) by $\mathcal{R}$ the class of all such finite structures $\mathbf{A}$ where $\leq_\mathbf{A}$ is a linear ordering of vertices.

It is clear that some structures of $\mathcal{R}$ correspond to partial (k, t)-Steiner systems with a linear ordering of its vertices. The function F assigns to a given t-tuple of distinct elements the unique set of k elements containing it, providing such a set exists. Thus the range of the function F is the set of edges of a partial (k, t)-Steiner system. However, not all structures in $\mathcal{R}$ correspond to $\vec{\mathcal{S}}_{k,t}$ because they may not satisfy conditions 3 and 4 given earlier. Conditions 3 and 4 are clearly local and induces a subclass $\mathcal{K}$ of $\mathcal{R}$ which can be clearly identified with $\vec{\mathcal{S}}_{k,t}$. Consequently, by application of Theorem 14.3 we get:

Theorem 14.5. *The class $\mathcal{K}$ (and consequently $\vec{\mathcal{S}}_{k,t}$ with strong embeddings) is a Ramsey class.*

Proof. In analogy to the proof of Theorem 14.4 we can show that $\mathcal{K}$ is an $\mathcal{R}$-multiamalgamation class. The completion property follows similarly, too: Given $\mathbf{A}, \mathbf{B} \in \mathcal{K}$ and $\mathbf{C}_0 \in \mathcal{R}$ put $n(\mathbf{B}, \mathbf{C}_0) = 0$. For $\mathbf{C} \longrightarrow (\mathbf{B})_2^\mathbf{A}$ with a homomorphism-embedding to $\mathbf{C}_0$ construct $\mathbf{C}'$ on vertex set C by completing $\leq_\mathbf{C}$ to a linear order $\leq_{\mathbf{C}'}$ and putting $F_{\mathbf{C}'}(\vec{t}) = E$ if and only if $F_\mathbf{C}(\vec{t}) = E$ and $\vec{t} \cup E \subseteq \tilde{B}$ for some $\tilde{\mathbf{B}} \in \binom{\mathbf{C}}{\mathbf{B}}$. $\qquad \square$

This theorem was proved in [3] by an explicit construction. A different proof (not using Theorem 14.3) is given in [6].

Note also that Steiner systems (such as Steiner Triple Systems) correspond to those structures $\mathbf{A} = (A, \leq_\mathbf{A}, F_\mathbf{A}) \in \mathcal{K}$ where $F_\mathbf{A}$ is a total (not partial) function and, equivalently, to the irreducible structures in $\mathcal{K}$. It is a nontrivial fact that nontrivial Steiner systems exist for all values (k, t) and that any partial (k, t)-Steiner systems can be extended to a (full) Steiner system, see [15, 30, 31, 32]. But this fact together with Theorem 14.4 implies that ordered

(k, t)-Steiner systems form a Ramsey class. Note also that the first nontrivial Ramsey property for Steiner systems was proved in [22].

14.6 Degenerate Graphs and Graphs with k-Orientations (Two Views)

Here we give yet another interpretation of Ramsey results for finite models. Given a positive integer k, a *k-orientation* is an oriented graph such that the out-degree of every vertex is at most k. (Thus a particular case of such graphs are degenerate graphs or graphs with maximal average degree bounded by k; see [19] for a comprehensive study of such "sparse" classes.) A k-orientation $G_1 = (V_1, E_1)$ is *successor closed* in a k-orientation $G_2 = (V_2, E_2)$ if G_1 is a subgraph of G_2 and moreover there is no edge oriented from V_1 to $V_2 \setminus V_1$.

Denote by $\mathcal{D}_k$ the class of all finite k-orientations. This is a hereditary class closed for free amalgamation over successor-closed subgraphs and thus the successor-closeness plays the role of embeddings and strong substructures.

We consider a language $L = \{F^1, F^2, \ldots, F^k\}$ consisting of function F^i of domain arity 1 and range arity i, for $1 \le i \le k$. Given an oriented graph $G = (V, E) \in \mathcal{D}_k$ denote by $\mathbf{G}^+$ the structure with vertex set V and unary functions $F^1, F^2, \ldots, F^k$. Function F^i, $1 \le i \le k$, is defined for every vertex of outdegree i and maps the vertex to all vertices in its out-neighborhood. Denote by $\mathcal{D}_k^+$ the class of all structures $\mathbf{G}^+$ for $G \in \mathcal{D}_k$. Because $\mathbf{G}_1^+$ is a substructure of $\mathbf{G}_2^+$ if and only if G_1 is successor closed in G_2 it follows that $\mathcal{D}_k^+$ is a free amalgamation class.

Denote by $\overrightarrow{\mathcal{D}}_k^+$ a class of all structures $\overrightarrow{\mathbf{A}} = (A, \le_\mathbf{A}, F_\mathbf{A}^1, F_\mathbf{A}^2, \ldots, F_\mathbf{A}^k)$ such that $\mathbf{A} = (A, \le_\mathbf{A}, F_\mathbf{A}^1, F_\mathbf{A}^2, \ldots, F_\mathbf{A}^k) \in \mathcal{D}_k$ and $\le_\mathbf{A}$ is a linear order of A. We immediately get:

Theorem 14.6. *Class $\overrightarrow{\mathcal{D}}_k^+$ is a Ramsey class.*

This can be seen as the most elementary use of Theorem 14.3. But there is more to this than meets the eye: The classes $\mathcal{D}_k$ and $\overrightarrow{\mathcal{D}}_k^+$ are building blocks for the analysis of the Hrushovski construction in the context of structural Ramsey theory [5]. At the end these examples provide the first example of an ω-categorical structure without a precompact Ramsey lift [5].

It is interesting that the class $\mathcal{D}_k$ allows a different lift (or expansion) that makes it also a Ramsey class. Let $G = (V, E)$ be a k-oriented graph. For every vertex v let us fix some arbitrary enumeration of arcs starting from v: $e_1(v), e_2(v), \ldots, e_{k'}(v)$ for $k' \le k$. The the set E is decomposed into sets

$E_1, E_2, \ldots, E_k$, where each set E_i is an oriented graph with outdegrees ≤ 1 and thus each E_i can be viewed as a mapping F_i (by adding idempotent elements at vertices where there are no outgoing arcs).

Thus we get the structures with language $L = \{F^1, F^2, \ldots, F^k\}$ and L-models $\mathbf{A}$ representing $(A, F_\mathbf{A}^1, F_\mathbf{A}^2, \ldots, F_\mathbf{A}^k)$ where $F_\mathbf{A}^i$ are functions $A \to A$. In this situation we can apply Theorem 14.2 directly. The additional condition is that the functions $A \to A$ have (except for idempotents) disjoint sets of arcs. However, this is clearly a local condition. Denote by $\mathcal{F}_L$ the class of all such L-models together with embeddings (defined in Section 14.2). It is clear that the class $\mathcal{F}_L$ is locally finite and that it has free amalgamation. Again denote by $\overrightarrow{\mathcal{F}}_L$ the class of all structures $\overrightarrow{\mathbf{A}}$ extending structures $\mathbf{A} \in \mathcal{F}_L$ by binary relation $\leq_\mathbf{A}$ defining a linear ordering of vertices. Thus we have:

Theorem 14.7. *The class $\overrightarrow{\mathcal{F}}_L$ is a Ramsey class. In the other words: The class of all ordered oriented graphs which are decomposable into k partial mappings is a Ramsey class.*

By different more elaborate methods this has been proved in [27]. An even simpler direct proof is given in [12]. Note that the closed sets in classes $\overrightarrow{\mathcal{F}}_L$ and $\overrightarrow{\mathcal{D}}_k^+$ (which are "expansions" of the same class $\mathcal{D}_k$) are in a correspondence. However the objects differ in their "labeling". The formulation of a Ramsey expansion by means of decomposition will be further pursued in the next section.

14.7 Resolvable Block Designs

We can combine the results of previous two sections. Here is a particular combinatorially interesting example (with illustrious history [30, 31, 32, 25, 20]).

Let $k > 2$. A *Balanced Incomplete Block Design* (shortly BIBD) with parameters $(v, k, 1)$ is a k-uniform hypergraph $(X, \mathcal{B})$ where $\mathcal{B}$ is a system of k-element subsets of X, $|X| = v$, such that every two element subset of X is contained in exactly one subset (called *block*) in $\mathcal{B}$. (In fact this is the same definition as Steiner systems: our BIBD are $(v, k, 1)$-BIBD. We have chosen this term here as it is more standard in this area of resolvability. For $\lambda > 1$ it is possible to generalise our argument. For the sake of brevity we do not do so in this paper.) A partial incomplete balanced block design (PBIBD) contains every pair of elements of X in at most one block. A *Resolvable Balanced Incomplete Block Design* (shortly RBIBD) is block design with the additional property that $\mathcal{B}$ is partitioned into equivalences on X with block as equivalence classes (thus $|X|$ has to be divisible by k). In this case we denote blocks of these equivalences as $\mathcal{B}_1, \mathcal{B}_2, \ldots, \mathcal{B}_m$ (where of course $t = (|X| - 1)/(k - 1)$). A *partial* RBIBD (shortly PRBIBD) is just a partition of blocks into equivalences (not necessarily covering the whole X).

If $(X, \mathcal{B}) = (X, \mathcal{B}_1, \mathcal{B}_2, \ldots, \mathcal{B}_m)$ and $(X', \mathcal{B}') = (X', \mathcal{B}_1, \mathcal{B}_2', \ldots, \mathcal{B}_{m'}')$ are two (partial) RBIBD then isomorphism, embedding, substructure are defined as strong mappings (Section 14.5) and moreover preserving equivalence (i.e., partition into sets $\mathcal{B}_1, \mathcal{B}_2, \ldots, \mathcal{B}_m$).

Denote by $\mathcal{PRBIBD}$ the class of all partial resolvable BIBD and all their embeddings and denote by $\overrightarrow{\mathcal{PRBIBD}}$ the class of all linearly ordered PRBIBD and all their embeddings. One can prove that $\overrightarrow{\mathcal{PRBIBD}}$ is a $\mathcal{R}$-multiamalgamation class (such complicated abbreviations seem to by typical for this area). In fact, this is a combination of Sections 14.5 and 14.6. This can be outlined as follows:

The class $\overrightarrow{\mathcal{PRBIBD}}$ may be interpreted in our scheme (in order to apply Theorem 14.3) as follows. The language is $L = \{\leq, S, F^1, F^2\}$ with arities $d(F^1) = 2, r(F^1) = k, d(F^2) = k, r(F^2) = 1, a(R^S) = 1$. We consider the following class $\mathcal{R}$ of L-structures $\mathbf{A} = (A, \leq_\mathbf{A}, S_\mathbf{A}, F_\mathbf{A}^1, F_\mathbf{A}^2)$:

1. $F_\mathbf{A}^1$ is a partial function from $\binom{A}{2}$ to $\binom{A}{k}$ (function $F_\mathbf{A}^1$ will represent the blocks of $\mathbf{A}$).
2. $F_\mathbf{A}^2$ is a partial function from $\binom{A}{k}$ to A (function $F_\mathbf{A}^2$ will represent the equivalences).

(We want the functions $F_\mathbf{A}^1$ and $F_\mathbf{A}^2$ to be on tuples of distinct elements and moreover symmetric. We indicated this by choosing the domains $\binom{A}{2}$ and $\binom{A}{k}$.) The embeddings are defined in the usual way. Essentially these are strong maps (see Section 14.5) that preserve partition of blocks into equivalences.) Clearly, again by Theorem 14.2, $\mathcal{R}$ is a Ramsey class.

However, our class $\overrightarrow{\mathcal{PRBIBD}}$ is the class of structures that satisfies a few more axioms which relate relations and functions together. They can be formulated as follows (for brevity we put $S = \{x, (x) \in S\}$):

1. $\leq_\mathbf{A}$ is a linear ordering of A.
2. For every $v \in \vec{t} \in \mathrm{Dom}(F_\mathbf{A}^2)$ it holds that $v \notin S$.
3. $\left(\bigcup \mathrm{Rg}(F_\mathbf{A}^2)\right) \subseteq S$.
4. For every $\{x, y\} \in \mathrm{Dom}(F_\mathbf{A}^1)$ it holds that both x and y are in $F_\mathbf{A}^1(\vec{t})$.
5. For every $\{x, y\} \in \mathrm{Dom}(F_\mathbf{A}^1)$ it holds that every pair x', y' of distinct vertices of $F_\mathbf{A}^1(x, y)$ is in $\mathrm{Dom}(F_\mathbf{A}^1)$ and $F_\mathbf{A}^1(x', y') = F_\mathbf{A}^1(x, y)$.
6. the domain of $F_\mathbf{A}^2$ is equal to the range of $F_\mathbf{A}^1$ (i.e. $\mathrm{Dom}(F_\mathbf{A}^2)$ are all blocks).
7. if $F_\mathbf{A}^2(K) = F_\mathbf{A}^2(K')$ then K and K' are disjoint sets.

Denote by $\mathcal{K}$ the subclass of $\mathcal{R}$ of all L-structures with these properties. Observe that these are all local conditions and one can prove that with this the class $\mathcal{K}$ is an $\mathcal{R}$-multiamalgamation class and thus Theorem 14.3 can be applied. Moreover one sees easily that this axiomatization describes the class $\overrightarrow{\mathcal{PRBIBD}}$. As a consequence we have the following

Theorem 14.8. *The classes $\mathcal{K}$ and $\overrightarrow{\mathcal{PRBIBD}}$ are Ramsey classes.*

The previous construction may be further exploited to other partitioned and, say, $\mathcal{H}$ decomposable structures. In the next section we give such an example. However, we do not aim for generality here.

14.8 H-Factorization Theorem

In this example we consider undirected graphs with embeddings (i.e., induced subgraphs). Suppose that H is a fixed graph (see remark at the end concerning possible generalizations). Recall that $\binom{G}{H}$ the set of all induced subgraphs of G isomorphic to H. An H-*matching* in a graph G is a collection $H_1, H_2, \ldots, H_t$ (vertex disjoint) induced subgraphs of G (i.e., we assume that $V(H_i) \cap V(H_j) = \emptyset$ for $i \neq j$) such that all H_i are isomorphic to H. An H-matching will be denoted by M_H, thus in our case $M_H = (H_1, H_2, \ldots, H_m)$. An H-*factorization* of G is a collection of H-matchings that partition the set $\binom{G}{H}$. *Partial H-factorization* of G is a collection of disjoint H-matching in the set $\binom{G}{H}$.

Graphs with a partial H-factorization will be considered with an embedding preserving the factorization. Given $G = (V, E)$ and $G' = (V', E')$ with partial factorizations $(M_H^1(G), M_H^2(G), \ldots, M_H^m(G))$ and $(M_H^1(G'), M_H^2(G'), \ldots, M_H^{m'}(G'))$ we say that a pair f, ι is an H-factorization embedding if:

1. f is an embedding G into G'.
2. ι is a monotone injection $\{1, \ldots, m\} \rightarrow \{1, \ldots, m'\}$.
3. for every two graphs H', H'' isomorphic to H we have $H', H'' \in M_H^i(G)$ if and only if $f(H'), f(H'') \in M_H^{\iota(i)}(G)$.

The last condition of course means that two graphs H', H'' isomorphic to H belong to the same class of partial H-factorizations of G if and only if their images $f(H'), f(H'')$ belong to the same class of partial H-factorizations of G'.

Denote by $\text{Fact}(H)$ the class of all finite graphs endowed with partial H-factorizations. By $\overrightarrow{\text{Fact}(H)}$ we denote the corresponding class of linearly ordered partially H-factorized graphs endowed with monotone embedding. We shall see that $\overrightarrow{\text{Fact}(H)}$ can be always interpreted as an multiamalgamation class. For this it is convenient to consider structures in Fact(H) as structures with two mappings, and two relations as follows:

Again, we consider the language $L = \{R, \leq, R^S, F^1, F^2\}$. Function F^1 has domain arity 2 and range arity $|H|$. The function F^2 has domain of arity $|H|$ and range of arity 1, R^S has arity 1.

In this situation we consider the class $\mathcal{R}$ of all L-structures $\mathbf{A} = (A, R_\mathbf{A}, \leq_\mathbf{A}, R^S_\mathbf{A}, F^1_\mathbf{A}, F^2_\mathbf{A})$: Again applying Theorem 14.2 we know that $\mathcal{R}$ is a Ramsey class. However the class Fact(H) has some more properties and they will be reflected by the subclass $\mathcal{K}$ of $\mathcal{R}$ of all structures $\mathbf{A} = (A, R_\mathbf{A}, \leq_\mathbf{A}, S_\mathbf{A}, F^1_\mathbf{A}, F^2_\mathbf{A})$ that satisfy the following (for brevity, we put again $S = \{x, (x) \in R\}$):

1. $\leq_\mathbf{A}$ is a linear ordering of A.
2. For every $v \in \vec{t} \in \mathrm{Dom}(F^2_\mathbf{A})$ it holds that $v \notin S$.
3. $\left(\bigcup \mathrm{Rg}(F^2_\mathbf{A})\right) \subseteq S$.
4. $(A, R_\mathbf{A})$ is an undirected graph (denoted by $G_\mathbf{A}$).
5. For every t-tuple $\vec{t} \in \mathrm{Dom}(F^2_\mathbf{A})$ the graph $G_\mathbf{A}$ induces on $\vec{t}$ a graph isomorphic to H.
6. $F^1_\mathbf{A}$ is a partial function from A^2 to $A^{|H|}$.
7. $F^2_\mathbf{A}$ is a partial function from $A^{|H|}$ to A.
8. $F^1_\mathbf{A}$ is symmetric; explicitely $F^1_\mathbf{A}(u, v) = F^1_\mathbf{A}(v, u)$.
9. $F^2_\mathbf{A}$ is symmetric and defined on tuples of distinct elements.
10. For every pair $u, v \in \mathrm{Dom}(F^1_\mathbf{A})$ it holds that both vertices u, v belong also to the image $F^1_\mathbf{A}(u, v)$.
11. Every pair u', v' of distinct vertices such that $F^1_\mathbf{A}(u, v) = \{u', v'\}$ is in $\mathrm{Dom}(F^1_\mathbf{A})$ and $F^1_\mathbf{A}(u, v) = F^1_\mathbf{A}(u', v')$.
12. The domain of $F^2_\mathbf{A}$ is equal to the range of $F^1_\mathbf{A}$.
13. $F^2_\mathbf{A}$ is "idempotent" meaning that $\mathrm{Dom}(F^2_\mathbf{A}) \cap \mathrm{Rg}(F^2_\mathbf{A}) = \emptyset$.
14. If $F^2_\mathbf{A}(K) = F^2_\mathbf{A}(K')$ then K and K' are disjoint sets.

All the conditions on functions $F^1_\mathbf{A}$ and $F^2_\mathbf{A}$ are local and one can prove that with this interpretation the class $\mathcal{K}$ is a $\mathcal{R}$-multiamalgamation class and thus Theorem 14.3 can be applied. As a consequence we have the following.

Theorem 14.9. *Then $\mathcal{K}$ is a Ramsey class.*

14.9 Final Comments

1. The key constructions of this chapter may be iterated. In this paper we presented iterations of at most two functions. Examples given here illustrate two main features of applications (and influence) of inclusion of functions in our vocabularies: we have trees and equivalences, functions and partitions.

2. The calculus of functions and partitions leads to abstract data types and Ramsey classes which may have some computer science consequences. There are some precedents, see influential [33].

3. We may have more arities. For example, we may consider $\mathcal{H}$-factorization for a family $\mathcal{H}$ of graphs. The arities are then the set $\{|H|, H \in \mathcal{H}\}$. This simply means that the domain and range can contain any arity in the set. We may even have an infinite set of arities. In dealing with Ramsey classes this does not lead to difficulties as we always deal with finite structures of sizes given in advance. (Recall the basic scheme: Given **A**, **B**, we look for **C**.) One can clearly generalize here.

We decided to stop here.

Acknowledgments

The Computer Science Institute of Charles University (IUUK) is supported by grant ERC-CZ LL-1201 of the Czech Ministry of Education and CE-ITI P202/12/G061 of GAČR.

References

1. Fred G. Abramson and Leo A. Harrington. Models without indiscernibles. *J. Symbolic Logic* **43** (1978), 572–600.
2. Andres Aranda, David Bradley-Williams, Jan Hubička, Miltiadis Karamanlis, Michael Kompatscher, Matěj Konečný, and Micheal Pawliuk. Ramsey expansions of metrically homogeneous graphs. arXiv:1707.02612, 2017.
3. Vindya Bhat, Jaroslav Nešetřil, Christian Reiher, and Vojtěch Rödl. A Ramsey class for Steiner systems. arXiv:1607.02792, 2016 to appear in *J. Comb. Th. A.*
4. Gregory Cherlin, Saharon Shelah, and Niandong Shi. Universal graphs with forbidden subgraphs and algebraic closure. *Adv. Appl. Math.* **22**, no. 4, (1999), 454–491.
5. David M. Evans, Jan Hubička, and Jaroslav Nešetřil. Automorphism groups and Ramsey properties of sparse graphs. arXiv:1801.01165.
6. David M. Evans, Jan Hubička, and Jaroslav Nešetřil. Ramsey properties and extending partial automorphisms for classes of finite structures. arXiv:1705.02379, 2017.
7. Ronald L. Graham, Klaus Leeb, and Bruce L. Rothschild. Ramsey's theorem for a class of categories. *Adv. Math.* **8**, no. 3 (1972), 417–433.
8. Ronald L. Graham and Bruce L. Rothschild. Ramsey's theorem for n-parameter sets. *Trans. Amer. Math. Soc.* **159** (1971), 257–292.
9. James D. Halpern and Hans Läuchli. A partition theorem. *Trans. Amer. Math. Soc.* **124**, no. 2 (1966), 360–367.
10. Wilfrid Hodges. *Model Theory*, Vol. 42. Cambridge University Press, Cambridge, 1993.

11. Jan Hubička and Jaroslav Nešetřil. Bowtie-free graphs have a Ramsey lift. arXiv:1402.2700, 2014.
12. Jan Hubička and Jaroslav Nešetřil. All those Ramsey classes (Ramsey classes with closures and forbidden homomorphisms). arXiv:1606.07979, 2016.
13. Aleksander Ivanov. An ω-categorical structure with amenable automorphism group. *Math. Logic Quart.* **61**, no. 4–5 (2015), 307–314.
14. Alexander S. Kechris, Vladimir G. Pestov, and Stevo Todorčević. Fraïssé limits, Ramsey theory, and topological dynamics of automorphism groups. *Geometr. Funct. Anal.* **15**, no. 1 (2005), 106–189.
15. Peter Keevash. The existence of designs. arXiv:1401.3665, 2014.
16. Péter Komjáth. Some remarks on universal graphs. *Discrete Math.* **199**, no. 1 (1999), 259–265.
17. Keith R. Milliken. A Ramsey theorem for trees. *J. Combin. Theory, Ser. A* **26**, no. 3 (1979), 215–237.
18. Jaroslav Nešetřil. Ramsey theory. In R. L. Graham, M. Grötschel, and L. Lovász, editors, *Handbook of Combinatorics*, Vol. 2, pp. 1331–1403. MIT Press, Cambridge, MA, 1995.
19. Jaroslav Nešetřil and Patrice Ossona de Mendez. *Sparsity: Graphs, Structures, and Algorithms*, Vol. 28. Springer Science + Business Media, New York, 2012.
20. Jaroslav Nešetřil and Helena Nešetřilová. A remark on Kirkman's school girls problem (in Czech, English abstract). *Dějiny vědy a techniky (History of Science and Technology)* **5**, no. 3–4 (1971), 171–173.
21. Jaroslav Nešetřil and Vojtěch Rödl. The Ramsey property for graphs with forbidden complete subgraphs. *J. Combinat. Theory, Ser. B* **20**, no. 3 (1976), 243–249.
22. Jaroslav Nešetřil and Vojtěch Rödl. Strong Ramsey theorems for Steiner systems. *Trans. Amer. Math. Soc.* **303**, no. 1 (1987), 183–192.
23. Jaroslav Nešetřil and Vojtěch Rödl. Partitions of finite relational and set systems. *J. Combinat. Theory, Ser. A* **22**, no. 3 (1977), 289–312.
24. Lionel Nguyen Van Thé. More on the Kechris–Pestov–Todorcevic correspondence: Precompact expansions. *Fundamenta Mathematicae* **222** (2013), 19–47.
25. Dwijendra K. Ray-Chaudhuri and Richard M. Wilson. Solution of Kirkmans schoolgirl problem. In Theodore S. Motzkin, editor, *Proceedings of the Symposia in Pure Mathematics*, Vol. 19 of *Combinatorics*, pp. 187–203. American Mathematical Society, 1971.
26. Norbert W. Sauer. Distance sets of Urysohn metric spaces. *Canad. J. Math.* **65**, no. 1 (2013), 222–240.
27. Miodrag Sokić. Unary functions. *Eur. J. Combin.* **52** (2016), 79–94.
28. Sławomir Solecki. Direct Ramsey theorem for structures involving relations and functions. *J. Combinat. Theory, Ser. A* **119**, no. 2 (2012), 440–449.
29. Sławomir Solecki. Monoid actions and ultrafilter methods in Ramsey theory. arXiv:1611.06600, 2016.
30. Richard M. Wilson. An existence theory for pairwise balanced designs I: Composition theorems and morphisms. *J. Combinat. Theory, Ser. A* **13**, no. 2 (1972), 220–245.
31. Richard M. Wilson. An existence theory for pairwise balanced designs II: The structure of PBD-closed sets and the existence conjectures. *J. Combinat. Theory, Ser. A* **13**, no. 2 (1972), 246–273.

32. Richard M. Wilson. An existence theory for pairwise balanced designs, III: Proof of the existence conjectures. *J. Combinat. Theory, Ser. A* **18**, no. 1 (1975), 71–79.

33. Andrew Chi-Chih Yao. Should tables be sorted? *JACM* **28**, no. 3 (1981), 615–628.

15

Borsuk and Ramsey Type Questions
in Euclidean Space

Peter Frankl, János Pach, Christian Reiher, and Vojtěch Rödl

Abstract

We give a short survey of problems and results on (1) diameter graphs and hypergraphs, and (2) geometric Ramsey theory. We also make some modest contributions to both areas. Extending a well-known theorem of Kahn and Kalai that disproved Borsuk's conjecture, we show that for any integer $r \geqslant 2$, there exist $\varepsilon = \varepsilon(r) > 0$ and $d_0 = d_0(r)$ with the following property. For every $d \geqslant d_0$, there is a finite point set $P \subset \mathbb{R}^d$ of diameter 1 such that no matter how we color the elements of P with fewer than $(1 + \varepsilon)^{\sqrt{d}}$ colors, we can always find r points of the same color, any two of which are at distance 1.

Erdős, Graham, Montgomery, Rothschild, Spencer, and Strauss called a finite point set $P \subset \mathbb{R}^d$ *Ramsey* if for every $r \geqslant 2$, there exists a set $R = R(P, r) \subset \mathbb{R}^D$ for some $D \geqslant d$ such that no matter how we color all of its points with r colors, we can always find a monochromatic congruent copy of P. If such a set R exists with the additional property that its diameter is the same as the diameter of P, then we call P *diameter-Ramsey*. We prove that, in contrast to the original Ramsey property, (1) the condition that P is diameter-Ramsey is not hereditary, and (2) not all triangles are diameter-Ramsey. We raise several open questions related to this new concept.

15.1 Introduction

The aim of this chapter is twofold. In the spirit of Graham-Yao [15], we give a "whirlwind tour" of two areas of Geometric Ramsey Theory, and make some modest contributions to them.

The *diameter* of a finite point set P, denoted by $\operatorname{diam}(P)$, is the largest distance that occurs between two points of P. Borsuk's famous conjecture [2],

259

restricted to finite point sets, states that any such set of unit diameter in $\mathbb{R}^d$ can be colored by $d + 1$ colors so that no two points of the same color are at distance 1. This conjecture was disproved in a celebrated paper of Kahn and Kalai [20]. We extend the theorem of Kahn and Kalai as follows.

Theorem 15.1. *For any integer $r \geqslant 2$, there exist $\varepsilon = \varepsilon(r) > 0$ and $d_0 = d_0(r)$ with the following property. For every $d \geqslant d_0$, there is a finite point set $P \subset \mathbb{R}^d$ of diameter 1 such that no matter how we color the elements of P with fewer than $(1 + \varepsilon)^{\sqrt{d}}$ colors, we can always find r points of the same color, any two of which are at distance 1.*

In a seminal paper of Erdős, Graham, Montgomery, Rothschild, Spencer, and Strauss [7], the following notion was introduced. A finite set P of points in a Euclidean space is a *Ramsey configuration* or, briefly, is *Ramsey* if for every $r \geqslant 2$, there exists an integer $d = d(P, r)$ such that no matter how we color all points of $\mathbb{R}^d$ with r colors, we can always find a monochromatic subset of $\mathbb{R}^d$ that is congruent to P. In two follow-up articles [8, 9], Erdős, Graham, and their coauthors established many important properties of these sets.

In the present chapter, we introduce a related notion.

Definition 15.1. *A finite set P of points in a Euclidean space is* diameter-Ramsey *if for every integer $r \geqslant 2$, there exist an integer $d = d(P, r)$ and a finite subset $R \subset \mathbb{R}^d$ with $\mathrm{diam}(R) = \mathrm{diam}(P)$ such that no matter how we color all points of R with r colors, we can always find a monochromatic subset of R that is congruent to P.*

Obviously, every diameter-Ramsey set is Ramsey, but the converse is not true. For example, we know that all triangles are Ramsey, but not all of them are diameter-Ramsey.

Theorem 15.2. *All acute and all right-angled triangles are diameter-Ramsey.*

Theorem 15.3. *No triangle that has an angle larger than $150°$ is diameter-Ramsey.*

There is another big difference between the two notions: By definition, every subset of a Ramsey configuration is Ramsey. This is not the case for diameter-Ramsey sets.

Theorem 15.4. *The seven-element set consisting of a vertex of a six-dimensional cube and its six adjacent vertices is not diameter-Ramsey.*

We will see that the vertex set of a cube (in fact, the vertex set of any brick) is diameter-Ramsey; see Lemma 15.1. Therefore, the property that a set is diameter-Ramsey is not hereditary.

It appears to be a formidable task to characterize all diameter-Ramsey simplices. It easily follows from the definition that all regular simplices are diameter-Ramsey; see Proposition 15.1. We will show that the same is true for "almost regular" simplices.

Theorem 15.5. *For every integer $n \geqslant 2$, there exists a positive real number $\varepsilon = \varepsilon(n)$ such that every n-vertex simplex whose side lengths belong to the interval $[1 - \varepsilon, 1 + \varepsilon]$ is diameter-Ramsey.*

This chapter is organized as follows: In Section 15.2, we give a short survey of problems and results on the structure of diameters and related coloring questions. In Section 15.3, we prove Theorem 15.1. In Section 15.4, we establish some simple properties of diameter-Ramsey sets and prove Theorems 15.2, 15.3, and 15.4, in a slightly stronger form. The proof of Theorem 15.4 is presented in Section 15.5. The last section contains a few open problems and concluding remarks.

15.2 A Short History

15.2.1 The Number of Edges of Diameter Graphs and Hypergraphs

Hopf and Pannwitz [18] noticed that in any set P of n points in the plane, the diameter occurs at most n times. In other words, among the $\binom{n}{2}$ distances between pairs of points from P at most n are equal to $\mathrm{diam}(P)$. This bound can be attained for every $n \geqslant 3$. For odd n this is shown by the vertex set of a regular n-gon, and for even n it is not hard to observe that one may add a further point to the vertex set of a regular $(n-1)$-gon so as to obtain such an example. In fact, all extremal configurations were characterized by Woodall [36].

The same question in $\mathbb{R}^3$ was raised by Vázsonyi, who conjectured that the maximum number of times the diameter can occur among $n \geqslant 4$ points in 3-space is $2n - 2$. Vázsonyi's conjecture was proved independently by Grünbaum [16], by Heppes [17], and by Straszewicz [33]; see also [34] for a simple proof. The extremal configurations were characterized in terms of ball polytopes by Kupitz, Martini, and Perles [25].

In dimensions larger than 3, the nature of the problem is radically different.

Theorem 15.6. (Erdős [6]) For any integer $d > 3$, the maximum number of occurrences of the diameter (and, in fact, of any fixed distance) in a set of n points in $\mathbb{R}^d$ is $\frac{1}{2} \left(1 - \frac{1}{\lfloor d/2 \rfloor} + o(1) \right) n^2$.

More recently, Swanepoel [35] determined the exact maximum number of appearances of the diameters for all $d > 3$ and all n that are sufficiently large depending on d.

The *diameter graph* associated with a set of points P is a graph with vertex set P, in which two points are connected by an edge if and only if their distance is $\mathrm{diam}(P)$. Erdős noticed that there is an intimate relationship between the foregoing estimates for the number of edges of diameter graphs and the following attractive conjecture of Borsuk [2]: Every (finite) d-dimensional point set can be decomposed into at most $d + 1$ sets of smaller diameter. If it were true, this bound would be best possible, as demonstrated by the vertex set of a regular simplex in $\mathbb{R}^d$.

One can generalize the notion of diameter graph as follows. Given a point set $P \subset \mathbb{R}^d$ and an integer $r \geqslant 2$, let $H_r(P)$ denote the hypergraph with vertex set P whose hyperedges are all r-element subsets $\{p_1, \ldots, p_r\} \subseteq P$ with $|p_i - p_j| = \mathrm{diam}(P)$ whenever $1 \leqslant i \neq j \leqslant r$. Obviously, $H_2(P)$ is the diameter graph of P, and $H_r(P)$ consists of the vertex sets of all r-*cliques* (complete subgraphs with r vertices) in the diameter graph. Note that every r-clique corresponds to a regular $(r - 1)$-dimensional simplex with side length $\mathrm{diam}(P)$. We call $H_r(P)$ the r-*uniform diameter hypergraph* of P.

It was conjectured by Schur that the Hopf–Pannwitz theorem mentioned at the beginning of this subsection can be extended to higher dimensions in the following way: For any $d \geqslant 2$ and any d-dimensional n-element point set P, the hypergraph $H_d(P)$ has at most n hyperedges. This was proved for $d = 3$ by Schur, Perles, Martini, and Kupitz [32]. Building on work of Morić and Pach [30], the case $d = 4$ was resolved by Kupavskii [23], and the general case of Schur's conjecture was subsequently settled by Kupavskii and Polyanskii [24].

However, for $2 < r < d$ we know very little about the number of edges of the diameter hypergraphs $H_r(P)$ and it would be interesting to investigate this matter further.

15.2.2 The Chromatic Number of Diameter Graphs and Hypergraphs

Erdős [5] pointed out that if we could prove that the number of edges of the diameter graph of every n-element point set $P \subset \mathbb{R}^d$ is smaller than $\frac{d+1}{2} n$, then this would imply that there is a vertex of degree at most d. Hence, the chromatic number of the diameter graph would be at most $d + 1$, and the color classes of any proper coloring with $d + 1$ colors would define a decomposition of P into at most $d + 1$ pieces of smaller diameter, as required by Borsuk's conjecture. For $d = 2$ and 3, this is the case. However, as is shown by Theorem 15.6, in higher dimensions the number of edges of an n-vertex diameter graph can grow quadratically in n. Based on this, Erdős later suspected that Borsuk's conjecture may be false (personal communication). This was verified only in 1993 by Kahn and Kalai [20].

Using a theorem of Frankl and Wilson [13], Kahn and Kalai established the following much stronger statement.

Theorem 15.7. (Kahn–Kalai) *For any sufficiently large d, there is a finite point set P in the d-dimensional Euclidean space such that no matter how we partition it into fewer than $(1.2)^{\sqrt{d}}$ parts, at least one of the parts contains two points whose distance is* $\operatorname{diam}(P)$.

In other words, the chromatic number of the diameter graph of P is at least $(1.2)^{\sqrt{d}}$. Today Borsuk's conjecture is known to be false for all dimensions $d \geqslant 64$; cf. [19].

Definition 15.2. *The* chromatic number *of a hypergraph H is the smallest number $\chi = \chi(H)$ with the property that the vertex set of H can be colored with χ colors such that no hyperedge of H is monochromatic.*

Clearly, we have

$$\chi(H_r(P)) \leqslant \chi(H_{r-1}(P)) \leqslant \ldots \leqslant \chi(H_2(P)),$$

for every P and $r \geqslant 2$. Moreover,

$$\chi(H_r(P)) \leqslant \left\lceil \frac{\chi(H_2(P))}{r - 1} \right\rceil.$$

To see this, take a proper coloring of the diameter graph $H_2(P)$ with the minimum number, $\chi = \chi(H_2(P))$, of colors and let $P_1, \ldots, P_\chi$ be the corresponding color classes. Coloring all elements of

$$P_{(i-1)(r-1)+1} \cup P_{(i-1)(r-1)+2} \cup \ldots \cup P_{i(r-1)}$$

with color i for $1 \leqslant i \leqslant \frac{\chi}{r-1}$, we obtain a proper coloring of the hypergraph $H_r(P)$. (Here we set $P_s = \varnothing$ for all $s > \chi$.)

Using the foregoing notation, the Kahn–Kalai theorem states that for any sufficiently large integer d, there exists a set $P \subset \mathbb{R}^d$ with $\chi(H_2(P)) \geqslant (1.2)^{\sqrt{d}}$.

According to a result of Schramm [31], we have $\chi(H_2(P)) \leqslant (\sqrt{3/2} + \varepsilon)^d$ for every $\varepsilon > 0$, provided that d is sufficiently large.

In the next section, we prove Theorem 15.1 stated in the Introduction. It extends the Kahn–Kalai theorem to r-uniform diameter hypergraphs with $r \geqslant 2$. Using the preceding notation, we will prove the following.

Theorem 15.8. *For any integer $r \geqslant 2$, there exist $\varepsilon = \varepsilon(r) > 0$ and $d_0 = d_0(r)$ with the following property. For every $d \geqslant d_0$, there is a finite point set $P \subset \mathbb{R}^d$ of diameter 1 such that*

$$\chi(H_r(P)) \geqslant (1 + \varepsilon)^{\sqrt{d}}.$$

That is, for any partition of P into fewer than $(1 + \varepsilon)^{\sqrt{d}}$ parts at least one of the parts contains r points any two of which are at distance 1.

15.2.3 Geometric Ramsey Theory

Recall from the Introduction that, according to the definition of Erdős, Graham et al. [7], a finite set of points in some Euclidean space is said to be *Ramsey* if for every $r \geqslant 2$, there exists an integer $d = d(P, r)$ such that no matter how we color all points of $\mathbb{R}^d$ with r colors, we can always find a monochromatic subset of $\mathbb{R}^d$ that is congruent to P. Erdős, Graham et al. proved, among many other results, that every Ramsey set is *spherical*, i.e., embeddable into the surface of a sphere. Later Graham [14] conjectured that the converse is also true: every spherical configuration is *Ramsey*. An important special case of this conjecture was settled by Frankl and Rödl.

Theorem 15.9. [11] *Every simplex is Ramsey.*

It was shown in [7] that the class of all Ramsey sets is closed both under taking subsets and taking Cartesian products. This implies.

Corollary 15.1. [7] *All* bricks, *i.e., Cartesian products of finitely many two-element sets, are Ramsey.*

Further progress in this area has been rather slow. The first example of a planar Ramsey configuration with at least *five* elements was exhibited by Kříž, who showed that every regular polygon is Ramsey. He also proved that the same is true for every Platonic solid. Actually, he deduced both of these statements from the following more general theorem.

Theorem 15.10. [22] *If there is a soluble group of isometries acting on a finite set of points P in $\mathbb{R}^d$, which has at most two orbits, then P is Ramsey.*

Graham's conjecture is still widely open. In fact, it is not even known whether all quadrilaterals inscribed in a circle are Ramsey.

An alternative conjecture has been put forward by Leader, Russell, and Walters [28]. They call a point set *transitive* if its symmetry group is transitive. A subset of a transitive set is said to be *subtransitive*. Leader et al. conjecture that a set is Ramsey if and only if it is subtransitive. It is not obvious a priori that this conjecture is different from Graham's, that is, if there exists any spherical set which is not subtransitive. However, this was shown to be the case in [28]. In [27] the same authors showed further that not all quadrilaterals inscribed in a circle are subtransitive.

The "compactness" property of the chromatic number, established by Erdős and de Bruijn [3], implies that for every Ramsey set P and every positive integer r, there exists a *finite* configuration $R = R(P, r)$ with the property that *no matter how we color the points of R with r colors, we can find a congruent copy of P which is monochromatic*. Following the (now standard) notation introduced by Erdős and Rado, we abbreviate this property by writing

$$R \longrightarrow (P)_r.$$

In Section 15.4, we address the problem how small the diameter of such a set R can be. In particular, we investigate the question whether there exists a set R with $\mathrm{diam}(R) = \mathrm{diam}(P)$ such that $R \longrightarrow (P)_r$. If such a set exists for every r, then according to Definition 15.1, P is called diameter-Ramsey.

15.3 Proof of Theorem 15.1

The proof of Theorem 15.1, reformulated as Theorem 15.8, is based on the construction used by Kahn and Kalai in [20].

Suppose for simplicity that $d = \binom{2n}{2}$ holds for some *even* integer n and set $[2n] = \{1, 2, \ldots, 2n\}$. The construction takes place in $\mathbb{R}^d$ and in the following we will index the coordinates of this space by the two-element subsets of $[2n]$.

To each partition $[2n] = X \cup Y$ of $[2n]$ into two n-element subsets X and Y, we assign the point $p(X, Y) = p(Y, X) \in \mathbb{R}^d$ whose coordinate $p_T(X, Y)$ corresponding to some unordered pair $T \subseteq [2n]$ is given by

$$p_T(X, Y) = \begin{cases} 1 & \text{if } |T \cap X| = |T \cap Y| = 1, \\ 0 & \text{otherwise.} \end{cases}$$

Let $P \subseteq \mathbb{R}^d$ be the set of all such points $p(X, Y)$. We have $|P| = \frac{1}{2}\binom{2n}{n}$.

Each point $p(X, Y) \in P$ has precisely $|X| \, |Y| = n^2$ nonzero coordinates. The squared Euclidean distance between $p(X, Y)$ and $p(X', Y')$, for two different partitions of $[2n]$, is equal to the number of coordinates in which $p(X, Y)$ and $p(X', Y')$ differ. The number of coordinates in which both $p(X, Y)$ and $p(X', Y')$ have a 1 is equal to

$$|X \cap X'| \, |Y \cap Y'| + |X \cap Y'| \, |X' \cap Y| \, .$$

Denoting $|X \cap X'| = |Y \cap Y'|$ by t, the last expression is equal to $t^2 + (n - t)^2$. Thus, we have

$$\|p(X, Y) - p(X', Y')\|^2 = 2n^2 - 2(t^2 + (n - t)^2) \, ,$$

which attains its maximum for $t = \frac{n}{2}$. The maximum is n^2, so that $\mathrm{diam}(P) = n$.

Fact 15.1. *An r-element subset $\{p(X_1, Y_1), \dots, p(X_r, Y_r)\} \subseteq P$ is a hyperedge of $H_r(P)$, the r-uniform diameter hypergraph of P, if and only if*

$$|X_i \cap X_j| = \frac{n}{2} \quad \text{for all } 1 \leqslant i \neq j \leqslant r. \qquad \square$$

We need the following important special case of a result of Frankl and Rödl [10] from extremal set theory. The set of all n-element subsets of $[2n]$ is denoted by $\binom{[2n]}{n}$.

Theorem 15.11. *[10] For every integer $r \geqslant 2$, there exists $\gamma = \gamma(r) > 0$ with the following property. Every family of subsets $\mathcal{F} \subseteq \binom{[2n]}{n}$ with $|\mathcal{F}| \geqslant (2 - \gamma)^{2n}$ has r members, $F_1, \dots, F_r \in \mathcal{F}$, such that*

$$|F_i \cap F_j| = \left\lfloor \frac{n}{2} \right\rfloor \quad \text{for all } 1 \leqslant i \neq j \leqslant r.$$

To establish Theorem 15.8, fix a subset Q of the set P defined earlier. The elements of Q are points $p(X, Y) \in \mathbb{R}^d$ for certain partitions $[2n] = X \cup Y$. Let $\mathcal{F}(Q) \subseteq \binom{[2n]}{n}$ denote the family of all sets X and Y defining the points in Q. Notice that $|\mathcal{F}(Q)| = 2 \, |Q|$.

By definition, $\chi = \chi(H_r(P))$ is the smallest number for which there is a partition

$$P = Q_1 \cup \dots \cup Q_\chi$$

such that no Q_k contains any hyperedge belonging to $H_r(P)$. According to Fact 15.1, this is equivalent to the condition that $\mathcal{F}(Q_k)$ does not contain r members such that any two have precisely $\frac{n}{2}$ elements in common. Now Theorem 15.11 implies that

$$|\mathcal{F}(Q_k)| = 2 \, |Q_k| < \left(2 - \gamma(r)\right)^{2n} \quad \text{whenever} \quad 1 \leqslant k \leqslant \chi \, .$$

Thus, we have

$$|P| = \sum_{k=1}^{\chi} |Q_k| < \frac{\chi}{2} \left(2 - \gamma(r)\right)^{2n} .$$

Comparing the last inequality with the equation $|P| = \frac{1}{2}\binom{2n}{n}$, we obtain

$$\chi = \chi(H_r(P)) > \frac{\binom{2n}{n}}{\left(2 - \gamma(r)\right)^{2n}} > \left(1 + \frac{\gamma(r)}{3}\right)^{\sqrt{2d}} .$$

This completes the proof of Theorem 15.8.

The proof of Theorem 15.8 gives the following result. The *regular* simplex S_r with r vertices and *unit side length* is not only a Ramsey configuration, but for every k there exists set $P(k) \subseteq \mathbb{R}^d$ of *unit diameter* with $d \leqslant c(r) \log^2 k$ such that no matter how we color $P(k)$ with k colors, it contains a monochromatic congruent copy of S_r. (Here $c(r) > 0$ is a suitable constant that depends only on r.)

15.4 Diameter-Ramsey Sets: Proofs of Theorems 15.2, 15.3, and 15.5

According to Definition 15.1, a finite point set P is diameter-Ramsey if for every $r \geqslant 2$, there exists a finite set R in some Euclidean space with $\operatorname{diam}(R) = \operatorname{diam}(P)$ such that no matter how we color all points of R with r colors, we can always find a monochromatic subset of R that is congruent to P. Before proving Theorems 15.2, 15.3, and 15.5, we make some general observations about diameter-Ramsey sets.

Proposition 15.1. *Every regular simplex is diameter-Ramsey.*

Proof. Let P be (the vertex set of) a d-dimensional regular simplex. For a fixed integer $r \geqslant 2$, let R be an rd-dimensional regular simplex of the same side length. By the pigeonhole principle, no matter how we color the vertices of R with r colors, at least $d + 1$ of them will be of the same color, and they induce a congruent copy of P. □

Recall that a *brick* is the vertex set of the Cartesian product of finitely many two-element sets.

Lemma 15.1. *If P and Q are diameter-Ramsey sets, then so is their Cartesian product $P \times Q$. Consequently, any brick is diameter-Ramsey.*

Proof. It was shown in [7] that for any Ramsey sets P and Q, their Cartesian product,

$$P \times Q = \{p \times q \mid p \in P, q \in Q\},$$

is also a Ramsey set. Their argument, combined with the equation

$$\text{diam}^2(P \times Q) = \text{diam}^2(P) + \text{diam}^2(Q),$$

proves the lemma.   □

Proof of Theorem 15.2. Consider a right-angled triangle T whose legs are of length l_1 and l_2. Let P (resp., Q) be a set consisting of two points at distance l_1 (resp., l_2) from each other, so that we have $T \subseteq P \times Q$. By Lemma 15.1, $P \times Q$ is diameter-Ramsey. Since $\text{diam}(T) = \text{diam}(P \times Q)$, we also have that T is diameter-Ramsey.

Now let T be an acute triangle with sides a, b, and c, where $a \leqslant b \leqslant c$. Set

$$l_1 = \sqrt{c^2 - a^2}, \quad l_2 = \sqrt{c^2 - b^2}, \quad \text{and} \quad x = \sqrt{a^2 + b^2 - c^2}.$$

Since T is acute, we have $a^2 + b^2 - c^2 > 0$. Therefore, x is well defined. We have $l_1 \geqslant l_2 \geqslant 0$. Suppose first that $l_1 \geqslant l_2 > 0$. Let T_0 be a right-angled triangle with legs l_1 and l_2, and let S be an equilateral triangle of side length x. We have $a^2 = l_2^2 + x^2, b^2 = l_1^2 + x^2$, and $c^2 = l_1^2 + l_2^2 + x^2$. Thus,

$$T \subseteq T_0 \times S \quad \text{and} \quad \text{diam}(T) = \text{diam}(T_0 \times S) = c.$$

By Proposition 15.1 and Lemma 15.1, we conclude that T is diameter-Ramsey. In the remaining case, we have $l_2 = 0$. Now T_0 degenerates into a line segment or a point. It is easy to see that the foregoing proof still applies.   □

We will prove Theorem 15.3 in a more general form. For this, we need a definition.

Definition 15.3. *Let t be a positive integer. A finite set of points P in some Euclidean space is said to be t-degenerate if it has a point $p \in P$ such that for the vertex set S of any regular t-dimensional simplex with $p \in S$ and $\text{diam}(S) = \text{diam}(P)$, we have*

$$\text{diam}(P \cup S) > \text{diam}(P).$$

Theorem 15.12. *Let $t \geqslant 1$ and let P be a finite t-degenerate set of points in some Euclidean space, which contains the vertex set of a regular t-dimensional simplex of side length $\text{diam}(P)$. Then P is not diameter-Ramsey.*

Proof. Suppose for contradiction that P is diameter-Ramsey. This implies that there exists a set R with $\mathrm{diam}(R) = \mathrm{diam}(P)$ such that no matter how we color it by two colors, it always contains a monochromatic congruent copy of P.

Color the points of R with red and blue, as follows. A point is colored *red* if it belongs to a subset $S \subset R$ that spans a t-dimensional simplex of side length $\mathrm{diam}(R)$. Otherwise, we color it blue. Let P' be a monochromatic copy of P. By the assumptions, P' contains the vertices of a regular t-dimensional simplex of side length $\mathrm{diam}(P)$, and all of these vertices are red. Since P is t-degenerate, the point of P' corresponding to p is blue, which is a contradiction. $\qquad\square$

Theorem 15.3 is an immediate corollary of Theorem 15.12 and the following statement.

Lemma 15.2. *Every triangle that has an angle larger than $150°$ is 1-degenerate.*

With no danger of confusion, for any two points p and p', we write pp' to denote both the segment connecting them and its length.

To establish Lemma 15.2, it is sufficient to verify the following.

Lemma 15.3. *Let $T = \{p_1, p_2, p_3\}$ be the vertex set of a triangle and q another point in some Euclidean space such that*

$$\max(p_2 q, p_3 q) \leqslant p_1 q \leqslant p_2 p_3 .$$

Then the angle of T at p_1 is at most $150°$.

First, we show why Lemma 15.3 implies Lemma 15.2. Let $T = \{p_1, p_2, p_3\}$ be a triangle whose angle at p_1 is larger than $150°$, so that $\mathrm{diam}(T) = p_2 p_3$. Suppose without loss of generality that $\mathrm{diam}(T) = 1$. To prove that T is 1-degenerate, it is enough to show that for any unit segment $S = p_1 q$, we have $\mathrm{diam}(T \cup S) > 1$. Suppose not. Then we have

$$\max(p_2 q, p_3 q) \leqslant p_1 q = p_2 p_3 = 1 .$$

Hence, by Lemma 15.3, the angle of T at p_1 is at most $150°$, which is a contradiction.

Proof of Lemma 15.3. Proceeding indirectly, we assume that

$$\sphericalangle p_2 p_1 p_3 > 150° . \qquad (15.1)$$

Let Π denote a (two-dimensional) plane containing T, and let q' denote the orthogonal projection of q to Π. In the plane Π, let g and h denote the perpendicular bisectors of the segments $p_1 p_2$ and $p_2 p_3$, respectively.

Since $p_1 q \geqslant p_2 q$, we have $p_1 q' \geqslant p_2 q'$. Thus, q' belongs to the closed half-plane of Π bounded by g where p_2 lies. By symmetry, q' belongs to the

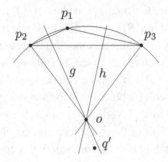

Fig. 15.1. Diagram for Lemma 15.1.

half-plane bounded by h that contains p_3. This implies that the intersection of these two half-planes is nonempty. In particular, p_1 cannot be an interior point of $p_2 p_3$ and, by (15.1), it follows that the triangle T must be nondegenerate. Hence, g and h must meet at a point o, the circumcenter of T (see Fig. 15.1).

Due to the inscribed angle theorem, we have

$$\measuredangle p_2 p_1 p_3 + \tfrac{1}{2}\measuredangle p_2 o p_3 = 180°$$

and hence $\measuredangle p_2 o p_3 < 60°$ by (15.1). This, in turn, implies that $p_2 o$, $p_3 o > p_2 p_3$. Thus, we have

$$p_2 q' \leqslant p_2 q \leqslant p_2 p_3 < p_2 o$$

and, in particular, $q' \neq o$. If one side of a triangle is smaller than another, then the same is true for the opposite angles. Applying this to the triangle $p_2 q' o$, we obtain that $\measuredangle q' o p_2 < 90°$. Analogously, we have $\measuredangle q' o p_3 < 90°$, which contradicts the position of q' described in the previous paragraph. □

We have been unable to answer the following question.

Question 15.1. Does there exist any obtuse triangle that is diameter-Ramsey?

We would like to remark, however, that the answer would be affirmative if we would just consider colorings with two colors. This is shown by the following example.

Example 15.1. Let R be the vertex set of a regular heptagon $p_1 p_2 \ldots p_7$ and let $P = \{p_1, p_2, p_4\}$. Clearly, P is the vertex set of an obtuse triangle having an angle of size $\frac{4}{7} \cdot 180° > 90°$ and $\operatorname{diam}(R) = \operatorname{diam}(P)$. Moreover, we have $R \longrightarrow (P)_2$, because the triple system with vertex set R whose edges are all sets of the form $\{p_i, p_{i+1}, p_{i+3}\}$ (the addition being performed modulo 7) is known to be isomorphic to the Fano plane, which in turn is known to have chromatic number 3.

It seems to be quite difficult to characterize all diameter-Ramsey simplices. According to Proposition 15.1, every regular simplex is diameter-Ramsey. Theorem 15.5 states that this remains true for "almost regular" simplices. It is a direct corollary of the following statement.

Lemma 15.4. *Every simplex S with vertices* $p_1, p_2, \ldots, p_n$ *satisfying*

$$\sum_{1 \leqslant i < j \leqslant n} (p_i p_j)^2 \geqslant \left(\binom{n}{2} - 1 \right) \operatorname{diam}^2(S)$$

is diameter-Ramsey.

Proof. Suppose without loss of generality that $\operatorname{diam}(S) = p_1 p_2 = 1$. Our strategy is to embed S into the Cartesian product R of $1 + \binom{n}{2}$ regular simplices, some of which might degenerate to a point. We will be able to achieve this, while making sure that $\operatorname{diam}(R) = 1$. Thus, in view of Proposition 15.1 and Lemma 15.1, we will be done.

Set

$$a = \sqrt{\sum_{i<j}(p_i p_j)^2 - \binom{n}{2} + 1} \quad \text{and} \quad x_{ij} = \sqrt{1 - (p_i p_j)^2}$$

for every $i < j$. Let T_0 be a regular simplex of side length a with n vertices. Let S_{ij} be a regular simplex of side length x_{ij} with $n - 1$ vertices, $1 \leqslant i < j \leqslant n$. For the Cartesian product of these simplices,

$$R = T_0 \times \prod_{i<j} S_{ij},$$

we have

$$\operatorname{diam}^2(R) = a^2 + \sum_{i<j} x_{ij}^2 = 1,$$

as required.

Let $\pi_0 \colon R \longrightarrow T_0$ and $\pi_{ij} \colon R \longrightarrow S_{ij}$ denote the canonical projections. Choose n points, $q_1, \ldots, q_n \in R$ such that

$$T_0 = \{\pi_0(q_1), \ldots, \pi_0(q_n)\}, \quad S_{ij} = \{\pi_{ij}(q_1), \ldots, \pi_{ij}(q_n)\}$$
$$\text{and} \quad \pi_{ij}(q_i) = \pi_{ij}(q_j),$$

for $1 \leqslant i < j \leqslant n$. It remains to check that the simplex $\{q_1, \ldots, q_n\}$ is congruent to S. However, this is obvious, because

$$(q_k q_\ell)^2 = a^2 + \sum_{i<j} x_{ij}^2 - x_{k\ell}^2 = 1 - x_{k\ell}^2 = (p_k p_\ell)^2,$$

for every $1 \leqslant k < \ell \leqslant n$. □

15.5 Proof of Theorem 15.4

Throughout this section, let $d \geqslant 6$, let p_0 denote the origin of $\mathbb{R}^d$, and let $S = \{p_0, p_1, p_2, p_3\} \subset \mathbb{R}^d$ be the vertex set of a regular tetrahedron of side length $\sqrt{2}$. Further, let $P \subset \mathbb{R}^d$ denote the seven-element set consisting of the origin $p_0 \in \mathbb{R}^d$ and the (endpoints of the) first 6 unit coordinate vectors $q_1 = (1, 0, 0, 0, 0, 0, \ldots)$, $q_2 = (0, 1, 0, 0, 0, 0, \ldots)$, ..., $q_6 = (0, 0, 0, 0, 0, 1, \ldots)$. Obviously, we have $\operatorname{diam}(S) = \operatorname{diam}(P) = \sqrt{2}$.

In view of Theorem 15.12, in order to establish Theorem 15.4, it is sufficient to prove that P is 3-degenerate. That is, we have to show that $\operatorname{diam}(P \cup S) > \sqrt{2}$. In other words, we have to establish

Claim 15.1. *There exist integers i and j $(1 \leqslant i \leqslant 3, 1 \leqslant j \leqslant 6)$ with $p_i q_j >$* $\sqrt{2}$.

The rest of this section is devoted to the proof of this claim.

For $i = 1, 2, 3$, decompose p_i into two components: the orthogonal projection of p_i to the subspace induced by the first 6 coordinate axes and its orthogonal projection to the subspace induced by the remaining coordinate axes. That is, if $p_i = (x_i(1), \ldots, x_i(d))$, let $p_i = p'_i + p''_i$, where

$$p'_i = (x_i(1), \ldots, x_i(6), 0, \ldots, 0) \quad \text{and} \quad p''_i = (0, \ldots, 0, x_i(7), \ldots, x_i(d)) .$$

Obviously, we have

$$|p_i|^2 = |p'_i|^2 + |p''_i|^2 = 2 . \tag{15.2}$$

The proof of Claim 15.1 is indirect. Suppose, for the sake of contradiction, that

$$\operatorname{diam}\{p_0, p_1, p_2, p_3, q_1, \ldots, q_6\} = \sqrt{2} .$$

Since q_j and p_0 differ only in their jth coordinate and $p_i q_j \leqslant p_i p_0$, the points p_i and q_j lie on the same side of the hyperplane perpendicularly bisecting the segment $p_0 q_j$. That is,

$$x_i(j) \geqslant \frac{1}{2} \quad \text{for every } i, j \ (1 \leqslant i \leqslant 3, 1 \leqslant j \leqslant 6). \tag{15.3}$$

Hence, we have $|p'_i|^2 = \sum_{j=1}^{6} x_i^2(j) \geqslant \frac{3}{2}$ and, by (15.2),

$$|p''_i|^2 = |p_i|^2 - |p'_i|^2 \leqslant \frac{1}{2} \quad \text{for every } i \ (1 \leqslant i \leqslant 3). \tag{15.4}$$

Moreover, if $i, i' \in \{1, 2, 3\}$ are distinct, then

$$\langle p_i, p_{i'} \rangle = \tfrac{1}{2}(|p_i|^2 + |p_{i'}^2| - |p_i - p_{i'}|^2) = \tfrac{1}{2}(2 + 2 - 2) = 1,$$

whence (15.3) implies

$$\langle p_i'', p_{i'}'' \rangle = 1 - \sum_{j=1}^{6} x_i(j)x_{i'}(j) \leqslant -\tfrac{1}{2}.$$

In view of (15.4) it follows that

$$|p_1'' + p_2'' + p_3''|^2 = |p_1''|^2 + |p_2''|^2 + |p_3''|^2 + 2\left(\langle p_1'', p_2'' \rangle + \langle p_1'', p_3'' \rangle + \langle p_2'', p_3'' \rangle\right)$$
$$\leqslant -\tfrac{3}{2},$$

which is a contradiction. This concludes the proof of Claim 15.1 and, hence, also the proof of Theorem 15.4.

15.6 Concluding Remarks

15.6.1 Kneser Graphs and Hypergraphs

Let $d = rn + (k - 1)(r - 1)$, where $r, k \geqslant 2$ are integers. Assign to each n-element subset $X \subseteq [d]$ the characteristic vector of X. That is, assign to X the point $p(X) \in \mathbb{R}^d$, whose ith coordinate is

$$p_i(X) = \begin{cases} 1 \text{ if } i \in X, \\ 0 \text{ if } i \notin X. \end{cases}$$

Let $P \subseteq \mathbb{R}^d$ be the set of all points $p(X)$. We have $|P| = \binom{d}{n}$ and $\mathrm{diam}(P) = \sqrt{2n}$.

For $r = 2$, we have $P \subset \mathbb{R}^{2n+k-1}$, and the diameter graph $H_2(P)$ is called a *Kneser graph*. It was conjectured by Kneser [21] and proved by Lovász [26] that $\chi(H_2(P)) > k$. On the other hand, if $k \leqslant n$, we have $H_3(P) = \varnothing$.

This was generalized to any value of r by Alon, Frankl, and Lovász [1], who showed that $\chi(H_r(P)) > k$, while $H_{r+1}(P) = \varnothing$, provided that $(k - 1)(r - 1) < n$. In other words, the fact that the chromatic number of the r-uniform diameter hypergraph of a point set is high does not imply that the same must hold for its $(r + 1)$-uniform counterpart.

For any integers $r, d \geqslant 2$, let $\chi_r(d)$ denote the maximum chromatic number which an r-uniform diameter hypergraph of a point set $P \subseteq \mathbb{R}^d$ can have.

Question 15.2. Is it true that for every $r \geqslant 2$, we have $\chi_{r+1}(d) = o(\chi_r(d))$, as d tends to infinity?

15.6.2 Relaxations of the Diameter-Ramsey Property

Diameter-Ramsey configurations seem to constitute a somewhat peculiar subclass of the class of all Ramsey configurations. We suggest classifying all Ramsey configurations P according to the growth rate of the minimum diameter of a point set R with $R \longrightarrow (P)_r$, as $r \to \infty$.

Definition 15.4. *Given a Ramsey configuration P and an integer r, we define*

$$d_P(r) = \inf \{\operatorname{diam}(R) \mid R \longrightarrow (P)_r\}.$$

We have $d_P(r) \geqslant \operatorname{diam}(P)$, for any Ramsey set P and any integer r, and this holds with equality if and only if for every $\varepsilon > 0$ there exists a configuration R with $R \longrightarrow (P)_r$ and $\operatorname{diam}(R) \leqslant (1 + \varepsilon)\operatorname{diam}(P)$. Certainly, all diameter-Ramsey sets P satisfy $d_P(r) = \operatorname{diam}(P)$ for all r, but perhaps the configurations with the latter property form a broader class.

Definition 15.5. *We call a Ramsey set P, lying in some Euclidean space,*

(1) almost diameter-Ramsey if $d_P(r) = \operatorname{diam}(P)$ holds for all integers $r \geqslant 1$;
(2) diameter-bounded if there is $C_P > 0$ such that $d_P(r) < C_P$ holds for every positive integer r;
(3) diameter-unbounded if $d_P(r)$ tends to infinity, as $r \to \infty$.

We do not know whether there exists any almost diameter-Ramsey configuration that fails to be diameter-Ramsey. Thus, we would like to ask the following

Question 15.3. Is it true that every almost diameter-Ramsey set is diameter-Ramsey?

To establish the diameter-boundedness of certain sets, we may utilize a result of Matoušek and Rödl [29]. They showed that, given a simplex S with circumradius ϱ, any number of colors r, and any $\varepsilon > 0$, there exists an integer d such that the d-dimensional sphere of radius $\varrho + \varepsilon$ contains a configuration R with $R \longrightarrow (S)_r$. In particular, this implies the following.

Corollary 15.2. *Every simplex is diameter-bounded Ramsey.*

Consequently, every diameter-unbounded Ramsey set must be affinely dependent. We cannot decide whether there exists any diameter-unbounded Ramsey set, but the regular pentagon may serve as a good candidate. Kříž's proof establishing that the regular pentagon is Ramsey [22] does not seem to imply that it is also diameter-bounded.

Question 15.4. Is the regular pentagon diameter-unbounded?

Finally we mention that one can also define these notions for families of configurations and ask, e.g., whether they be uniformly diameter-bounded Ramsey. As an example, we remark that a slight modification of a coloring appearing in [7] shows that no bounded subset of any Euclidean space can simultaneously arrow all triangles whose diameter is 2 with 8 colors. To see this, one may color each point x with the residue class of $\lfloor 2\|x\|^2 \rfloor$ modulo 8. Given any $K > 1$ we set $\xi = \frac{1}{17K^2}$ and consider the isosceles triangle with legs of length $1 + \xi$ and base of length 2. Assume for the sake of contradiction that there is a monochromatic copy abc of this triangle with apex vertex b and with $\|a\|, \|b\|, \|c\| \leqslant K$. Let m denote the midpoint of the segment ac and observe that $bm = \sqrt{\xi}$. The triangle inequality yields

$$\sqrt{\xi} = \|b - m\| \geqslant \big| \|b\| - \|m\| \big|$$

and, hence, we have

$$\sqrt{\xi} \cdot \big(\|b\| + \|m\| \big) \geqslant \big| \|b\|^2 - \|m\|^2 \big|.$$

Multiplying by 4, and applying triangle inequality to the left-hand side and the parallelogram law to the right-hand side we infer

$$
\begin{aligned}
2\sqrt{\xi} \cdot \big(\|a\| + 2\|b\| + \|c\| \big) &\geqslant \big| 4\|b\|^2 - \|a + c\|^2 \big| \\
&= \big| 4\|b\|^2 - 2\|a\|^2 - 2\|c\|^2 + \|a - c\|^2 \big| \\
&= \big| 4 + \big(2\|b\|^2 - 2\|a\|^2 \big) + \big(2\|b\|^2 - 2\|c\|^2 \big) \big|,
\end{aligned}
$$

which due to $\lfloor 2\|a\|^2 \rfloor \equiv \lfloor 2\|b\|^2 \rfloor \equiv \lfloor 2\|c\|^2 \rfloor$ (mod 8) leads to $8K\sqrt{\xi} \geqslant 2$, contrary to our choice of ξ.

Remark 15.1. While revising this chapter, we learned from Nora Frankl about some progress regarding Question 15.1 obtained jointly with Jan Corsten [4]. They proved that the bound of $150°$ appearing in Theorem 15.3 can be lowered to $135°$. Their elegant proof involves the spherical coloring and Jung's inequality.

Acknowledgments

A part of this work was carried out while Peter Frankl was visiting EPFL in May 2015. János Pach was Supported by Swiss National Science Foundation Grants 200020-144531 and 200021-137574. Vojtěch Rödl was supported by National Science Foundation grants DMS-1301698 and DMS-1102086.

References

1. N. Alon, P. Frankl, and L. Lovász. The chromatic number of Kneser hypergraphs. *Trans. Amer. Math. Soc.* **298**, no. 1 (1986), 359–370.
2. K. Borsuk. Drei Sätze über die n-dimensionale euklidische Sphäre. *Fund. Math.* **20** (1933), 177–190.
3. N. G. de Bruijn and P. Erdős. A colour problem for infinite graphs and a problem in the theory of relations. *Indag. Math.* **13** (1951), 369–373.
4. J. Corsten and N. Frankl. A note on Diameter-Ramsey sets. arXiv:1708.07373.
5. P. Erdős. On sets of distances of n points. *Amer. Math. Monthly* **53** (1946), 248–250.
6. P. Erdős. On sets of distances of n points in Euclidean space. *Magyar Tudom. Akad. Matem. Kut. Int. Közl. (Publ. Math. Inst. Hung. Acad. Sci.)* **5** (1960), 165–169.
7. P. Erdős, R. L. Graham, P. Montgomery, B. L. Rothschild, J. Spencer, and E. G. Straus. Euclidean Ramsey theorems. I. *J. Combin. Theory Ser. A* **14** (1973), 341–3638.
8. P. Erdős, R. L. Graham, P. Montgomery, B. L. Rothschild, J. Spencer, and E. G. Straus. Euclidean Ramsey theorems. II. In *Infinite and Finite Sets* (Colloq., Keszthely, 1973; dedicated to P. Erdős on his 60th birthday), Vol. I, *Colloq. Math. Soc. János Bolyai*, Vol. **10**, North-Holland, Amsterdam, 1975, 529–557.
9. P. Erdős, R. L. Graham, P. Montgomery, B. L. Rothschild, J. Spencer, and E. G. Straus. Euclidean Ramsey theorems. III. In *Infinite and Finite Sets* (Colloq., Keszthely, 1973; dedicated to P. Erdős on his 60th birthday), Vol. I, *Colloq. Math. Soc. János Bolyai*, Vol. **10**, North-Holland, Amsterdam, 1975, 559–583.
10. P. Frankl and V. Rödl. Forbidden intersections. *Trans. Amer. Math. Soc.* **300** (1987), 259–286.
11. P. Frankl and V. Rödl. A partition property of simplices in Euclidean space. *J. Amer. Math. Soc.* **3**, no. 1 (1990), 1–7.
12. P. Frankl and V. Rödl. Strong Ramsey properties of simplices. *Israel J. Math.* **139** (2004), 215-236.
13. P. Frankl and R. M. Wilson. Intersection theorems with geometric consequences. *Combinatorica* **1** (1981), 357–368.
14. R. L. Graham. Recent trends in Euclidean Ramsey Theory. *Discrete Math.* **136**, no. 1–3 (1994), 119–127.
15. R. L. Graham and F. Yao. A whirlwind tour of computational geometry. *Amer. Math. Monthly* **97**, no. 8 (1990), 687–701.
16. B. Grünbaum. A proof of Vázsonyi's conjecture. *Bull. Res. Council Israel Sect. A* **6** (1956), 77–78.
17. A. Heppes. Beweis einer Vermutung von A. Vázsonyi. *Acta Math. Acad. Sci. Hungar.* **7** (1956), 463–466.
18. H. Hopf and E. Pannwitz. Aufgabe Nr. 167. *Jahresbericht d. Deutsch. Math.-Verein.* **43** (1934), 114.
19. T. Jenrich and A. E. Brouwer. A 64-dimensional counterexample to Borsuk's conjecture. *Electron. J. Combin.* **21**, no. 4 (2014), Paper 4.29.
20. J. Kahn and G. Kalai. A counterexample to Borsuk's conjecture. *Bull. Amer. Math. Soc. (N.S.)* **29** (1993), 60–62.
21. M. Kneser. Aufgabe 300. *Jber. Deutsch. Math.-Verein.* **58** (1955), 27.

22. I. Kříž. Permutation groups in Euclidean Ramsey theory. *Proc. Amer. Math. Soc.* **112** (1991), no. 3, 899–907.
23. A. Kupavskii. Diameter graphs in $\mathbf{R}^4$. *Discrete Comput. Geom.* **51**, no. 4 (2014), 842–858.
24. A. B. Kupavskii and A. Polyanskii. Proof of Schur's conjecture in $\mathbf{R}^d$. arXiv:1402.3694v1.
25. Y. S. Kupitz, H. Martini, and M. A. Perles. Ball polytopes and the Vázsonyi problem. *Acta Math. Hungar.* **126**, no. 1–2 (2010), 99–163.
26. L. Lovász. Kneser's conjecture, chromatic number, and homotopy. *J. Combin. Theory Ser. A* **25**, no. 3 (1978), 319–324.
27. I. Leader, P. A. Russell, and M. Walters. Transitive sets and cyclic quadrilaterals. *J. Combin.* **2**, no. 3 (2011), 457–462.
28. I. Leader, P. A. Russell, and M. Walters. Transitive sets in Euclidean Ramsey Theory. *J. Combin. Theory Ser. A* **119**, no. 2 (2012), 382–396.
29. J. Matoušek and V. Rödl. On Ramsey sets in spheres. *J. Combin. Theory Ser. A* **70**, no. 1 (1995), 30–44.
30. F. Morić and J. Pach. Remarks on Schur's conjecture. *Comput. Geom.* **48**, no. 7 (2015), 520–527.
31. O. Schramm. Illuminating sets of constant width. *Mathematika* **35**, no. 2, 180–189.
32. Z. Schur, M. A. Perles, H. Martini, and Y. S. Kupitz. On the number of maximal regular simplices determined by n points in $\mathbb{R}^d$. In *Discrete and Computational Geometry, The Goodman-Pollack Festschrift*, eds. B. Aronov, S. Basu, J. Pach, and M. Sharir. Algorithms Combin. Vol. **25**, Springer, Berlin, 2003, 767–787.
33. S. Straszewicz. Sur un problème géométrique de P. Erdős. *Bull. Acad. Pol. Sci., Cl. III* **5** (1957), 39–40.
34. K. J. Swanepoel. A new proof of Vázsonyi's conjecture. *J. Combinat. Theory, Ser. A* **115** (2008), 888–892.
35. K. J. Swanepoel. Unit distances and diameters in Euclidean spaces. *Discrete Comput. Geom.* **41** (2009), 1–27.
36. D. R. Woodall. Thrackles and deadlock. In *Combinatorics, Proc. Conf. Comb. Math.* (D. Welsh, ed.), Academic Press, London, 1971, pp. 335–347.

16

Pick's Theorem and Sums of Lattice Points

Karl Levy and Melvyn B. Nathanson

Abstract

Pick's theorem is used to prove that if P is a lattice polygon (that is, the convex hull of a finite set of lattice points in the plane), then every lattice point in the h-fold sumset hP is the sum of h lattice points in P.

For sets $X, X_1, \ldots, X_h$ in $\mathbf{R}^n$, we define the *sumset*

$$X_1 + \cdots + X_h = \{x_1 + \cdots + x_h : x_i \in X_i \text{ for } i = 1, \ldots, h\}$$

and the *h-fold sumset*

$$hX = \underbrace{X + \cdots + X}_{h \text{ summands}}.$$

We also define the *dilation*

$$h * X = \{hx : x \in X\}.$$

Lemma 16.1. *For every convex subset of $\mathbf{R}^n$ and every positive integer h, the sumset equals the dilation, that is,*

$$hX = h * X.$$

Proof. If $x \in X$, then

$$hx = \underbrace{x + \cdots + x}_{h \text{ summands}} \in hX$$

and so $h * X \subseteq hX$.

If the set X is convex and if $x_1, \ldots, x_h \in X$, then X contains the convex combination

$$(1/h)x_1 + \cdots + (1/h)x_h$$

278

and so

$$x_1 + \cdots + x_h = h((1/h)x_1 + \cdots + (1/h)x_h) \in h * X.$$

Thus, if X is convex, then $hX \subseteq h * X$. This completes the proof. $\square$

A *lattice polytope* in $\mathbf{R}^n$ is the convex hull of a nonempty finite set of lattice points in $\mathbf{Z}^n$. A *lattice polygon* in $\mathbf{R}^2$ is the convex hull of a nonempty finite subset of $\mathbf{Z}^2$, that is, of lattice points in the plane.

Let $n \geq 2$ and $h \geq 2$. If $P, P_1, \ldots, P_h$ are lattice polytopes in $\mathbf{R}^n$, then

$$h(P \cap \mathbf{Z}^n) \subseteq (hP) \cap \mathbf{Z}^n \tag{16.1}$$

and

$$(P_1 \cap \mathbf{Z}^n) + \cdots + (P_h \cap \mathbf{Z}^n) \subseteq (P_1 + \cdots + P_h) \cap \mathbf{Z}^n. \tag{16.2}$$

These set inclusions can be strict. For example, if P_1 is the triangle in $\mathbf{R}^2$ whose vertices are $\{(0, 0), (1, 0), (1, -1)\}$, and if P_2 is the triangle whose vertices are $\{(0, 0), (1, 2), (2, 3)\}$, then $P_1 + P_2$ is the hexagon with vertices

$$\{(0, 0), (1, -1), (3, 2), (3, 3), (2, 3), (1, 2)\}.$$

We have

$$(1, 1) = (1/2, 0) + (1/2, 1) \in (P_1 + P_2) \cap \mathbf{Z}^2$$

but

$$(1, 1) \notin (P_1 \cap \mathbf{Z}^2) + (P_2 \cap \mathbf{Z}^2).$$

Therefore,

$$(P_1 \cap \mathbf{Z}^2) + (P_2 \cap \mathbf{Z}^2) \neq (P_1 + P_2) \cap \mathbf{Z}^2.$$

In $\mathbf{R}^3$, let P be the tetrahedron with vertices

$$\{(0, 0, 0), (1, 0, 0), (0, 1, 0), (1, 1, 2)\}$$

We have

$$(1, 1, 1) = (1/2, 1/2, 0) + (1/2, 1/2, 1) \in (2P) \cap \mathbf{Z}^3$$

but

$$(1, 1, 1) \notin 2(P \cap \mathbf{Z}^3).$$

Therefore,

$$2(P \cap \mathbf{Z}^3) \neq (2P) \cap \mathbf{Z}^3.$$

It is an only partially solved problem to determine the lattice polytopes $P, P_1, \ldots, P_h$ in $\mathbf{R}^n$ for which we have equalities and not inclusions in (16.1) and (16.2). This is usually discussed in the language of toric geometry [2, 3, 6]. It is known that in the plane we have

$$h(P \cap \mathbf{Z}^2) = (hP) \cap \mathbf{Z}^2 \tag{16.3}$$

for every lattice polygon P and every positive integer h (Koelman [5, 4]). In this note we apply Pick's theorem (Pick [7], Beck and Robins [1, pp. 38–40]) to obtain a simple proof of this result. Pick's theorem states that if P is a lattice polygon with area A and with I lattice points in its interior and B lattice points on its boundary, then

$$A = I + \frac{B}{2} - 1. \tag{16.4}$$

A triangle is the convex hull of three noncollinear points. A *primitive lattice triangle* is a lattice triangle whose only lattice points are its three vertices. By Pick's theorem, a lattice triangle in $\mathbf{R}^2$ is primitive if and only if its area is $1/2$.

Lemma 16.2. *Let T be a primitive lattice triangle with vertex set $\{0, u, v\}$, and let*

$$W = \{iu + jv : i = 0, 1, \ldots, h \text{ and } j = 0, 1, 2, \ldots, h - i\}.$$

Then

$$W = hT \cap \mathbf{Z}^2.$$

Proof. Because $u, v \in \mathbf{Z}^2$ and because the triangle T is convex and

$$hT = h * T = \{\alpha u + \beta v : \alpha \geq 0, \beta \geq 0, \text{ and } \alpha + \beta \leq h\}$$

it follows that $W \subseteq hT \cap \mathbf{Z}^2$. The linear independence of the vectors u and v implies that

$$|W| = \sum_{i=0}^{h} (h - i + 1) = \frac{(h + 1)(h + 2)}{2}.$$

Let $A(h)$ denote the area of the dilated triangle hT, and let $I(h)$ and $B(h)$ denote, respectively, the number of interior lattice points and boundary lattice points of hT. Because T is primitive, we have $A(1) = 1/2$ and $B(1) = 3$. It follows that

$$A(h) = A(1)h^2 = \frac{h^2}{2}$$

and

$$B(h) = B(1)h = 3h.$$

By Pick's theorem,

$$A(h) = I(h) + \frac{B(h)}{2} - 1$$

and so the number of lattice points in hT is

$$|hT \cap \mathbf{Z}^2| = I(h) + B(h) = A(h) + \frac{B(h)}{2} + 1 = \frac{(h+1)(h+2)}{2}.$$

Because W and $hT \cap \mathbf{Z}^2$ are finite sets with $W \subseteq hT \cap \mathbf{Z}^2$ and $|W| = |hT \cap \mathbf{Z}^2|$, it follows that $W = hT \cap \mathbf{Z}^2$. This completes the proof. $\qquad\square$

Theorem 16.1. *Let P be a lattice polygon. If w is a lattice point in the sumset hP, then there exist lattice points a, b, c in P and nonnegative integers i, j, k such that $h = i + j + k$ and*

$$w = ia + jb + kc \in h(P \cap \mathbf{Z}^2).$$

In particular,

$$h(P \cap \mathbf{Z}^2) = (hP) \cap \mathbf{Z}^2.$$

Proof. Every lattice polygon P can be triangulated into primitive lattice triangles. If w is a lattice point in $hP = h * P$, then $w/h \in P$ and so there is a primitive lattice triangle T' contained in P with $w/h \in T'$. Let $\{a, b, c\}$ be the set of vertices of T', and let $T = T' - c$ be the primitive lattice triangle with vertices 0, $u = a - c$, and $v = b - c$. We have $w/h - c \in T$ and so $w - hc \in hT$. By Lemma 16.2, there are nonnegative integers i and j such that $i + j = h - k \leq h$ and

$$w - hc = iu + jv.$$

This implies that

$$w = iu + jv + hc = ia + jb + kc \in 3(T' \cap \mathbf{Z}^2) \subseteq 3(P \cap \mathbf{Z}^2).$$

This completes the proof. $\qquad\square$

References

1. M. Beck and S. Robins. *Computing the Continuous Discretely.* Undergraduate Texts in Mathematics. Springer Science+Business Media, New York, 2007.

2. N. Fakhruddin. Multiplication maps of linear systems on smooth projective toric varieties. arXiv: 0208178, 2002.
3. C. Haase, B. Nill, A. Paffenholz, and F. Santos. Lattice points in Minkowski sums. *Electron. J. Combin.* **15**, no. 1 (2008), Note 11, 5.
4. R. J. Koelman. A criterion for the ideal of a projectively embedded toric surface to be generated by quadrics. *Beiträge Algebra Geom.* **34**, no. 1 (1993), 57–62.
5. R. J. Koelman. Generators for the ideal of a projectively embedded toric surface. *Tohoku Math. J.* (2) **45**, no. 3 (1993), 385–392.
6. T. Oda. Problems on Minkowski sums of convex lattice polytopes. arXiv:0812.1418, 2008.
7. G. A. Pick. Geometrisches zur Zahlenlehre. Sitzenber. Lotos (Prague) **19** (1899), 311–319.

17

Apollonian Ring Packings

Adrian Bolt, Steve Butler, and Espen Hovland

Abstract

Apollonian packings are a well-known object in mathematics formed by starting with a set of three mutually tangent circles and then repeatedly filling in the "holes" with maximally sized circles. There have been several generalizations of these packings over the years, and we introduce a new family of such packings based on a simple "ring" structure of circles and give the tools to construct and mathematically explore these packings.

17.1 Introduction

Circle packing traces its roots back to the the ancient Greeks, and in particular to the work of Apollonius, who posed the problem of finding circles that are tangent to three given circles with disjoint interiors. For the case in which the initial circles are mutually tangent there are two possible solutions.

When we start with three mutually tangent circles and then repeatedly add circles to any existing set of three mutually tangent circles we get the well-known Apollonian circle packing. This leads to beautiful pictures (see Fig. 17.1), but also interesting mathematics. This is a consequence of the Descartes Circle Theorem (see [8]) on bends of circles. A bend is the *signed curvature* of the circle. A bend of 0 indicates a straight line (which we consider as a circle with infinite radius), and a negative bend corresponds to a circle that has been turned inside out (e.g., all circle packing pictures in this chapter have negative bend for the outermost circle).

Theorem 17.1 (The Descartes Circle Theorem). *Given four mutually tangent circles having disjoint interiors and with bends b_1, b_2, b_3, b_4, then*

$$(b_1 + b_2 + b_3 + b_4)^2 = 2(b_1^2 + b_2^2 + b_3^2 + b_4^2).$$

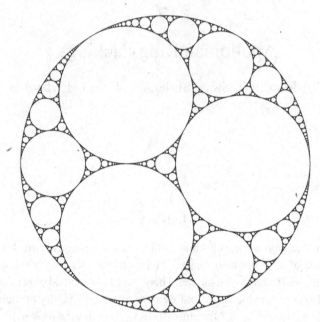

Fig. 17.1. Apollonian packing (ring packing for $n = 3$).

From Theorem 17.1 we can solve for the bends of the two new circles in terms of the existing circles. Given that the bends of the original circles are b_1, b_2, b_3 then the new bends are

$$b_4 = b_1 + b_2 + b_3 \pm 2\sqrt{b_1 b_2 + b_1 b_3 + b_2 b_3}.$$

In particular if b_1, b_2, b_3 are integers and $b_1 b_2 + b_1 b_3 + b_2 b_3$ is a square (i.e., $w := \sqrt{b_1 b_2 + b_1 b_3 + b_2 b_3}$ is an integer), then the new bends are also integers. Moreover, it can be shown that all further bends will also be integral. This has led to extensive study of group and number theoretic properties of these packings (see [3, 4, 5, 6]).

There are other variations of packings. One in particular that has exhibited similar properties was studied by Guettler and Mallows [7] and uses three circles instead of one in filling the holes.[1] This is done in the unique way so that the three new circles are mutually tangent and also each new circle is tangent to exactly two of the original circles. An example of this packing is shown in Fig. 17.2. This has a similar number theoretic property as the original Apollonian packing; namely, if the three initial bends are integral and w is *twice* a square number then all bends will be integral.

---

[1] Hints of this packing were given in a Japanese sangaku of Adachi Mitsuaki dating to 1821; see [9].

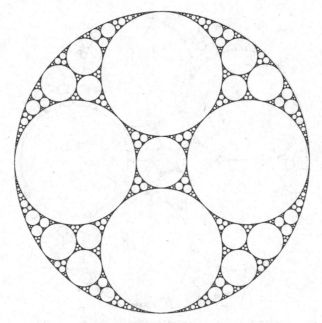

Fig. 17.2. Guettler–Mallows packing (ring packing for $n = 4$).

These are only two of infinitely many packing rules. A theoretical approach to packing rules that use *finitely* many circles to fill in the hole was given by Butler et al. [1]. One packing given there that will be relevant to our discussion will be a packing that was based on the skeleton of the icosahedron and is shown in Fig. 17.3. Every packing has some minimal ring wherein bends will lie, and for this packing we have for $\varphi = (1 + \sqrt{5})/2$, the golden ratio, that if the initial bends are in $\mathbb{Z}[\varphi]$ and if w is a square in this ring, then all bends are in $\mathbb{Z}[\varphi]$.

These three packings are the first in an infinite sequence of packings that are based on starting with a configuration of circles such that any circle is tangent to exactly n other circles and then using circle inversion to fill in any space between three mutually tangent circles. (To make it more precise for any three mutually tangent circles we perform an inversion through the unique circle that passes through the three points of tangency; a discussion on circle inversion can be found in the work of Hidetoshi and Rothman [9].) The packings we have already discussed correspond to $n = 3, 4, 5$ respectively. For $n \geq 6$ the situation will change dramatically, as there will be infinitely many circles to be put between three mutually tangent circles. Nevertheless we will see that there is a set of three basic tools (and corresponding matrices) that can be used to construct the packings and explore their mathematical properties.

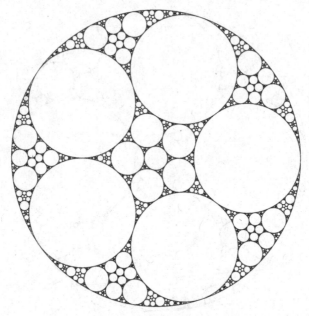

Fig. 17.3. Icosahedron packing (ring packing for $n = 5$).

17.2 Construction of Packings

Our starting point is to form a configuration of circles with high symmetry so that each circle is tangent to precisely n others. We achieve this by forming a tiling of congruent equilateral triangles so that at each vertex precisely n triangles come together. Given such a tiling the desired circle packing is easily recovered by placing a circle at each vertex and making the radius precisely half the length of the side of the triangle. These tiling are formed in three different ways:

- $n = 3, 4, 5$: In this case we use the Platonic solids tetrahedron, octahedron, and icosahedron respectively. We can embed the vertices of these solids on the surface of the sphere and then form triangles on the sphere (i.e., use *spherical geometry*). To find a configuration in the plane we then use projection.
- $n = 6$: In this case we use the familiar infinite tiling of the *Euclidean plane* with six equilateral triangles coming together at every point.
- $n = 7$: In this case we start by noting that our desired tiling exists in the *hyperbolic plane*. To find a configuration in the plane we then use some model of the hyperbolic plane; in particular we can use the Poincaré disk model that will embed into the plane and further circles become circles, giving our configuration.

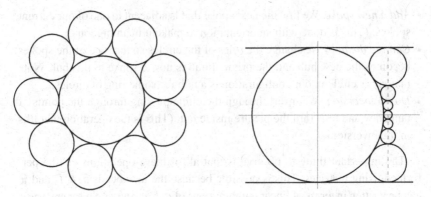

Fig. 17.4. An example of a ring and its Möbius transform.

For $n = 3, 4, 5$ the configuration is unique up to Möbius transformation. This is a consequence of the following theorem, which uses the tangency graph of a finite circle packing; i.e., the graph formed by letting each circle be a vertex and joining two vertices by an edge if the corresponding circles are tangent (see [10]).

Theorem 17.2 (Koebe–Andreev–Thurston). *If G is a finite maximal planar graph, then the circle packing whose tangency graph is isomorphic to G is unique up to Möbius transformations.*

For $n \geq 6$ we do not have a finite graph, which is why we have carefully ensured that our formation of our initial packing given earlier has high symmetry (i.e., since the preceding theorem does *not* apply). For our configuration of circles we have that by an appropriate Möbius transformation we can put any circle at the center with n equally sized circles that form a ring on the outside. (This is why we call these packing *ring packings*.) An example is shown on the left of Fig. 17.4, and result of an appropriate Möbius transformation (in particular inverting at a point of tangency between two outer circles) of this ring is shown on the right.

We are now ready to give the tools to form a packing. Instead of trying to add all new circles simultaneously (which is the historical approach), we will add circles one at a time. We start by making our fundamental unit a set of three (ordered) mutually tangent circles, one will act as the "hub" (the center circle on the left of Fig. 17.4) and the other two are adjacent "spokes" (two tangent outer circles on the left of Fig. 17.4). We now have three operations that we perform.

- *Add a new spoke.* We find the next spoke that is adjacent to one of our current spokes. (This is done with an orientation to make it unambiguous.)
- *Change the hub.* We change the roles of the circles so that one of the spokes becomes the new hub and the original hub is now a spoke to this hub. Note that *every* circle in our configuration is a hub for some ring of circles.
- *Do an inversion.* We invert through the circle passing through the points of tangency and thus turn the picture inside out. (This is the operation that fills in the inversions.)

The important thing to observe is that all of these operations can be performed using matrices. This is possible because the new bends $\widehat{a}, \widehat{b}, \widehat{c}$, and $\widehat{w}$ can be written in terms of linear combinations of a, b, c, and w. So for any triple of mutually tangent circles will be represented by a vector as

$$\begin{pmatrix} a \\ b \\ c \\ w \end{pmatrix} = \begin{pmatrix} \text{bend of the hub} \\ \text{bend of first spoke} \\ \text{bend of second spoke} \\ \sqrt{ab + ac + bc} \end{pmatrix}.$$

With this in mind two of the three operations are easy to describe by a matrix. For changing the hub we won't change w and we permute the roles so we permute the first three entries. This gives the matrix

$$B = \begin{pmatrix} 0 & 1 & 0 & 0 \\ 0 & 0 & 1 & 0 \\ 1 & 0 & 0 & 0 \\ 0 & 0 & 0 & 1 \end{pmatrix}.$$

When doing an inversion we will not change the circles with bends a, b and c but we will change their relative ordering so this is done by a permutation matrix. On the other hand if we treat w as a bend then doing inversion turns the circle inside out and we need to negate. This gives the matrix

$$C = \begin{pmatrix} 1 & 0 & 0 & 0 \\ 0 & 0 & 1 & 0 \\ 0 & 1 & 0 & 0 \\ 0 & 0 & 0 & -1 \end{pmatrix}.$$

That leaves adding a new spoke. For this we use the results in Butler et al. [1] on finding the bends of new circles. Namely we first take the known circles and through Möbius inversion change them to be a unit circle centered at the origin and the lines $y = \pm 1$ (this is known as a standard configuration; see the right in Fig. 17.4). We then determine the coordinates (x, y) and bend β of the corresponding circle which we want to find the new bend for (note these will be fixed). Finally we have that the new bend d will be

$$d = \frac{\beta^2(x^2 + y^2 + 1) - 1}{4\beta}(a + b) + \frac{1}{2}\beta(b - a)y + \beta c + \beta x w. \qquad (17.1)$$

In our case we have that the centers (x, y) and bends β are

$$(x, y) = \left(\sec\tfrac{\pi}{n}, \frac{\tan\tfrac{\pi}{n}\sin\frac{(2k+1)\pi}{n}}{2\sin\frac{(k+1)\pi}{n}\sin\frac{k\pi}{n}}\right), \qquad \beta = \frac{2\sin\frac{(k+1)\pi}{n}\sin\frac{k\pi}{n}}{\tan\tfrac{\pi}{n}\sin\tfrac{\pi}{n}} \qquad (17.2)$$

for $k = 1, \ldots, n - 2$.

On a side note these circles all have the same x coordinate and so are lined up on top of each other. Therefore the sum of the radii must be 1 (i.e., twice their sum is the distance between $y = \pm 1$). From this we can conclude for $n \geq 3$,

$$\sum_{k=1}^{n-2} \csc\tfrac{(k+1)\pi}{n}\csc\tfrac{k\pi}{n} = 2\cot\tfrac{\pi}{n}\csc\tfrac{\pi}{n},$$

a curious trigonometric identity!

Now using the information in (17.2) in (17.1) we can determine that the bend of the new spoke will satisfy

$$d = 4\cos^2\tfrac{\pi}{n}a + b + 4\cos^2\tfrac{\pi}{n}c + 4\cos\tfrac{\pi}{n}w.$$

With a little more work it can be shown that we also have that the value $\widehat{w}$ for the new triple satisfies

$$\widehat{w} = -2\cos\tfrac{\pi}{n}a - 2\cos\tfrac{\pi}{n}c - w.$$

Finally, we can conclude that the matrix for adding a spoke is

$$A = \begin{pmatrix} 1 & 0 & 0 & 0 \\ 0 & 0 & 1 & 0 \\ 4\cos^2\tfrac{\pi}{n} & 1 & 4\cos^2\tfrac{\pi}{n} & 4\cos\tfrac{\pi}{n} \\ -2\cos\tfrac{\pi}{n} & 0 & -2\cos\tfrac{\pi}{n} & -1 \end{pmatrix}.$$

Theorem 17.3. *For a fixed $n \geq 3$ and the corresponding Apollonian ring packing we can use A, B, and C to get from any set of three mutually tangent circles to any other set of three mutually tangent circles. In particular, A, B, and C can generate the packing.*

Proof. We illustrate the procedure using Fig. 17.5. We have a starting set of mutually tangent circles (triangle) and we have a desired set of mutually tangent circles (triangle) that is our destination (either because it is the final triple, or because our desired triple is in the interior of the corresponding hole).

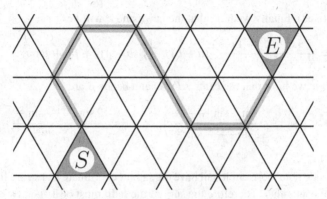

Fig. 17.5. Moving between sets of mutually tangent circles.

In the diagram, the operation A consists of holding one vertex (the hub) fixed and then "rotating" to an adjacent triangle; the operation B consists of changing which vertex acts as the hub. We now fix a path from a vertex on the starting triangle to a vertex on the end triangle. From the start triangle we now apply B until the hub is the starting vertex and then we apply A until the rotation takes us onto the first edge of the path. We now have a shorter path, so by induction using only A and B we can get to the end.

At the end of the path when we reach our destination, if the triple is in the interior we then apply C to carry out the inversion.

We continue alternating between these operations until we reach our desired triple (only finitely many iterations will be needed since otherwise the limit would involve a circle with infinite bend, i.e., a point).   □

We now have the tools to make any Apollonian ring packing and we show the results for $n = 6, 7, 8$ in Figs. 17.6, 17.7, and 17.8, respectively. It is interesting to note that for $n = 6$ this is a proper packing in that in the limit we cover all of the plane (i.e., the black holes in the middle are infinitely many circles getting smaller and smaller). For $n \geq 7$ the situation is dramatically different in that we will *not* cover all of the plane. In particular, the large holes in the center are not part of the packing, as are several additional holes.

17.3 Modular Considerations

One of the most active problems related to Apollonian packings has been determining what bends are possible in a given packing. Graham et al. noticed that in the original Apollonian packings some configurations had all integral bends,

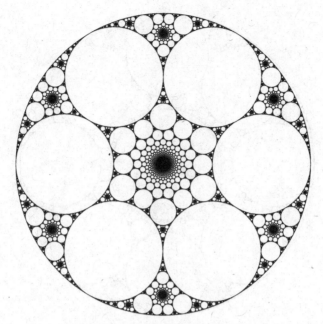

Fig. 17.6. Apollonian ring packing for $n = 6$.

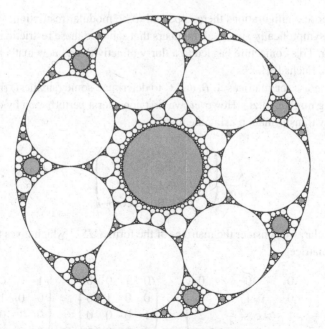

Fig. 17.7. Apollonian ring packing for $n = 7$.

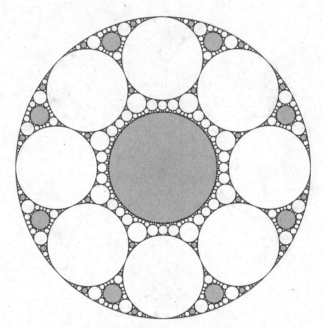

Fig. 17.8. Apollonian ring packing for $n = 8$.

and in those configurations there appeared to be modular restrictions modulo 24 and asymptotically almost all numbers that satisfied these restrictions seem to appear. This conjecture has led to a flurry of activity and new mathematical tools (see Fuchs [2]).

We can use the matrices A, B, and C to determine some (smallest) ring that a packing might contain. However, we first consider a perturbation by doing a similarity transformation using the matrix

$$S = \begin{pmatrix} 1 & 0 & 0 & 0 \\ 0 & 1 & 0 & 0 \\ 0 & 0 & 1 & 0 \\ 0 & 0 & 0 & 2\cos\frac{\pi}{n} \end{pmatrix}.$$

In particular, we consider the matrices of the form SMS^{-1} which gives the following matrices.

$$\begin{pmatrix} 1 & 0 & 0 & 0 \\ 0 & 0 & 1 & 0 \\ 4\cos^2\frac{\pi}{n} & 1 & 4\cos^2\frac{\pi}{n} & 2 \\ -4\cos^2\frac{\pi}{n} & 0 & -4\cos^2\frac{\pi}{n} & -1 \end{pmatrix} \quad \begin{pmatrix} 0 & 1 & 0 & 0 \\ 0 & 0 & 1 & 0 \\ 1 & 0 & 0 & 0 \\ 0 & 0 & 0 & 1 \end{pmatrix} \quad \begin{pmatrix} 1 & 0 & 0 & 0 \\ 0 & 0 & 1 & 0 \\ 0 & 1 & 0 & 0 \\ 0 & 0 & 0 & -1 \end{pmatrix}$$

This similarity transformation gives consistency in the form of the entries of the first matrix. The interpretation of this is we are changing $ab + ac + bc = (2\cos\frac{\pi}{n})^2 w^2$ from the previous $ab + ac + bc = w^2$. As an example for $n = 4$ (the Guettler–Mallows packing) this becomes $ab + ac + bc = 2w^2$. Guettler and Mallows [7] in general showed that if the initial bends of the circles satisfied $ab + ac + bc = 2d^2$ for $a, b, c, d \in \mathbb{Z}$ then all bends were integers. More generally the preceding matrices allows us to establish the following.

Theorem 17.4. *Consider an Apollonian ring packing with $n \geq 3$. If a, b, c are the bends of three mutually tangent circles and $ab + ac + bc = (2\cos\frac{\pi}{n})^2 d^2$ for $a, b, c, d \in \mathbb{Z}[4\cos^2\frac{\pi}{n}]$, then all bends are in $\mathbb{Z}[4\cos^2\frac{\pi}{n}]$.*

Corollary 17.1. *For the Apollonian ring packing with $n = 6$, if a, b, c are the bends of three mutually tangent circles and $ab + ac + bc = 3d^2$ for $a, b, c, d \in \mathbb{Z}$, then all bends are in $\mathbb{Z}$.*

We can even go further and note modular possibilities that can happen in this packing. If we consider what possible bends can happen modulo 6 for three mutually tangent circles we have two natural constraints: (1) they must be relatively prime (otherwise all bends have a common term and we can scale to remove); (2) $ab + ac + bc \equiv 0 \pmod{3}$ (i.e., it must be a multiple of 3). The possible triples are thus

$$\{1, 1, 1\}, \{1, 1, 4\}, \{1, 3, 3\}, \{1, 3, 6\}, \{1, 4, 4\}, \{1, 6, 6\}, \{2, 2, 5\}, \{2, 3, 3\},$$
$$\{2, 3, 6\}, \{2, 5, 5\}, \{3, 3, 4\}, \{3, 3, 5\}, \{3, 4, 6\}, \{3, 5, 6\}, \{5, 5, 5\}, \{5, 6, 6\}.$$

These can be further refined by placing them into six families so that if any triple occurs modulo 6 as the bends of three mutually adjacent circles in the packing, then the only triples that can occur modulo 6 of three mutually adjacent circles are in the family. These families are as follows (the proof can readily be established by using the foregoing matrices for $n = 6$ and some case analysis):

1. $\{1, 1, 1\}$
2. $\{5, 5, 5\}$
3. $\{1, 3, 3\}, \{3, 3, 5\}$
4. $\{1, 1, 4\}, \{1, 4, 4\}$
5. $\{2, 2, 5\}, \{2, 5, 5\}$
6. $\{1, 3, 6\}, \{1, 6, 6\}, \{2, 3, 3\}, \{2, 3, 6\}, \{3, 3, 4\}, \{3, 4, 6\}, \{3, 5, 6\}, \{5, 6, 6\}$

We note that if we set the outside circle to have bend -1 for the packing shown in Fig. 17.6 then all bends are integer, and moreover this falls into the $\{1, 3, 3\}, \{3, 3, 5\}$ family. We believe further refinements of this are possible and in addition modulo 6 might not tell the whole story.

17.4 Finding Centers with Bends

We have primarily been concerned with the bends of the circles in the packings, but for visualization we also need to have a way to find the centers. This can be done in a brute force manner by determining the bend of a new circle and then solving various quadratics to place the circle. However, there is a more efficient manner to carry this out given by Lagarias et al. [8]. Namely they noted that the linear relationships which hold for the bends also hold for the "scaled" centers.

So if we have circles with bends b_1, b_2, b_3 and corresponding centers (x_1, y_1), (x_2, y_2), (x_3, y_3) then the scaled centers will be (b_1x_1, b_1y_1), (b_2x_2, b_2y_2), (b_3x_3, b_3y_3). There is a *fourth* circle which will have as radius $w = \sqrt{ab + ac + bc}$ and scaled center $(w\widehat{x}, w\widehat{y})$. With all of this in place we now apply any of our matrices A, B, and C to

$$\begin{pmatrix} b_1x_1 & b_1y_1 & b_1 \\ b_2x_2 & b_2y_2 & b_2 \\ b_3x_3 & b_3y_3 & b_3 \\ w\widehat{x} & w\widehat{y} & w \end{pmatrix}$$

and read off any new row to find the new circle's bend and then, by division, the new circle's center.

The question remains how to find this new "circle." The answer is that this is the dual circle, i.e., the circle which passes through the three points of tangency of the other circles. (Note that there is still one of two possibilities for the sign of w, but this can be quickly determined by trial and error.) We can plot these circles overlaying our packing to see the relationship of these other circles (and to produce more interesting pictures).

For the illustrations of ring packings used in this paper our initial matrix for a starting triple of mutually tangent circles was

$$\begin{pmatrix} 0 & 0 & -1 \\ \csc\frac{\pi}{n} & 0 & 1 + \csc\frac{\pi}{n} \\ \csc\frac{\pi}{n} - 2\sin\frac{\pi}{n} & 2\cos\frac{\pi}{n} & 1 + \csc\frac{\pi}{n} \\ -\cot\frac{\pi}{n} & -1 & -\cot\frac{\pi}{n} \end{pmatrix}.$$

Of course many other starting configurations are possible and lead to other different and interesting pictures.

17.5 Concluding Remarks

In this paper we have introduced the Apollonian ring packings and given simple tools to construct these packings. There still remain many interesting problems

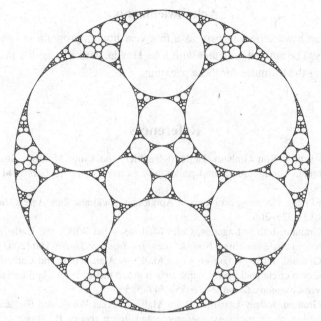

Fig. 17.9. Packing based off of the cube.

including looking at what bends are possible (particularly the modular restrictions for the hexagon packing), as well as determining the group generated by the matrices A, B, C. There are also many interesting symmetries to explore. For example, any Möbius transformation which takes three mutually tangent circles in the packing to three other mutually tangent circles will leave the image unchanged (i.e., any set of three mutually tangent circles will produce the same packing). Also what can be said about these "packings" which do not cover all of space.

There are other generalizations that could also be considered. These packings are based off of tilings by equilateral triangles, but tilings with other shapes could also be considered. For example one could start with the cube (a tiling using squares) and then to turn it into an Apollonian type packing we add a point in the center of each face connected to the vertices of the face. The resulting base configuration was unwittingly touched on in Butler et al. [1] and the resulting packing is shown in Fig. 17.9. On a side note the packings where three hexagons or four octagons come together at every vertex lead to the $n = 6$ and $n = 8$ Apollonian ring packings. And of course one can always look into investigating sphere packings in higher dimensions.

Acknowledgments

This research was done, in part, as a first year honors program at Iowa State University. The authors, together with John Hovda, worked together in a group on deriving the formulas for these packings.

References

1. Steve Butler, Ron Graham, Gerhard Guettler, and Colin Mallows. Irreducible Apollonian configurations and packings. *Discrete Computat. Geom.* **44** (2010), 487–507.
2. Elena Fuchs. Counting problems in Apollonian packings. *Bull. Amer. Math. Soc.* **50** (2013), 229–266.
3. Ron Graham, Jeffrey Lagarias, Colin Mallows, Alan Wilks, and Catherine Yan. Apollonian circle packings: Number theory. *J. Number Theory* **100** (2003), 1–45.
4. Ron Graham, Jeffrey Lagarias, Colin Mallows, Alan Wilks, and Catherine Yan. Apollonian circle packings: Geometry and group theory I. The Apollonian group. *Discrete Computat. Geom.* **34** (2005), 547–585.
5. Ron Graham, Jeffrey Lagarias, Colin Mallows, Alan Wilks, and Catherine Yan. Apollonian circle packings: Geometry and group theory II. Super-Apollonian group and integral packings. *Discrete Computat. Geom.* **34** (2006), 37–72.
6. Ron Graham, Jeffrey Lagarias, Colin Mallows, Alan Wilks, and Catherine Yan. Apollonian circle packings: Geometry and group theory III. Higher dimensions. *Discrete Computat. Geom.* **35** (2006), 1–36.
7. Gerhard Guettler, and Colin Mallows. A generalization of Apollonian packing of circles. *J. Combinat.* **1** (2010), 1–27.
8. Jeffrey Lagarias, Colin Mallows, and Alan Wilks. Beyond the Descartes circle theorem. *Amer. Math. Monthly* **109** (2002), 338–361.
9. Fukagawa Hidetoshi, and Tony Rothman. *Sacred Mathematics: Japanese Temple Geometry.* Princeton University Press, Princeton, NJ, 2008.
10. Kenneth Stephenson. Circle packing: A mathematical tale. *Notices* **50** (2003), 1376–1388.

18

Juggling and Card Shuffling Meet Mathematical Fonts

Erik D. Demaine and Martin L. Demaine

Abstract

We explore two of Ron Graham's passions – juggling patterns and perfect card shuffling – through one of our passions, mathematical fonts. First, for each letter of the English alphabet, we design a one-person three-ball juggling pattern where the balls trace out the letter (possibly rotated 90°). Second, using a deck of 26 cards labeled A through Z, we show how to perform a sequence of in/out perfect riffle shuffles to display any desired sequence of letters (forming a word, phrase, etc.), using algorithms already developed by Diaconis and Graham. Along the way, we pose some new open problems about perfect shuffling.

18.1 Introduction

A *mathematical typeface* is based on a mathematical theorem or open problem. In this way, the way that text is written, and not just the text itself, can engage the reader in mathematical thinking. A *puzzle typeface* hides the text from plain sight, requiring solving a puzzle to decipher the underlying text. By combining these two ideas, we can write secret messages encoded with various mathematical ideas, and readers who are enthusiastic about puzzles can learn about beautiful mathematics by solving mathematical puzzles to decipher text.

We have so far developed half a dozen of these mathematical/puzzle typefaces, based on mathematics ranging from hinged dissections to geometric tours to computational origami to the geometry of blown glass. See [4] for a survey, and visit our website to play with the fonts yourself.[1]

To celebrate Ron Graham's 80th birthday, we created two new typefaces based on two of Ron's (and our) favorite contexts where mathematics meet

---

[1] http://erikdemaine.org/fonts/

the unusual: juggling and card shuffling. Both typefaces are designed to be performed. The juggling typeface consists of a three-ball two-hand juggling pattern for each letter, enabling the expression of an n-letter word or phrase through n simultaneous jugglers or one juggler performing a sequence of patterns. The card shuffling typeface expresses an n-letter word or phrase through a sequence of $O(n)$ operations – in perfect shuffles, out perfect shuffles, and revealing the top card in a 26-card alphabet deck – enabling an expert card magician to pull out an audience-specified message.

18.2 Juggling Fonts

Juggling patterns are well studied mathematically, from the initial work of Shannon in 1981 [10] to the foundational paper by Buhler, Eisenbud, Graham, and Wright [2]; see [9]. One major revolution, embodied by *siteswap notation* used by many jugglers, is the abstraction of juggling patterns into a timing sequence for when objects land after being thrown. In addition to raising many clean mathematical questions about juggling patterns, this perspective enables easy communication of juggling patterns as well as computer simulation of these patterns. Some computer simulators, such as Juggling Lab [1], add modeling of performance aspects of juggling patterns beyond the mathematical abstraction, such as hand motion between object throws.

Our idea for representing letters as juggling patterns is visual: we design cyclic juggling patterns, with particular hand motions, such that the trajectories of the balls trace the letter, in some cases rotated 90°. Figure 18.1 shows a few such juggling patterns and their trajectories. One way to experience the trajectories is with persistence of vision: lit balls in a dark room will paint a longer time range onto the retina. But we prefer the puzzle of watching the juggling pattern and visualizing its projection down the time axis.

Figure 18.2 shows the full typeface in the (easy-to-read) trajectory font. To see the more puzzling animated font, and to really see the underlying juggling patterns, visit the webapp.[2] For a classic juggling feel, we kept all patterns to three balls, two hands, and toss juggling, though we plan to make other typefaces with other styles of juggling. Some letters come from classic juggling patterns: A is Half Shower, E is Columns, F is an Exchange out of two balls, I is Yo-Yo, O is Shower, U is Box, W is Double Box / Extended Box, and X is Reverse Cascade. Letters D, M, and N are all variations of Columns; N is perhaps new, while the others are known but perhaps unnamed. Letter B is

---

[2] http://erikdemaine.org/fonts/juggling/

Fig. 18.1. Expressing "RON GRAHAM" as juggling trajectories and animations. Time proceeds vertically. The letter R is rotated 90°. Animations produced with Juggling Lab [1].

essentially three balls out of a four-ball Reverse Fountain (in siteswap notation, 4440). Letter C is a known (but hard) variation of Cascade (sometimes called Arches), while S and its mirror image Z are new and extra-hard variations of Cascade involving very fast hand motions between throws. Several letters (H, K, P, Q, R, T, V, Y) are essentially two-ball juggling with a third ball held and moved to finish the letter, similar to classic tricks like Fake Columns and Yo-Yo (in siteswap notation, 42). Letter G and its mirror image J are challenging variations of Columns with a back-and-forth cross throw in between (in siteswap notation, 6411). Finally, L is a synchronous pair of cross throws mixed

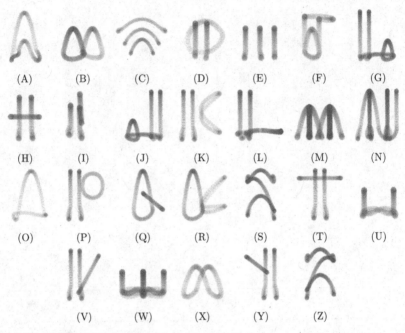

Fig. 18.2. Juggling typeface, trajectory font. Underlying animations produced with Juggling Lab [1].

with two-ball Columns (in extended siteswap notation, "(2,6)(2x,2x)"). Letters B, E, and R are rotated 90°, as it seems difficult to construct them upright.

This typeface is meant to be performed. With many jugglers arranged in a line, we can write a message by having each juggler perform one letter. Figure 18.3 shows a photograph from such a performance of Fig. 18.1. With only one juggler, we can transition from one letter to another. For the mathematical abstraction of juggling patterns, optimal transitions can be found algorithmically [3]; for the constant-size patterns of this typeface, performers can find suitable transitions through practice.

18.3 Card Shuffling Fonts

Perfect riffle shuffles (often called Fano shuffles) fascinate mathematicians and magicians alike. The trick is difficult to perform, requiring the deck to be perfectly divided in half, and cards to alternate perfectly from the two half-decks. In fact, there are two types of perfect shuffles, depending on which half-deck drops a card first; see Fig. 18.4. *Outside* perfect shuffles preserve the topmost

Fig. 18.3. RON GRAHAM (Fig. 18.1) performed by Colin Wright, Jeffrey Davis, Erik Demaine, Glenn Hurlbert, Jay Cummings, Jacob Landgraf, Peter Frankl, Pat Dragon, and M. Puck Rombach (from left to right) at the Connections in Discrete Mathematics conference on the occasion of Ron Graham's 80th birthday.

and bottommost cards (on the "outside" of the deck), while *inside* perfect shuffles bring the two middle cards to the outside.

When successful, perfect shuffles imbue a seemingly random operation (riffle shuffling) with powerful deterministic properties, which can be used to great effect. Perhaps most famously, eight outside perfect shuffles of a 52-card deck return it to its original order. On the other hand, if we repeatedly inside

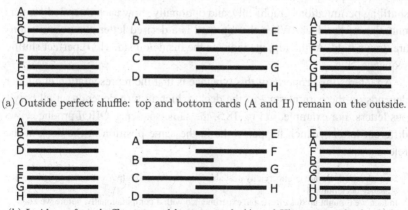

(a) Outside perfect shuffle: top and bottom cards (A and H) remain on the outside.

(b) Inside perfect shuffle: top and bottom cards (A and H) move inside the deck.

Fig. 18.4. The two types of Fano/perfect shuffles.

perfect shuffle a 52-deck card, it would take 52 iterations before the deck returns to its original order [5, 6]. These iteration counts, $k = 8$ and $k = 52$, come from finding the minimum values of k for which $2^k \equiv 1 \pmod{52 - 1}$ and $2^k \equiv 1 \pmod{52 + 1}$, respectively.

When Persi Diaconis was 13, he learned about the mathematical magic of perfect shuffles from Alex Elmsley, a British computer scientist [8]. In addition to the property of eight outside perfect shuffles, Elmsley showed Diaconis his technique for moving a card from the top (0th) position to the kth position in the deck using the binary representation of k: for each 0, perform an outside perfect shuffle (O), and for each 1, perform an inside perfect shuffle (I) [7, 5].

Half a century later, Diaconis and Graham [5] solved the reverse problem, posed by Elmsley [7]: how to move a card from the kth position up to the top of the deck. Amazingly, this feat can be achieved using at most $\lceil \lg n \rceil$ perfect shuffles in an n-card deck. This bound is information theoretically optimal: the trick provides a binary encoding (mapping outside/O to 0 and inside/I to 1) for each of the n cards. Thus, a magician who knows a card's location within a 52-card deck can extract it from the top of the deck after at most six perfect shuffles.

This trick suggests a performance font with a 26-card deck, one card for each letter of the alphabet.[3] The deck starts sorted A through Z. To spell a message (for example, provided by the audience), the magician performs at most five perfect shuffles to bring each letter to the top, reveals the top card, and continues to the next letter.[4]

This effect can also be illustrated by a variety of visual fonts. Figure 18.5 shows examples produced by the webapp.[5] Here we optionally represent the shuffling permutations graphically, and optionally show each intermediate permutation of the deck. When the top card is a desired letter, the permutations are drawn bold. At the top, we indicate the inside/outside (I/O) perfect shuffle sequence.

An interesting property of this typeface is that the representation of a letter depends on the current state of the deck, which in turn depends on all previous letters. For example, in Fig. 18.5, the same sequence OIIOI produces two different letters (which happen to be in the same position in the deck at the relevant times).

---

[3] Several such decks are commercially available. We recommend http://www.magictricks.com/ alphabet-deck.html.

[4] In fact, we thought up this effect, and experimented with it computationally, before we knew about Elmsley's Problem and its solution. Thanks to Persi and Ron for pointing us to their work during the Connections in Discrete Mathematics conference!

[5] http://erikdemaine.org/fonts/shuffle/

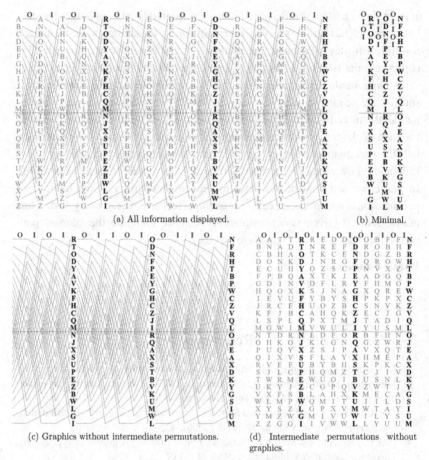

(a) All information displayed.

(b) Minimal.

(c) Graphics without intermediate permutations.

(d) Intermediate permutations without graphics.

Fig. 18.5. Representing "RON" by a sequence of Fano/perfect shuffles: OIOI | OIIOI | OIIOI |.

We can use the perfect shuffle sequence alone to encode messages in a puzzle font. Can you decode the following puzzles, where each "|" symbol indicates the reveal of the top card? (Solutions follow.)

Puzzle 1. OOOII | IOOOI |
Puzzle 2. IOII | OIOOI | IIOOI | IIIOI | OIOOI | OOOII | OOII | OIOI |
Puzzle 3. OOOOI | IIOOI | OOII | IIOOI | | III | OOOI |

Other perfect shuffling typefaces could use different decks, for example, a 52-card deck of uppercase and lowercase letters.

A natural open question is to find the shortest execution of a given letter sequence. Diaconis and Graham [5] point out that there can be two solutions of at most $\lceil \lg n \rceil$ perfect shuffles, and there are presumably more of longer lengths. Our approach has been to greedily repeatedly apply the shortest solution to each resulting Elmsley's Problem, but plausibly it could help to do more shuffles earlier in order to permute the cards so that later letters become easier to reach. ·

A more fundamental open question is to characterize how many perfect shuffles are needed to achieve each (reachable) permutation of a deck. Diaconis, Graham, and Kantor [6] characterized which permutations are reachable for each deck size, viewed as a group. For example, a 26-card deck can reach half the permutations, while a 52-card deck can reach them all. But we know little about how many perfect shuffles we need to represent each permutation. In particular, what is the diameter of the directed graph of reachable card permutations, which has an edge for each perfect shuffle (so every vertex has out-degree 2)? Given two permutations, can we compute the shortest path connecting them, or is this problem NP-hard?

Puzzle Solutions:

1. HI   •   2. BIRTHDAY   •   3. SHUFFLE

References

1. Jack Boyce et al. Juggling lab. http://jugglinglab.sourceforge.net/, 2014.
2. Joe Buhler, David Eisenbud, Ron Graham, and Colin Wright. Juggling drops and descents. *Amer. Math. Monthly* **101** (1984), 507–519.
3. Jean Cardinal, Steve Kremer, and Stefan Langerman. Juggling with pattern matching. *Theory Comput. Systems* **39**, no. 3 (2006), 425–437.
4. Erik D. Demaine and Martin L. Demaine. Fun with fonts: Algorithmic typography. *Theoreti. Comput. Sci.* **586** (June 2015), 111–119.
5. Persi Diaconis and Ron Graham. The solutions to Elmsley's problem. *Math Horizons* 14:22–27, February 2007.
6. Persi Diaconis, R. L. Graham, and William M. Kantor. The mathematics of perfect shuffles. *Adv. Appl. Math.* **4**, no. 2 (1983), 175–196.
7. Alex Elmsley. The mathematics of the weave shuffle. *The Pentagram* **11**, no. 9–11 (June–August 1957).
8. Jascha Hoffman. Q&A: The mathemagician. *Nature* **478** (October 2011), 457.
9. Burkard Polster. *The Mathematics of Juggling*. Springer-Verlag, Heidelberg, 2003.
10. Claude E. Shannon. Scientific aspects of juggling. In N. J. A. Sloane and A. D. Wyner, eds., *Claude Elwood Shannon: Collected Papers*, 850–864. IEEE Press, New York, 1993.

19

Randomly Juggling Backwards

Allen Knutson

Abstract

We recall the directed graph of *juggling states*, closed walks within which give juggling patterns, as studied by Ron in [CG08, BG10]. Various random walks in this graph have been studied before by several authors, and their equilibrium distributions computed. We motivate a random walk on the reverse graph (and an enrichment thereof) from a very classical linear algebra problem, leading to a particularly simple equilibrium: a Boltzmann distribution closely related to the Poincaré series of the b-Grassmannian in ∞-space.

We determine the most likely asymptotic state in the limit of many balls, where the probability of a 0-throw is kept fixed in the limit.

19.1 Walks on the Juggling Digraph

The "siteswap" theory of juggling patterns was invented in the early-mid 1980s by Paul Klimek of Santa Cruz, Bruce Tiemann and Bengt Magnusson at Caltech, and Mike Day and Colin Wright in Oxford. In 1988, having had some time to digest this theory, Jack Boyce and I (each also at Caltech) independently invented a directed graph of "juggling states" to study siteswaps. We recall this definition now.

19.1.1 The Digraph

Fix $b \in \mathbb{N}$ for the rest of the chapter; it is called the **number of balls**. (Of course one might want a theory in which b varies, but we won't vary it in this chapter.) A **juggling state** σ is just a b-element subset of $\mathbb{N}$, but we will draw it as a semi-infinite word in $\times$ and $-$ using only b many $\times$s, e.g., $- \times \times - - \times - - - - \ldots$. We won't generally write any of the infinitely many $-$s after the last $\times$. The physical interpretation of σ is as follows: If a juggler is making one throw each second, and we stop her mid-juggle and let the b balls fall to the ground, σ records the sound "wait, thump, thump, wait, wait, thump" (and thereafter, silence) that the balls make. In the standard "cascade" pattern (b odd) and asynchronous "fountain" pattern (b even), this state is always the **ground state** $\times \times \cdots \times$, but other patterns go through more interesting states.[1]

Put a directed edge $\sigma \to \tau$ if $\times\tau \supseteq \sigma$ (meaning, containment of the $\times$-locations). If the first letter of σ is $\times$, then there is one extra $\times$ in $\times\tau$ not in σ, in some position $t + 1$; call t the **throw** and label the edge with it. If the first letter of σ is $-$, the throw is conventionally taken to be 0 (even though it's not much of a throw; the juggler just waits for one beat). If $\sigma \to \tau$, then any two of (σ, τ, t) determine the third.

A closed walk in this digraph is called a **juggling pattern** and is determined by its sequence of throws, the **siteswap**. Here is the (excellent) siteswap 501 as a closed walk:

$$\times - \times \xrightarrow{\;5\;} \;-\times--\times \xrightarrow{\;0\;} \times--\times \xrightarrow{\;1\;} \times - \times$$

Perhaps the earliest nontrivial theorem about siteswaps is Ron et al.'s calculation that the number of patterns of length n with at most b balls is $(b + 1)^n$ [BEGW94]. Ron and his coauthors have also counted cycles that start from a given state [CG08, BG10].

The first book on the general subject is [Pol03].

In [Wa05] was studied the finite subgraph in which only throws of height $\leq n$ are allowed, giving $\binom{n}{b} = \binom{n}{n-b}$ states. This digraph for (b, n), with arrows reversed, is isomorphic to the one for $(n - b, n)$: we reverse the length-n states, and switch $\times$s and $-$s. This remark serves as foreshadowing for the second paragraph that follows, and the chapter.

19.1.2 A Markov Chain

In [Wa05, LV12, ELV15, Va14, ABCN, ABCLN] are studied Markov chains of juggling states, in which the possible throws from a state are given probabilities. (Sometimes probability 0, making them impossible; e.g., [Wa05] puts a bound on the highest throw.)

We now define a Markov chain that follows the edges *backwards*, using a coin with $p(\text{heads}) = 1/q$. **However, we will never write the arrows as reversed: $\tau \to \tau'$ will have a consistent meaning throughout the chapter.** Let σ be a juggling state.

---

[1] For example, the three-ball "shower" (juggling in a circle) alternates between the states $\times \times -\times$ and $\times - \times - \times$.

(1) Flip the coin at most b times, or until it comes up tails.
(2) If the coin never comes up tails, attach $-$ to the front of σ.
(3) If the coin comes up tails on the ith flip, move the ith *last* $\times$ in σ to the front, leaving a $-$ in its place.

Example: $\sigma = - - \times \times - \times$, so $b = 3$ and we flip at most three times. If the flips are

- Tails: we get $\times - - \times \times - \cancel{\times}$, i.e., $\times - - \times \times$.
- Heads, then tails: we get $\times - - \times \cancel{\times} - \times$, i.e., $\times - - \times - - \times$.
- Heads, heads, tails: we get $\times - - \cancel{\times} \times - \times$, i.e., $\times - - - \times - \times$.
- Heads, heads, heads: we get $- - - \times \times - \times$.

Note that the resulting juggling states are exactly those that point to σ in the digraph; the $\times$ that moves is the ball thrown (if any; in the all-heads case the "throw" is a 0).

Our main results in this chapter are

- a calculation of the (quite simple) stationary distribution of this chain,
- a motivation and solution of the chain from linear algebra considerations, and
- a study of the typical states in the $b \to \infty$ limit.

The limit $q \to \infty$ (always tails) is boring; after b throws we get to the ground state and stay there. The limit $q \to 1$ (b fixed) has no stationary distribution. In Section 19.3 we show the limit $b \to \infty, q \to 1$ is well-behaved if we keep fixed the all-heads probability $E = q^{-b}$, which acts as a sort of temperature. Specifically, we compute the typical ball density around position h to be $(1 - E)/(1 + E^{1-h/b} - E)$.

The linear algebra itself suggests in Section 19.4 a Markov chain on (the reverse of) a richer digraph with distinguishable balls that can bump one another out of position; we solve this one as well. (This digraph appeared first in [ABCLN], though we had been considering it already for a few years; as far as we can tell our motivations for studying it are different than theirs.)

19.2 The Linear Algebra Motivation

Let U_b be the space of $b \times \mathbb{N}$ matrices of full rank, over the field $\mathbb{F}$ with q elements. Define a map

$$\sigma : U_b \to \{\text{juggling states}\}$$

where there is a $\times$ in position i of $\sigma\left([\vec{c}_0 \vec{c}_1 \vec{c}_2 \cdots]\right)$ if $\vec{c}_i$ is not in the span of $\vec{c}_0, \ldots, \vec{c}_{i-1}$. Equivalently, $\sigma(M)$ records the pivot columns in M's reduced row-echelon form.

This σ is preserved by and is the complete invariant for {row operations} $\times$ {rightward column operations}. On the Grassmannian $GL(b)\backslash U_b$ of b-planes in $\mathbb{F}^N$, to which σ descends, σ records the (finite-codimensional) Bruhat cell of $rowspan(M)$.

We define a Markov chain on U_b, called "add a random column $\vec{c}$ on the left," meaning uniformly with respect to the counting measure on $\mathbb{F}^b$. Although this chain does not have an invariant probability distribution, it obviously preserves the counting measure on U_b.

Proposition 19.1. *Let $M \in U_b$, so $L = [\vec{c}\,M]$ is also in U_b. Then $\sigma(L) \to \sigma(M)$ in the juggling digraph. If we let τ range over the finite set of possible values of $\sigma(L)$, the probability of obtaining τ is given by the process described in. Section 19.1.2.*

Proof. For the first statement, we need only observe that if a column is pivotal in $[\vec{c}\,M]$, it is certainly pivotal in M.

For the second, choose j minimal such that $\vec{c}$ is in the span of M's left j pivot columns.

- $j = 0 \iff \vec{c} = 0 \iff \sigma(L)$ is $\sigma(M)$ with a $-$ in front. Otherwise,
- $\sigma(L)$ is $\sigma(M)$ with its jth $\times$ moved to the front. There are q^j $\vec{c}$s in the span of those j columns, of which q^{j-1} are in the span of the first $j-1$, for a probability of $\left(q^j - q^{j-1}\right)/q^b = \left(1 - \frac{1}{q}\right)/q^{b-j}$, also the probability of tails after $b - j$ heads.

Very similar results to the first statement appeared in [KLS13] and [Pos], but about rotating the columns of a finite matrix. $\square$

We now want to push this "measure" down to the set of juggling states, i.e., for each juggling state τ, we want to define the probability that $M \in U_b$ has $\sigma(M) = \tau$.

Proposition 19.2. *Let τ be a juggling state, and pick $N >$ the last $\times$-position of τ. Then the fraction of $b \times N$ matrices with pivots in position τ is*

$$\frac{|GL_b(q)|}{q^{b^2}} \bigg/ q^{\ell(\tau)} \qquad \text{where } \ell(\tau) := \#\{\text{"inversion" pairs } \ldots - \ldots \times \ldots \text{ in } \tau\}$$

independent of N.

Proof. Each such M is row-equivalent, by a unique element of $GL_b(q)$, to a unique one in reduced row-echelon form. The pivotal columns in that are fixed (an identity matrix), accounting for the q^{b^2} factor. With those columns erased, the remaining $b \times (N - b)$ matrix has a partition's worth of 0s in the lower left, and the complementary partition of free variables in the upper right. Each 0 entry corresponds to a pair $\ldots - \ldots \times \ldots$ in τ. $\qquad\square$

Rewriting the prefactor $|GL_b(q)|/q^{b^2}$, we get

Corollary 19.1. *The mapping* $\tau \mapsto \prod_{i=1}^{b}(1 - q^{-i})/q^{\ell(\tau)}$ *is a probability measure on the space of juggling states (meaning it sums to* 1, *summing over all* τ).

Proof. Summing over only those τ with last $\times$ in position $\leq N$, we get the fraction of $b \times N$ matrices that are full rank. This goes quickly to 1 as $N \to \infty$. $\qquad\square$

Put another way $\prod_{i=1}^{b}(1 - q^{-i})^{-1} = \sum_{\tau} q^{-\ell(\tau)}$, which is easily justified for q a formal variable: each side[2] is a sum over Young diagrams with columns of height at most b, of $q^{-\text{area}}$. The left-hand side computes this by counting how many columns of height i there are, for each i.

The Weil conjectures relate counting points over $\mathbb{F}$ to homology, and each side of this equation is computing the Poincaré series of the Grassmannian of b-planes in $\mathbb{C}^N$. This is closely related to Bott's formula for the Poincaré series of the *affine* Grassmannian, which also was related to juggling in [ER96].

Theorem 19.1. *This distribution in Corollary 19.1 is stationary for the Markov chain in Section 19.1.2.*

We could likely justify this for q a prime power through some limiting procedure in N, but it's easy enough to check for a formal variable q (i.e., $q^{-\infty} = 0$), so we do that now.

Proof. Stationarity at τ says

$$\sum_{\tau \to \tau'} p(\tau')p(\tau', \tau) = p(\tau)$$

(here $\to$ indicates the edge in the usual juggling digraph, whereas $p(\tau', \tau)$ is the transition probability calculated in Proposition 19.1). If τ begins with $-$, then the only τ' is τ with that $-$ removed, and stationarity says

$$p(\tau')q^{-b} = p(\tau)$$

$$\prod_{i=1}^{b}(1 - q^{-i})^{-1} q^{-\ell(\tau')}q^{-b} = \prod_{i=1}^{b}(1 - q^{-i})^{-1} q^{-\ell(\tau)}$$

or $\ell(\tau') + b = \ell(\tau)$, which is obvious from the definition of ℓ.

If τ begins with $\times$ (i.e., we have a ball to throw), then there are infinitely many τ' it could throw to. We group τ's $-$s into b many groups $j \in [1, b]$, where

[2] Matt Szczesny points out that this is a "partition function." If you don't get that joke be grateful.

j is the number of $\times$ in τ that the throw skips past (counting itself, hence $j \geq 1$). Let λ_j be the position of the jth $\times$ in τ, with $\lambda_1 = 0$, $\lambda_{b+1} := \infty$. Then

$$\prod_{i=1}^{b}(1 - q^{-i})^{-1} \sum_{\tau \to \tau'} p(\tau')p(\tau', \tau) = \sum_{\tau \to \tau'} q^{-\ell(\tau')}p(\tau', \tau)$$

$$= \sum_{j=1}^{b} \sum_{\substack{t \in (\lambda_j, \lambda_{j+1}) \\ \tau' := \tau \text{ after } t\text{-throw}}} q^{-\ell(\tau')}p(\tau', \tau)$$

$$= \sum_{j=1}^{b} \sum_{\substack{t \in (\lambda_j, \lambda_{j+1}) \\ \tau' := \tau \text{ after } t\text{-throw}}} q^{-\ell(\tau')}q^{j-b}(1 - q^{-1})$$

$$= q^{-b} \sum_{j=1}^{b} q^{j}(1 - q^{-1}) \sum_{\substack{t \in (\lambda_j, \lambda_{j+1}) \\ \tau' := \tau \text{ after } t\text{-throw}}} q^{-\ell(\tau')}.$$

To make τ', we remove the $\times$ from the front of τ (preserving ℓ), and put it in position $t - 1 \in \mathbb{N}$, which places it right of $t - 1$ other letters of which $j - 1$ are $\times$. That creates $t - j$ of the $-\times$ inversions to its left, while destroying $b - j$ inversions from its right:

$$q^{-b} \sum_{j=1}^{b} q^{j}(1 - q^{-1}) \sum_{\substack{t \in (\lambda_j, \lambda_{j+1}) \\ \tau' := \tau \text{ after } t\text{-throw}}} q^{-\ell(\tau')} = q^{-b} \sum_{j=1}^{b} q^{j}(1 - q^{-1})$$

$$\times \sum_{\substack{t \in (\lambda_j, \lambda_{j+1}) \\ \tau' := \tau \text{ after } t\text{-throw}}} q^{-(\ell(\tau)+(t-j)-(b-j))}$$

$$= q^{-\ell(\tau)} \sum_{j=1}^{b} q^{j}(1 - q^{-1}) \sum_{-t \in (-\lambda_{j+1}, -\lambda_j)} q^{-t}$$

which telescopes as $= q^{-\ell(\tau)} \sum_{j=1}^{b} q^{j}(q^{-\lambda_j - 1} - q^{-\lambda_{j+1}})$

which in turn telescopes as $= q^{-\ell(\tau)}(q^{1}q^{-\lambda_1 - 1} - q^{b}q^{-\lambda_{b+1}})$

$$= q^{-\ell(\tau)}(q^{1}q^{-0-1} - q^{b}q^{-\infty}) \quad = q^{-\ell(\tau)}.$$

$\square$

Two comments. Another way we could have made rigorous the probability measure on matrices, and then pushed it down to the set of states, would be to group matrices into equivalence classes where $M \sim M'$ if they have the same pivot columns, and agree in the columns up to and including their bth pivot column.

Also, instead of working with full-rank matrices we could have worked with all matrices, allowing $b' < b$ pivot columns. On the level of juggling states, this amounts to having the remaining $b - b'$ many $\times$s sitting in abeyance at the (infinite) right end of the state; in short order those $\times$s move to finite positions and never go back. This larger Markov chain is not ergodic, and the $b' < b$ states have 0 probability, so we just left them out for ease of exposition.

We record for later use this function

$$s_n = s_n(q) := \prod_{i=1}^{n} (1 - q^{-i})$$

whose reciprocal is $\sum_{n\text{-state } \tau} q^{-\ell(\tau)}$, and connect it to other well-known Poincaré series:

Proposition 19.3. *The Poincaré series of the flag manifold $Fl(n)$, in the non-traditional variable q^{-1}, is $s_n/s_1^n = \sum_{\pi \in S_n} q^{-\ell(\pi)}$. The Poincaré series of the Grassmannian $Gr(j, h)$ is $s_h/s_j s_{h-j} = \sum_{\sigma \in \binom{h}{j}} q^{-\ell(\sigma)}$. The second sum is over j-ball juggling states with no $\times$s in position h or later.*

19.3 The $b \to \infty$ Limit, with Fixed Probability of Placing an Initial "$-$"

How many $\times$s are we most likely to have in the first h spots?

The trick we use to measure such states is to look at concatenations $\sigma_L \sigma_R$ where σ_L is a finite string with multiplicities $\times^c -^{h-c}$, and σ_R an infinite one with multiplicities $\times^{b-c} -^{\infty}$. Then

$$\ell(\sigma_L \sigma_R) = \ell(\sigma_L) + \ell(\sigma_R) + (b - c)(h - c)$$

as the third term counts the inversions of the $(h - c)$ $-$s in σ_L with the $(b - c)$ $\times$s in σ_R.

Using this and Proposition 19.3, we calculate the probability P_c of having exactly $c \in [0, \min(h, b)]$ $\times$s in $[0, h - 1]$ as

$$\sum_{\sigma_L \in \binom{[0,h-1]}{c}} \sum_{\sigma_R} s_b q^{-\ell(\sigma_L) - \ell(\sigma_R) - (b-c)(h-c)} = s_b q^{-(b-c)(h-c)} \sum_{\sigma_L \in \binom{[0,h-1]}{c}} q^{-\ell(\sigma_L)} \sum_{\sigma_R} q^{-\ell(\sigma_R)}$$

$$= s_b q^{-(b-c)(h-c)} \frac{s_h}{s_c s_{h-c}} \frac{1}{s_{b-c}}$$

which is maximized at the c where P_c/P_{c-1} crosses from > 1 to < 1. That ratio is

$$P_c/P_{c-1} = \frac{q^{-(b-c)(h-c)}}{q^{-(b-c+1)(h-c+1)}} \frac{S_{c-1}}{S_c} \frac{S_{h-c+1}}{S_{h-c}} \frac{S_{b-c+1}}{S_{b-c}}$$

$$= q^{-c+h+b-c+1}(1 - q^{-c})^{-1}(1 - q^{-(h-c+1)})(1 - q^{-(b-c+1)})$$

$$= (q^c - 1)^{-1}(q^h - q^{c-1})(q^{b-c+1} - 1)$$

and setting it to 1 gives

$$q^c - 1 = (q^h - q^{c-1})(q^{b-c+1} - 1)$$

$$(q^c - 1) / (q^{b-c+1} - 1) = q^h - q^{c-1}$$

$$q^{c-1} + \frac{q^c - 1}{q^{b-c+1} - 1} = q^h$$

Toward considering the $b \to \infty$ limit, let $\lambda := c/b \in [0, 1]$, $\mu := h/b \in [\lambda, \infty)$, and $E = q^{-b}$:

$$q^{-1}E^{-\lambda} + \frac{E^{-\lambda} - 1}{qE^{\lambda-1} - 1} = E^{-\mu}$$

This E (for "empty hand") is the probability of never flipping tails, thereby putting a $-$ at the front of the state. The limit $q \to 1, E \to 1$ of backwards juggling is thus not interesting. Instead, we consider simultaneous limits $q \to 1, b \to \infty$ in such a way that E has a limit in $(0, 1)$, e.g., $q = 1 - 1/b$. Then

$$E^{-\mu} = E^{-\lambda} + \frac{E^{-\lambda} - 1}{E^{\lambda-1} - 1} = E^{-\lambda}\left(1 + \frac{1 - E^{\lambda}}{E^{\lambda-1} - 1}\right) = E^{-\lambda}\frac{E^{\lambda-1} - E^{\lambda}}{E^{\lambda-1} - 1}$$

$$= E^{-\lambda}\frac{1 - E}{1 - E^{1-\lambda}}$$

$$\mu = \lambda + \log_{E^{-1}}\frac{1 - E}{1 - E^{1-\lambda}} \qquad \text{note } E^{-1} > 1 > E^{1-\lambda} > E > 0, \text{ so } \mu > \lambda$$

$$\sim \lambda + \frac{E^{1-\lambda}}{\log(E^{-1})} \qquad \text{as } E \to 0, \lambda \text{ fixed, or}$$

$$\frac{\log(1 - \lambda)^{-1}}{1 - E} \qquad \text{as } E \to 1, \lambda \text{ fixed}$$

To recap: if we consider the limit $b \to \infty$ of many balls, and don't control E, then in the $E \to 0$ limit we get $\mu = \lambda$, the ground state. The limit $E \to 1$ doesn't exist. But if $E \in (0, 1)$, then the foregoing function $\mu(\lambda)$ says how far

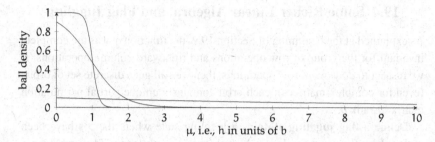

Fig. 19.1. The ball density functions for $E = 0.00001$ (the sigmoidal curve), $E = 0.9$ (the very flat one at the bottom), and $E = 0.1$ in between; each y-intercept is $1 - E$. Recall that the fraction in the tail $\mu \geq 1$ is about $\ln 2 / \ln(E^{-1})$, rather a lot, which is why we need E so very small to get a sigmoidal-looking curve.

out (as a multiple of b) one should look to find the first λ balls (as a fraction of b).

We can invert this relation to find λ in terms of μ:

$$E^{-\mu}(1 - E^{1-\lambda}) = E^{-\lambda}(1 - E)$$
$$E^{-\mu} - E^{1-\mu}E^{-\lambda} = E^{-\lambda}(1 - E)$$
$$E^{-\mu} = E^{-\lambda}(1 - E + E^{1-\mu})$$
$$E^{-\mu}(1 - E + E^{1-\mu})^{-1} = E^{-\lambda}$$
$$\mu - \log_{E^{-1}}(1 + E^{1-\mu} - E) = \lambda \qquad \text{note } E^{1-\mu} > E, \text{ so } \lambda < \mu.$$

For example, as $E \to 0$ the fraction λ of balls in the first b slots is $1 - \log_{E^{-1}}(2 - E) \sim 1 - \ln 2 / \ln(E^{-1})$, i.e., all but $\ln 2 / \ln(E^{-1})$.

The *ball density* at μ is the derivative of this $\lambda(\mu)$ with respect to μ,

$$1 - \frac{\frac{d}{d\mu}\log(1 + E^{1-\mu} - E)}{\log(E^{-1})}$$

$$= 1 - \frac{\frac{d}{d\mu}(1 + E^{1-\mu} - E)}{\log(E^{-1})(1 + E^{1-\mu} - E)} = 1 - \frac{\frac{d}{d\mu}(E^{1-\mu})}{\log(E^{-1})(1 + E^{1-\mu} - E)}$$

$$= 1 - \frac{\log(E^{-1})(E^{1-\mu})}{\log(E^{-1})(1 + E^{1-\mu} - E)} = 1 - \frac{E^{1-\mu}}{1 + E^{1-\mu} - E}$$

$$= \frac{1 - E}{1 + E^{1-\mu} - E}$$

which is $1 - E$ at $\mu = 0$ (as befits the definition of E) and decreases thereafter. As another sanity check, consider the limit $E \to 0$ with μ fixed: for $\mu < 1$ we get $\frac{1-0}{1+0-0} = 1$, whereas for $\mu > 1$ we get $\frac{1-0}{1+\infty-0} = 0$. See Fig. 19.1.

19.4　Some Richer Linear Algebra, and Flag Juggling

As explained at the beginning of Section 19.2, the function σ was the complete
invariant for the group of row operations and rightward column operations. If
we restrict to *downward* row operations, then we still get a discrete set of orbits
(even for complex matrices); each orbit contains a unique partial permutation
matrix of rank b.

Define a **flag juggling state** as a juggling state where the $\times$s have been
replaced by the numbers $1, \ldots, b$, each used exactly once. Then we have a
unique map $\tilde{\sigma} : U_b \to$ {flag juggling states} that takes a partial permutation
matrix of rank b with $m_{ij} = 1$ to a state with an i in the jth position, and such
that σ is invariant under downward row and rightward column operations.

To give the analogue of Proposition 19.2 requires us to extend the definition
of ℓ to flag juggling patterns: it should also count any pair $\ldots i \ldots j \ldots$ with
$i > j$ as an inversion, e.g., $\ell(-3-1\,2) = 7$. It is then reasonable to consider
"$-$" as $+\infty$ for this inversion count.

Proposition 19.4. *Let $\tilde{\tau}$ be a flag juggling state, and pick $W >$ the last $\times$-
position of τ.*

*Then the fraction of $b \times W$ matrices with $\tilde{\sigma}(M) = \tilde{\tau}$ is $(1 - q^{-1})^b / q^{\ell(\tilde{\tau})}$,
independent of W.*

Proof. We're computing the size of the $B_- \times N_+$-orbit through the partial permutation matrix M with $\tilde{\sigma}(\pi) = \tilde{\tau}$, where B_- is lower triangular $b \times b$ matrices
and N_+ is upper triangular $W \times W$ matrices with 1s on the diagonal.

The N_+-stabilizer of M consists of matrices R with $R_{ij} = 0$ unless $\tilde{\tau}$ has $-$ in
its ith position. The B_--stabilizer is trivial. However, the $(B_- \times N_+)$-stabilizer
of M is slightly larger than the product of the stabilizers; some row operations
can be canceled by some column operations, one such pair for each inversion
$\ldots i \ldots j \ldots$ with $i > j$.

The order of $B_- \times N_+$ is $(q-1)^b q^{\binom{b}{2}} q^{\binom{W}{2}}$; dividing by the stabilizer order
gives the size of the orbit, then by q^{bW} gives the fraction claimed. $\qquad\square$

Corollary 19.2. *The mapping $\tilde{\tau} \mapsto (1 - q^{-1})^b / q^{\ell(\tilde{\tau})}$ is a probability measure
on the space of flag juggling states.*

Side note. The corresponding equation $(1 - q^{-1})^{-b} = \sum_{\tilde{\tau}} q^{-\ell(\tilde{\tau})}$ gives two
formulae for the Poincaré series of the manifold $B_- \backslash U_b$ of partial flags $(V^1 <
V^2 < \ldots < V^b < \mathbb{C}^{\mathbb{N}})$. Since $B_- \backslash U_b$ is a Leray–Hirsch-satisfying bundle over
$GL(b) \backslash U_b$ with fiber $B_- \backslash GL(b)$, the Poincaré series $b_{B_- \backslash U_b}$ of this bundle factors

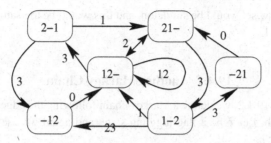

Fig. 19.2. The $b = 2$ flag juggling digraph with only throws ≤ 3 drawn.

as

$$b_{B_- \backslash U_b} = b_{GL(b) \backslash U_b} \, b_{B_- \backslash GL(b)}$$

where

$$b_{B_- \backslash U_b} = (1 - q^{-1})^{-b} = \sum_{\tilde{\tau}} q^{-\ell(\tilde{\tau})}$$

$$b_{GL(b) \backslash U_b} = \prod_{i=1}^{b} (1 - q^{-i})^{-1} = \sum_{\tau} q^{-\ell(\tau)}$$

$$b_{B_- \backslash GL(b)} = \prod_{i=1}^{b} \frac{1 - q^{-i}}{1 - q^{-1}} = \sum_{\pi \in S_n} q^{-\ell(\pi)}$$

the three sums on the right being derivable from the respective Bruhat decompositions.

We now define the edges out of a flag juggling state $\tilde{\tau}$, again making the set of states into the vertices of a digraph which appeared already in [ABCLN]. (As far as we can tell our motivations for studying this digraph are different than theirs.) A small example is shown in Fig. 19.2.

If $\tilde{\tau}$ begins with $-$, then there is a unique outgoing edge, to $\tilde{\tau}$ with the $-$ removed. Otherwise we pick up the number that $\tilde{\tau}$ starts with and begin walking east.

(1) At any $-$, we can replace the $-$ with the carried number and be done.
(2) At any strictly larger number, we can pick up that larger number, drop the number we were carrying in its place, and go back to (1).

Define the **throw set** for the transition $\tilde{\tau} \to \tilde{\tau}'$ as the places a number is dropped. Neither drop is required; we can continue walking east instead (though not forever). Note that if we used the label 1 b times instead of $[1, b]$ each once, then each throw set would be singleton, and this would be the same digraph as in Section 19.1.

19.4.1 Another Markov Chain

As in Section 19.1.2, we define a Markov chain following the edges backwards in this digraph. Let $\tilde{\tau}$ be a flag juggling state, and again we use a coin with $p(\text{heads}) = 1/q$.

(1) Hold a $-$, and point at the rightmost number in $\tilde{\tau}$.
(2) Flip the coin. If tails, put down what we're holding and pick up the number we're pointing at. If heads, do nothing.
(3) Move leftwards; stop when we meet a number smaller than what we're holding (interpreting $-$ as $+\infty$, jibing with our definition of $\ell(\tilde{\tau})$).
(4) If we meet such a number, go back to (2). Otherwise we've fallen off the left end of (the now modified) $\tilde{\tau}$; drop whatever we're holding, there.

For example, start with $\tilde{\tau} = -- 3\ 1 - 2$, holding a $-$, pointing at the 2.

- Tails: pick up the 2, leaving the $-$ in its place. Point at the 1.
  - Tails: drop the 2 for the 1, then carried all the way left to give $1 -- 3\ 2 - -$.
  - Heads: the 2 gets carried all the way left to give $2 -- 3\ 1 - -$.
- Heads: skip the 2 and point at the 1.
  - Tails: pick up the 1, leaving the $-$, and carry the 1 all the way left to give $1 -- 3 -- 2$.
  - Heads: leave the 1 and proceed to the 3.
    - Tails: pick up the 3 and carry it left to give $3 --- 1 - 2$.
    - Heads: leave the 3 and drop the $-$ on the left, giving $--- 3\ 1 - 2$.

$$\text{In all, } \tilde{\tau} \mapsto (1 - q^{-1})^2[1 --3\ 2] + (1 - q^{-1})q^{-1}([2 --3\ 1 --]$$
$$+ [1 --3 --2]) + q^{-2}(1 - q^{-1})[3 --- 1 - 2]$$
$$+ q^{-3}[--- 3\ 1 - 2]$$

This is again motivated by the "add a random column on the left" Markov chain on U_b:

Proposition 19.5. *Let M be the partial permutation matrix in U_b with $\tilde{\sigma}(M) = \tilde{\tau}$, and $\vec{c}$ a random column vector. Then the probability of $\tilde{\sigma}([\vec{c}M])$ being a particular state τ' is the probability of reaching τ' in the foregoing process.*

In particular, the possible τ' are the ones such that $\tau' \to \tau$ in the digraph defined in this section.

Proof sketch. Rightward column-reduction of $[\vec{c}\,M]$ corresponds to doing the coin flips, in reverse order. At each coin-flipping step, we determine that a certain entry of $\vec{c}$ is zero (heads) or nonzero (tails). We leave the details to the reader. □

We have the corresponding theorem, but skip the corresponding formal derivation:

Theorem 19.2. *The vector $\tilde{\tau} \mapsto (1 - q^{-1})^b q^{-\ell(\tilde{\tau})}$ is the stationary distribution of the Markov chain (1)–(4) defined earlier.*

19.4.2 The Linear Algebra of Repeated Labels

Given a finite multiset S of numbers, we can redefine **flag juggling states** to bear those labels, and the Markov chain in this section extends without changing a word. If the elements of S are all equal, or all different, we get the digraphs from Section 19.1 and Section 19.4 respectively. So one can ask for a corresponding linear algebra problem and, hopefully thereby, calculation of the stationary distribution.

To interpolate between *all* row operations and *downward* row operations, we consider the rows as coming in contiguous groups, and only allow row operations within a group or downward. For example if $S = \{1, 1, 5, 7, 7, 7\}$, then we have three groups, which are two rows above one row above three rows.

In the analogue of reduced row-echelon form, the b pivot columns are still arbitrary, but the pivots within a given group run northwest/southeast. In the analogue of Corollaries 19.1 and 19.2, the prefactor is

$$\prod^{\text{groups}} \prod_{i=1}^{\text{group size}} (1 - q^{-i})$$

which computes the probability that a *block* lower triangular matrix is invertible. These prefactors, times $q^{-\ell(\tilde{\tau})}$, again give the stationary distribution.

19.5 Drawing Out the Process

Since the transitions in the (flag) juggling Markov chain involve repeated flipping of coins, it seems natural to ask for an alternative version with intermediate states, such that only one coin is flipped at each transition.

19.5.1 The Digraph of Ordinary and Hatted States

Define a **hatted flag juggling state** τ_+ as a flag juggling state τ with one position hatted, as in $\hat{5}$ or $\hat{-}$. The hat is not allowed over any of the $-$s occurring after τ's last number, e.g., $3 - 4 - - \ldots$ can only be hatted as $\hat{3} - 4$, $3\hat{-}4$, or $3 - \hat{4}$.

The vertices of the digraph will be the usual unaugmented states, plus these new, intermediate, states. Make directed edges as follows:

- If τ is unhatted, then it has only one arrow out, hatting the 0th label.
  Example: $\quad 3 - 2\, 1 \quad \rightarrow \quad \hat{3} - 2\, 1.$
- If τ_+ has a hat, then there are one or two arrows out of τ_+.
  - We can move the hat and label one step rightward, switching places with the unhatted label just beyond, unless that involves switching a $\hat{-}$ with the last number in τ. Example: $\quad \hat{3} - 2\, 1 \quad \rightarrow \quad - \hat{3}\, 2\, 1.$
  - If the hatted label is the last number in τ, we can remove the hat. If it isn't, and the next label after the hatted label is larger (counting $-$ as ∞), then the hat can jump one step rightward to that larger label, with the labels not moving. Example: $\quad \hat{3} - 2\, 1 \quad \rightarrow \quad 3 \hat{-} 2\, 1 \not\rightarrow 3 - \hat{2}\, 1.$

It's easy to see that if τ, τ' are unhatted states, then $\tau \rightarrow \tau'$ in the digraph from Section 19.4 if and only if there is a directed path in this digraph from τ to τ' through only hatted states.

19.5.2 The Last Markov Chain

As twice before, we put probabilities on the reversed edges.

If τ is an unhatted state, the only edge into it has a hat on the last number in τ. Whereas if τ_+ is τ with a hat at position 0, the only edge into τ_+ comes from τ. In either case the unique edge gets probability 1.

In the remaining case, τ_+ has a hat at position $i > 0$, and we must move the hat left one step. If the label to the left is smaller than the hatted label, then we move the hat left with probability 1 and are done. But if that label is larger, then with probability $1/q$, we move the label bearing the hat, not just the hat. From the state $3 - 2\, \hat{1}$:

$$3 - \hat{2}\, 1 \quad \xrightarrow{1-q^{-1}} \quad 3 - 2\, \hat{1} \quad \xleftarrow{q^{-1}} \quad 3 - \hat{1}\, 2$$

(Don't forget that the Markov chain runs backwards along the arrows.)

Again, we claim that if τ, τ' are unhatted states, and we start this Markov chain at τ and stop when we next meet an unhatted state, the probability that it is τ' is the same as that given by the Markov chain from Section 19.4.

19.6 Questions

What is a linear algebra interpretation of the model in Section 19.5?

Is there an analogue of [BEGW94] for the flag juggling digraph?

The stationary distributions computed here live on $S_{b+\mathbb{N}}/(S_b \times S_{\mathbb{N}})$ and $S_{b+\mathbb{N}}/((S_1)^b \times S_{\mathbb{N}})$, and easily generalize to other W/W_P coset spaces ($\tau \mapsto q^{-\ell(\tau)}$ times a prefactor). What is a Markov chain on W/W_P for which they are the stationary distribution?

What is an analogue of the ball density function calculated in Section 19.3, when the balls are colored as in Section 19.4?

Is there a version of Section 19.1 in which b varies, whose stationary distribution is still $q^{-\mathrm{area}}$, up to overall scale? Does that have a linear algebra interpretation?

Is the $q \to 1$ limit related in any substantive way to the mythical field of one element? Can that theoretical theory include the $b \to \infty$ limit?

It is easy to define a version of the digraph from Section 19.5 with multiple hatted labels percolating through independently. Is this of any use?

Acknowledgments

A. Knutson was supported by National Science Foundation grant 0303523.

The author is very grateful to Ron Graham for many conversations about mathematics and juggling, and most especially for sending the author to speak about these subjects in his stead at the Secondo Festival Della Matematica in Rome.[3] Many thanks also to Jack Boyce, Svante Linusson, Harri Varpanen, and Greg Warrington. Some related linear algebra was developed with David Speyer and Thomas Lam in [KLS13].

References

ABCN. Arvind Ayyer, Jérémie Bouttier, Sylvie Corteel, and François Nunzi. Multivariate juggling probabilities. Preprint 2014. http://arxiv.org/abs/1402.3752

ABCLN. Arvind Ayyer, Jérémie Bouttier, Sylvie Corteel, Svante Linusson, and François Nunzi. Bumping sequences and multispecies juggling. Preprint 2015. http://arxiv.org/abs/1504.02688

BEGW94. Joe Buhler, David Eisenbud, Ron Graham, and Colin Wright. Juggling drops and descents. *Amer. Math. Monthly* **101** (1994), 507–519. Reprinted with new appendix in the Canadian Mathematical Society, Conference

---

[3] This lecture is available at https://www.youtube.com/playlist?list=PL3C8EC6BA111662D4 in both English and Italian.

Proceedings, Vol. 20, 1997. http://www.math.ucsd.edu/~ronspubs/94_01a_juggling.pdf

BG10. Steve Butler and Ron Graham. Enumerating (multiplex) juggling sequences. *Ann. Combinat.* **13** (2010), 413–424. http://www.math.ucsd.edu/~ronspubs/10_03_multiplex.pdf

CG08. F. R. K. Chung and Ron Graham. Primitive juggling sequences. *Amer. Math. Monthly* **115** (March 2008), 185–194. http://www.math.ucsd.edu/~ronspubs/08_03_primitive_juggling.pdf

ER96. Richard Ehrenborg and Margaret Readdy. Juggling and applications to q-analogues. *Discrete Math.* **157** (1996), 107–125. http://www.ms.uky.edu/~jrge/Papers/Juggle.pdf

ELV15. Alexander Engström, Lasse Leskelä, and Harri Varpanen. Geometric juggling with q-analogues. *Discrete Math.* **338** (2015), pp. 1067–1074. http://arxiv.org/abs/1310.2725

KLS13. Allen Knutson, David E Speyer, and Thomas Lam. Positroid varieties: Juggling and geometry. *Compositio Mathematica* **149**, no. 10 (October 2013), pp. 1710–1752. http://arxiv.org/abs/1111.3660

LV12. Lasse Leskelä and Harri Varpanen. Juggler's exclusion process. *J. Appl. Prob.* **49**(1), 2012, pp. 266–279. http://arxiv.org/abs/1104.3397

Pol03. Burkhard Polster, The mathematics of juggling. Springer-Verlag, New York, 2003.

Pos. Alexander Postnikov, Total positivity, Grassmannians, and networks. http://arxiv.org/abs/math/0609764

Va14. Harri Varpanen, Toss and spin juggling state graphs. *Proceedings of Bridges* 2014, pp. 301–308, 2014. http://arxiv.org/abs/1405.2628

Wa05. Greg Warrington, Juggling probabilities. *Amer. Math. Monthly* **112**, no. 2 (2005), 105–118. http://arxiv.org/abs/math/0302257

20

Explicit Error Bounds for Lattice Edgeworth Expansions

Joe P. Buhler, Anthony C. Gamst, Ron Graham,
and Alfred W. Hales

Abstract

Motivated, roughly, by comparing the mean and median of an IID sum of bounded lattice random variables, we develop explicit and effective bounds on the errors involved in the one-term Edgeworth expansion for such sums.

Let X be a bounded integer-valued random variable (these will occasionally be referred to in the text that follows as "dice"), and let $X[n]$ denote the sum of n independent and identically distributed (IID) copies of X (sometimes referred to as "rolls"). If p_x denotes $\mathbf{Pr}(X = x)$, then we say that x is a *value* of X if $p_x > 0$. The mean of X is

$$\mathbf{E}X = \mu_1(X) = \mu_1 = \sum_x x\, p_x,$$

the higher (central) moments of X are $\mu_k(X) = \mu_k = \mathbf{E}(X - \mu_1)^k$, and the standard deviation $\sigma = \sigma(X)$ is the square root of μ_2. (Here, and in the text that follows, the random variable X is omitted from the notation if it can be inferred easily from context.)

The *tilt* of X is

$$T(X) := \mathbf{Pr}(X > \mu_1) - \mathbf{Pr}(X < \mu_1).$$

Modulo the annoying question of the precise definition of the median, the sign of T measures whether the median is to the right or left of the mean. The primary focus of this chapter is on the sign of the tilt of $X[n]$ for lage n. Let

$$T_n = T_n(X) = T(X[n]).$$

As n goes to infinity the central limit theorem says that the distribution of $X[n]$, suitably shifted and scaled, converges to the standard normal distribution. This implies that the tilt goes to zero as n goes to infinity, and this will require us to accurately estimate T_n in order to say anything at all about its sign. This will be done by using the one-term Edgeworth expansion (an asymptotic refinement of the central limit theorem) for lattice random variables. The key goal of this chapter is to prove explicit formulas for the error in these approximations sufficient to enable the determination of the sign of T_n, for *all* n.

It turns out that the sign of T_n for large n depends on the third moment μ_3 (associated with the "skew" or "tilt" of the distribution) but also, perhaps more surprisingly, on the congruence class of n modulo the so-called "span" of X. We will see that for large n the sign of the tilt is (almost always) completely determined by these two pieces of data.

To make this more precise, it is convenient to make some definitions. The *span* of a die X is the largest integer b such that all values of X are contained in a single congruence class modulo b, i.e., an arithmetic progression of the form $a + bn$, for $n \in \mathbf{Z}$. The integer a is called a *shift* of X, and it is only well-defined modulo b. It isn't hard to see that the span is the gcd (greatest common divisor) of all $x - a$ as x ranges over the values of X. (For the span to be nonzero, X has to have at least two values, and we will always assume that this is the case.)

If x is an integer and b is any positive integer, let $x \bmod b$ denote the unique integer congruent to x modulo b that is in the interval $[0, b)$.

Theorem 20.1. *Let X, σ, a, b, and μ_3 be as earlier. Then for positive integers n,*

$$T_n = \frac{L(na)}{\sqrt{2\pi n}} + E$$

where

$$L(c) = \frac{(-c) \bmod b - c \bmod b}{\sigma} - \frac{\mu_3}{3\sigma^3},$$

and

$$E = o(1/\sqrt{n}).$$

Note that $L(na)$ depends only on the congruence class of n modulo b, and that if n goes to infinity in a fixed congruence class then the limit of $\sqrt{2\pi n}\, T_n$ exists and is equal to $L(na)$. The error E will turn out to be bounded by terms that are, roughly, constant multiples of $1/n$ and $\exp(-c\sqrt{n})/n^p$ for various c and p.

Proofs of the lattice Edgeworth expansions in the literature do not seem to include error bounds on the error E, and our goal is to exhibit such bounds, for bounded lattice random variables. Such bounds are necessary if one wants to find an n_0 together with a *proof* that

$$n \geq n_0, \quad an \equiv c \bmod b \quad \text{imply that} \quad \text{sign}(T_n) = \text{sign}(L(c)).$$

Briefly, one could say that "asymptopia" has arrived when the sign of T_n is equal to its asymptotic sign.

This question arose for us in [2] where the existence of "maximally intransitive" dice was shown. We[1] felt that it should be possible to determine when asymptopia arrives, i.e., when the desired dominance relation between the dice constructed in [2] was absolutely guaranteed for $n \geq n_0$ (see that article for details).

It is possible for $L(c)$ to be 0, though "unlikely" if X is not symmetric. In this case, higher order Edgeworth expansions are necessary, and this case will be left to the motivated and energetic reader.

No prior understanding of Edgeworth expansions is required to read this chapter, and we consider only a specific case. For a broader perspective, the reader could consult [3, 4, 5]. The techniques described here should be applicable more generally.

Section 20.1 develops some of the basic ideas needed to approximate $\mathbf{Pr}(X[n] < 0)$. Section 20.2 proves such an approximation, together with explicit error bounds. Section 20.3 uses the results of Section 20.2 to prove a refined version of Theorem 20.1, and looks at examples.

20.1 Preliminaries

It is convenient to focus on the case of real-valued X with mean 0 and span 1. If Y is a die with span b then

$$X := \frac{Y - \mu_1(Y)}{b}$$

is a bounded lattice random variable with mean 0 and span 1, which says that the values of X lie in a lattice $a + \mathbf{Z}$ but are not contained in a larger lattice $a + d\mathbf{Z}$, $d > 1$. The tilt is invariant under affine transformations $X \to bX + a$ so $T_n(Y) = T_n(X)$. It is convenient to fix this situation from now on: X will be a random variable with span 1, mean 0, and shift a.

If X arises from dice as earlier then the shift a is a rational number. In this case it might make sense to take a limit as n goes to infinity through a set of values where $\{na\}$ is fixed. Here $\{x\}$ denotes the *fractional part* of x, i.e., the unique y such that $x = y + j$ for some integer j, and $0 \leq y < 1$. However, the explicit estimates apply for irrational a and an arbitrary n, and may be useful in other contexts.

---

[1] Well, especially RLG.

The central limit theorem says that the *cumulative probability distribution* of the normalized random variable $Z_n := X[n]/(\sigma\sqrt{n})$ approaches that of a standard normal random variable in the sense that

$$\lim_{n\to\infty} \mathbf{Pr}\,(Z_n \leq s) = \frac{1}{\sqrt{2\pi}} \int_{-\infty}^{s} e^{-t^2/2} dt, \qquad \text{for all } s \in \mathbf{R}.$$

Our interest in the tilt suggests focusing on the mean $s = \mu_1(Z_n) = 0$. Then $\lim_{n\to\infty} \mathbf{Pr}\,(Z_n \leq 0) = 1/2$. The Berry–Esseen Theorem gives an explicit bound on the error, i.e., in the case $s = 0$,

$$|\mathbf{Pr}\,(Z_n \leq 0) - 1/2| \leq \frac{c}{\sqrt{n}}$$

for a small constant c, e.g., see [4]. However, it is easy to show that the tilt T_n is $O(1/\sqrt{n})$, so the Berry–Esseen level of accuracy is insufficient for saying anything about the tilt. The central result of this chapter is of the form

$$\left| \mathbf{Pr}\,(Z_n \leq 0) - \left(1/2 + \frac{\ell}{\sqrt{n}} \right) \right| \leq E(n).$$

Here ℓ depends on on the second, third, and fourth moments of X and the fractional part $\{na\}$. The error E is bounded by an expression whose principal term is of the form d/n, with about seven other terms that are each of the form $\lambda e^{-\tau n^\gamma}/n^\rho$ for various constants λ, τ, ρ, and γ. Although this is a very special case of the central limit theorem, the techniques should apply more generally.

As will be seen, this explicit bound on the error in the simplest nontrivial Edgeworth expansion allows us to prove theorems about the sign of the tilt.

Readers might remember that the skewness of the distribution of $X[n]$ depends on the third moment of X. This is reflected in the term $\mu_3/3\sigma^3$ of $L(c)$ in Theorem 20.1. One intuitive way to see that the third moment and asymptotic tilt might have opposite signs is that for large n the distribution of $X[n]$ should be approximately normal, and if the median is slightly negative then the positive values have to be somewhat larger to make the mean equal to 0, so the third moment will be positive.

The other term in $L(c)$, which becomes $(\{-na\} - \{na\})/\sigma$ in the span 1 case, shows that for lattice random variables the third moment does not give the full story. This term is sometimes called the lattice correction term. To get an intuitive feel for this term, consider a lattice random variable X of span 1 and shift a with mean and third moment equal to 0 (a simple linear algebra exercise shows that these are easy to come by). Since $\mu_3 = 0$ the lattice correction term is the only term. The support of the probability distribution of $X[n]$ is contained in the set of real numbers x that are congruent to na modulo 1. For large n, the probability distribution is close to a rescaled version of the standard normal distribution. To first order, it seems reasonable to suspect that if $c = na$ is less than $1/2$ then the sum of the probabilities $p_x = \mathbf{Pr}(X = x)$ for $\{x\} = c, x > 0$, will be slightly greater than the corresponding sum for $x < 0$ since the latter contribute more heavily to making the mean 0, which tends to make the tilt positive.

Of course the explicit formula for $L(c)$ emerges cleanly from the calculations that follow, and perhaps this supersedes all of these heuristic remarks!

20.1.1 The Characteristic Function

As earlier, fix X with mean 0, span 1, shift a, central moments μ_k, and standard deviation σ defined by $\sigma^2 = \mu_2$. The goal of this section is to give a formula for $\mathbf{Pr}(X[n]) < 0$ in terms of an integral. This formula can be proved fairly directly using Fourier series, but we will take a somewhat more leisurely approach that starts with a contour integral.

The probability generating function (PGF) of X is a function of a complex variable z defined by

$$F(z) = \mathbf{E}\, z^X = \sum_x p_x z^x = z^a \sum_j p_{a+j} z^j,$$

using the fact that values of X can be written $x = a + j$ for some integer j. Note that $z^{-a} F(z)$ is a finite Laurent series:

$$z^{-a} F(z) = \sum_x p_x z^{x-a} = \sum_j p_{a+j} z^j.$$

Applying Cauchy's Theorem gives

$$p_{a+j} = \mathbf{Pr}(X = a + j) = [z^j]\, z^{-a} F(z) = \frac{1}{2\pi i} \oint_\gamma \frac{z^{-a} F(z)}{z^j} \frac{dz}{z}$$

where $[z^j]z^{-a}F(z)$ denotes the coefficient of z^j in the polynomial $z^{-a}F(z)$, and the contour γ can be chosen to be a counterclockwise circle around the origin.

With an eye to ultimately applying this to the tilt, we use this integral to find a useful expression for $\mathbf{Pr}(X < 0)$. The set of negative values $x = a + j$ is the set of $\{a\} + j$ as j ranges over negative integers (where $\{a\}$ is the fractional part of a, defined earlier; this might reasonably be denoted $a \bmod 1$). Therefore,

$$\mathbf{Pr}(X < 0) = \sum_{j<0} p_{\{a\}+j} = \frac{1}{2\pi i} \oint (z + z^2 + z^3 \ldots)\, z^{-\{a\}} F(z) \frac{dz}{z}$$

$$= \frac{1}{2\pi i} \oint \frac{z^{1-\{a\}} F(z)}{1-z} \frac{dz}{z}$$

where the radius is less than 1 to ensure that the geometric series converges.

If n is a positive integer and $X[n] := \sum X_i$, where the X_i are n independent random variables, then independence implies that the PGF of $X[n]$ is $F(z)^n$. Applying the preceding formula to $X[n]$ gives

$$\mathbf{Pr}(X[n] < 0) = \frac{1}{2\pi i} \oint \frac{z^{1-\{na\}} F(z)^n}{1-z} \frac{dz}{z}$$

where the contour is a counterclockwise circle around the origin of radius slightly less than 1.

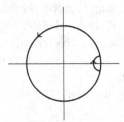

Move the contour outward to the unit circle except for an infinitesimal semicircular divot centered at, and to the left of, 1. In other words, the contour follows the unit circle counterclockwise from $z = e^{i\varepsilon}$ to $z = e^{-i\varepsilon}$ followed by a clockwise small circular arc back to $e^{i\varepsilon}$. For very small ε the integrand is close to $-1/(z-1)$, and the contour is basically a small clockwise semicircle; Cauchy's Theorem implies that the value of the integral over the divot is very close to $1/2$. Taking the limit as ε goes to zero gives

$$\mathbf{Pr}(X[n] < 0) = \frac{1}{2} + \frac{1}{2\pi i} \oint \frac{z^{1-\{na\}} F(z)^n}{1-z} \frac{dz}{z}$$

where the contour is the unit circle punctured at $z = 1$, with the "principal value interpretation" at the puncture. With an eye to changing variables by $z = e^{it}$, let

$$f(t) = F(e^{it}) = \mathbf{E} \, e^{itX} = \sum_x p_x e^{itx}$$

be the *characteristic function* (CF) of X. Then

$$\mathbf{Pr}(X[n] < 0) = \frac{1}{2} - \frac{1}{2\pi i} \int_{-\pi}^{\pi} e^{i\alpha t} f(t)^n D(t) \frac{dt}{t}$$

where $\alpha = \alpha_n = 1/2 - \{na\}$ and

$$D(t) = (t/2)/\sin(t/2).$$

The principal value interpretation of the integral at $t = 0$ will always be used, which means that

$$\int_{-\pi}^{\pi} := \lim_{\varepsilon \to 0} \left(\int_{-\pi}^{-\varepsilon} + \int_{\varepsilon}^{\pi} \right).$$

The following result summarizes the preceding discussion.

Theorem 20.2. *With the foregoing notation,*

$$\mathbf{Pr}(X[n] < 0) = 1/2 - I_0, \quad \text{where} \quad I_0 := \frac{1}{2\pi i} \int_{-\pi}^{\pi} e^{i\alpha t} f(t)^n D(t) \frac{dt}{t}.$$
(20.1)

In addition, if $\alpha' = 1/2 - \{-na\}$, then

$$T_n = \frac{1}{2\pi i} \int_{-\pi}^{\pi} e^{i\alpha t} f(t)^n D(t) \frac{dt}{t} - \frac{1}{2\pi i} \int_{-\pi}^{\pi} e^{i\alpha' t} f(-t)^n D(t) \frac{dt}{t}.$$

The last statement of the theorem follows easily from the first, using several obvious facts: (1) $\mathbf{Pr}(X[n] > 0) = \mathbf{Pr}((-X)[n] < 0)$, (2) α' is the shift of $-X$, and (3) the CF of $-X$ is $f(-t)$.

For later use, we remark that $D(t)$ is even, has $D(0) = 1$, $D(\pi) = \pi/2$, and has power series coefficients that can be expressed in terms of Bernoulli numbers and are positive. From that, or alternately by a simple calculus exercise, it follows $D(t)$ is increasing on $[0, \pi]$ so that

$$D(t) \leq D(\pi) = \pi/2 \quad \text{on} \quad [-\pi, \pi]. \tag{20.2}$$

20.1.2 Span

As earlier, X has span 1, mean 0, shift a, and CF $f(t)$.

Lemma 20.1. *There are integers c_x, one for each value x of X, such that*

$$\sum_x c_x = 0, \quad \text{and} \quad \sum_x c_x x = 1.$$

Proof. Let y be a value of X. If $b := \gcd(\{x - y\})$ is larger than 1 then the values of X are contained in $y + b\mathbf{Z}$ which contradicts the fact that X has span 1. Therefore, the gcd is 1 and there are integers c_x, for $x \neq y$, such that

$$\sum c_x(x - y) = 1.$$

Set $c_y = - \sum_{x \neq y} c_x$. The stated properties are easily verified. $\square$

A set $\{c_x\}$ as in the lemma is said to be a *certificate* of the fact that X has span 1.

Lemma 20.2. *The function $|f(t)|$ has period 2π, and $|f(t)| < 1$ for $t \in (0, 2\pi)$.*

Proof. We can assume that the shift a is a value of X. Since

$$f(t) = e^{iat} \sum_x p_x e^{it(x-a)}$$

and the $x - a$ are all integers it follows that $|f(t + 2\pi)| = |f(t)|$, and that the period of $|f(t)|$ is of the form $2\pi/b$ for some positive integer b. Then

$$1 = |f(2\pi/b)| = \left| e^{i2\pi a/b} \sum_x p_x e^{2\pi i(x-a)/b} \right| \le \sum_x p_x = 1.$$

Equality in this use of the triangle inequality implies that all $e^{\pi i(x-a)/b}$ are equal to 1, i.e., that all $x - a$ are multiples of b. If $b > 1$ then this contradicts the fact that the span of X is equal to 1. $\square$

These lemmas show that if the span is 1 then $\gcd(\{x - x'\}) = 1$, X has a certificate, and $|f(t)|$ has period 2π. It is not hard to show that any of these implies that the span is 1, so all four conditions are equivalent.

20.1.3 Bounding CF Power Series Tails

The power series of the CF for X converges for all real t, and has the form

$$f(t) = \sum_x p_x e^{itx} = \sum_{k \ge 0} \mu_k \frac{(it)^k}{k!} = 1 - \frac{\mu_2 t^2}{2} - i\frac{\mu_3 t^3}{6} + \frac{\mu_4 t^4}{24} + \dots.$$

$$(20.3)$$

The tail of this power series has an especially tight bound, saying that the remainder after k terms is at most the absolute value of the next term with the moment replaced by the corresponding absolute moment.

Lemma 20.3. *Let $\overline{\mu}_k = \mathrm{E}\,|X|^k$ denote the kth absolute moment of X. Then*

$$\left| \sum_{j \ge k} \frac{\mu_j(it)^j}{j!} \right| \le \frac{\overline{\mu}_k |t|^k}{k!}.$$

Proof. The expansion

$$e^{it} = \sum_{0 \le j < k} \frac{(it)^j}{j!} + \theta(t)\frac{(it)^k}{k!}, \qquad |\theta(t)| \le 1,$$

follows from several standard integral forms of the remainder in Taylor's Theorem. Replace t by tx, where x is a value of X, multiply by p_x, and sum over all values x to arrive at

$$f(tx) = \sum_{j<k} \sum_x p_x \frac{(itx)^j}{j!} + \sum_x \theta(tx) p_x \frac{(itx)^k}{k!}$$

$$= \sum_{j<k} \mu_j \frac{(it)^j}{j!} + \sum_x \theta(tx) p_x \frac{(itx)^k}{k!}.$$

Taking the absolute value of the remainder gives the bound

$$\left| \sum_x \theta(tx) p_x \frac{(itx)^k}{k!} \right| \le \sum_x p_x |x|^k \frac{|t|^k}{k!} = \overline{\mu}_k |t|^k / k!$$

as claimed. □

20.1.4 Bounding the CF

Fix a certificate $\{c_x\}$, as earlier, and let $C = \sum_x |c_x|$ be its ℓ_1 norm. Note that $C \ge 2$ since at least two of the integers c_x are nonzero. Before proving a bound on the CF $f(t)$ outside a neighborhood of 0, a preliminary lemma is needed.

Lemma 20.4. *If* $0 < t < \pi$ *then no interval on the circle of arc length less than* $2t/C$ *contains* e^{itx} *for all values* x *of* X.

Proof. Suppose, to the contrary, that all e^{itx} lie in the interior of the arc from e^{iu} to $e^{i(u+2t/C)}$ on the circle. (Since $2t/C < \pi$ the interior is well-defined — it is the smaller of the arcs into which those two points divide the circle.) Then there are integers j_x such that

$$u < tx + 2\pi j_x < u + 2t/C .$$

Subtract $u + t/C$ to get

$$-t/C < tx + 2\pi j_x - u - t/C < t/C.$$

Multiply by c_x and sum, noting that the inequalities reverse if c_x is negative, to get

$$-t = \frac{-\sum |c_x| t}{C} < t + 2\pi M < \frac{\sum |c_x| t}{C} = t$$

for some integer M. If M is nonnegative the right inequality is false, and if M is negative the left inequality is false. This finishes the proof of the lemma. □

Theorem 20.3. *Let X be a random variable as earlier, and $f(t)$ its CF. Let $m := \min\{p_x\}$ be the minimum probability of a value. If $|t| \leq \pi$ then*

$$|f(t)| \leq 1 - \frac{8mt^2}{\pi^2 C^2}.$$

Proof. By Lemma 20.4 there are two values x, y such that $x < y$ and the "arc-length" distance between e^{itx} and e^{ity} on the unit circle is at least $2t/C$ and at most π, i.e.,

$$\frac{2t}{C} \leq ty - tx \leq \pi.$$

Then

$$f(t) = \sum_u p_u e^{itu} = T + m e^{itx} + m e^{ity},$$

where T is a trigonometric sum with nonnegative coefficients that sum to $1 - 2m$, and therefore

$$
\begin{aligned}
|f(t)| &\leq 1 - 2m + \left| m e^{it(x+y)/2} \left(e^{it(x-y)/2} + e^{it(y-x)/2} \right) \right| \\
&= 1 - 2m + 2m \cos(t(y-x)/2) \\
&\leq 1 - 2m + 2m \cos(t/C) = 1 - 2m(1 - \cos(t/C)) \\
&= 1 - 4m \sin^2(t/2C).
\end{aligned}
$$

Since $x/\sin(x)$ is increasing on $[0, \pi/4]$ it follows that

$$\sin(t/2C) \geq \frac{t}{2C} \frac{\sin(\pi/4)}{\pi/4} = \frac{\sqrt{2}t}{\pi C}.$$

Thus

$$1 - 4m \sin^2(t/2C) \leq \frac{8mt^2}{\pi^2 C^2},$$

finishing the proof. □

A similar bound $|f(t)| \leq 1 - dt^2$ can be found in [1], for a completely different constant d (not consistently better or worse than the constant in Theorem 20.3). In addition a technique is given in [1] to improve the bound when there are several independent certificates, and that idea also applies to our bound. Note that from (20.3) it is clear that any such constant has to be strictly smaller than $\mu_2/2$. Later we will see how to, for practical purposes, find bounds that, roughly, say that in a practical context the constant can be made as close to $\mu_2/2$ as desired.

20.1.5 Facts about the Gamma Function

Several facts about values of the gamma function (and its upper and lower variants) at integers and half-integers will be needed in the text that follows. To slightly complicate matters, these will arise here as integrals of functions of the form of $t^x \exp(-ct^2/2)$. It is convenient to collect these in one place for the sake of future reference. In Proposition 20.1 the "double factorial" $x!!$ of a nonnegative integer denotes the product of all positive integers up to x that have the same parity as x, i.e.,

$$x!! = \prod_{0 \le k < x/2} (x - 2k).$$

Proposition 20.1. *Let c and s be positive real numbers, and k a positive integer.*

(1) $\int_0^\infty t^{2k} e^{-ct^2/2} \frac{dt}{t} = c^{-k} (2k - 2)!!,$

     $\int_0^\infty t^{2k-1} e^{-ct^2/2} \frac{dt}{t} = c^{-k+1} (2k - 3)!! \sqrt{\dfrac{\pi}{2c}}$

(2) $\int_0^s t^2 e^{-ct^2/2} \frac{dt}{t} = \dfrac{1}{c} \left(1 - e^{-cs^2/2}\right)$

     $\int_0^s t^4 e^{-ct^2/2} \frac{dt}{t} = \dfrac{2}{c^2} \left(1 - e^{-cs^2/2} \left(1 + \dfrac{cs^2}{2}\right)\right)$

(3) $\int_s^\infty e^{-ct^2/2} \frac{dt}{t} \le \dfrac{e^{-cs^2/2}}{cs^2}, \quad \int_s^\infty t e^{-ct^2/2} \, dt = \dfrac{e^{-cs^2/2}}{cs}$

     $\int_s^\infty t^2 e^{-ct^2/2} \, dt < \dfrac{e^{-cs^2/2}}{c^2 s} (1 + cs^2).$

From now on, WGFFx ("well-known gamma function fact x") will refer to some statement in part (x) of this proposition. The only nontrivial part in the entire proposition is the case $k = 1$ of the second part of (1), which is a famous integral. Everything else follows from elementary integration or a sufficiently cunning application of the integration by parts formula

$$\int t^x e^{-ct^2/2} \frac{dt}{t} = \frac{-t^{x-2} e^{-ct^2/2}}{c} + \frac{x - 2}{c} \int t^{x-2} e^{-ct^2/2} \frac{dt}{t}.$$

For instance, the first WGFF3 follows from

$$\int_s^\infty e^{-ct^2/2} \frac{dt}{t} = \left[\frac{-e^{-ct^2/2}}{ct^2}\right]_s^\infty - \frac{2}{c} \int_s^\infty \frac{e^{-ct^2/2}}{t^3} \, dt. \tag{20.4}$$

20.2 Main Theorem

The main technical theorem of this chapter, Theorem 20.10, says that

$$\mathbf{Pr}(X[n] < 0) = \frac{1}{2} - \frac{L_-}{\sqrt{2\pi n}} + E(n), \quad L_- = L_-(\{na\}) = \frac{1/2 - \{na\}}{\sigma} - \frac{v_3}{6}$$

where explicit bounds on $E(n)$ are given. As alluded to earlier, the formula for L_- is "well-known" from Edgeworth expansions; the point of the theorem is of course the bound on E. The precise statement of this will be deferred until the end of this section. This can be used to give an analogous statement for the tilt; our aim is for this to be good enough to use in practice to find an n_0 such that the sign of T_n is constant for $n \geq n_0$.

The following subsections (1) introduce a "scale-invariant" version of the earlier notation and results, (2) outline the steps of the proof of the theorem, (3) methodically work through those steps, and then (4) finally give a full statement of the theorem.

20.2.1 Scale-Invariance

It is convenient to modify the notation slightly by introducing *scale-invariant* quantities where possible. Note that T_n is scale-invariant: it is unchanged if X is replaced by a multiple of X. We introduce scale-invariant versions of μ_k, α, $f(t)$, and $D(t)$ by advancing alphabetically:

$$v_k := \mu_k/\sigma^k \qquad\qquad g(t) := f(t/\sigma) = \sum v_k(it)^k/k!$$
$$E(t) := D(t/\sigma) = (t/(2\sigma))/\sin(t/2\sigma) \qquad \beta := \alpha/\sigma = (1/2 - \{na\})/\sigma$$

The key results of the preceding section using this notation are:

- The value of the cumulative distribution function of $X[n]$ at 0, in terms of an integral, Theorem 20.2, becomes

$$\mathbf{Pr}(X[n] < 0) = 1/2 - \frac{1}{2\pi i} \int_{-\pi\sigma}^{\pi\sigma} e^{i\beta t} g(t)^n E(t) \frac{dt}{t}. \qquad (20.5)$$

- The bound on the tail of the power series of the CF, Lemma 20.3, becomes

$$\left| \sum_{j \geq k} \frac{v_j(it)^j}{j!} \right| \leq \frac{\bar{v}_k t^k}{k!}. \qquad (20.6)$$

- Finally, the bound on the CF, Theorem 20.3, becomes (introducing a factor of 2 with an eye to the earlier gamma function facts)

$$|g(t)| \le 1 - \frac{rt^2}{2}, \quad \text{where } r = \frac{16m}{(\pi C \sigma)^2}, \quad \text{for } |t| \le \pi \sigma. \quad (20.7)$$

Since the power series for $g(t)$ starts out with $1 - t^2/2$, the constant r measures how "tractable" X is; if r is small then the tail integral estimate will be weak, and the closer r is to 1 the better the estimate will be.

20.2.2 Proof Outline

Throughout the proof, X is a bounded lattice random variable with mean 0, span 1, shift a, and scale-invariant CF function $g(t) = \mathbf{E} \, e^{itX/\sigma}$. The standard deviation is $\sigma = \sqrt{\mu_2}$, and scale-invariant central moments are $\nu_k = \mu_k/\sigma^k$. Moreover, when n is given, $\beta_n = \beta = (1/2 - \{na\})/\sigma$.

Fix a positive integer n and let s be a positive real number. During the proof various upper bounds will be placed on s, and it will be assumed throughout that they hold. The bounds on the various approximation errors are in practice smallest when s is as large as possible. As will be seen, s will in practice be a constant (that depends on X) times $n^{-1/4}$.

For the sake of a (reasonably) simple statement of error bounds, no attempt will be made to optimize various constants that arise. This seems appropriate in the motivating example of determining the sign of the tilt by the fact that nowadays a computer can calculate T_n for large n. Thus the point of the estimates is to enable a proof of when the asymptotic tilt has arrived so that, when combined with computation, the sign is known for all T_n. Thus finding the absolute best possible n_0 may not be important.

The technique for proving Theorem 20.10 is as follows. From (20.5) we know that $\mathbf{Pr}(X[n] < 0) = 1/2 - I_0$, where

$$I_0 := \frac{1}{2\pi i} \int_{-\pi\sigma}^{\pi\sigma} e^{i\beta t} g(t)^n E(t) \frac{dt}{t}.$$

This integral will be approximated by defining a sequence of further integrals I_k, $1 \le k \le 5$. Each I_k will be a reasonable approximation to I_{k-1}, and the error between them will be explicitly bounded. The last integral I_5 can be evaluated directly, and is equal to $L_-/\sqrt{2\pi n}$. The difference between I_0 and I_5 is of course bounded by the sum of bounds on the differences between consecutive integrals, leading to a bound on E_-.

The core idea of these approximations is that the dominant contribution to I_0 should come from a small interval around 0 whose size depends on n.

The *modus operandi* of the proof here is then summarized by the following sequence of approximations; the subscript on each approximation gives the number of the subsection in which the corresponding error is bounded

$$I_0 = \frac{1}{2\pi i} \int_{-\pi\sigma}^{\pi\sigma} e^{i\beta t} g(t)^n E(t) \frac{dt}{t}$$

$$\simeq_3 \frac{1}{2\pi i} \int_{-s}^{s} e^{i\beta t} g(t)^n E(t) \frac{dt}{t} \qquad \text{tail integral bound}$$

$$\simeq_4 \frac{1}{2\pi i} \int_{-s}^{s} e^{iu} e^{-nt^2/2} E(t) \frac{dt}{t}, \qquad u = \beta t - \frac{nv_3 t^3}{6}$$

$$= \frac{1}{2\pi} \int_{-s}^{s} \sin(u) e^{-nt^2/2} E(t) \frac{dt}{t}, \qquad \frac{\cos(u)}{t} \text{ is odd}$$

$$\simeq_5 \frac{1}{2\pi} \int_{-s}^{s} \sin(u) e^{-nt^2/2} \frac{dt}{t}, \qquad E(t) \simeq 1$$

$$\simeq_6 \frac{1}{2\pi} \int_{-s}^{s} u e^{-nt^2/2} \frac{dt}{t}, \qquad \sin(u) \simeq u$$

$$\simeq_7 \frac{1}{2\pi} \int_{-\infty}^{\infty} (\beta - nv_3 t^2/6) e^{-nt^2/2} dt, \quad \text{another tail bound.}$$

20.2.3 Bounding the Tail Integral

For large n the dominant contribution to the integral

$$I_0 = \frac{1}{2\pi i} \int_{-\pi\sigma}^{\pi\sigma} e^{i\beta t} g(t)^n E(t) \frac{dt}{t},$$

should come from a small neighborhood of the origin. In fact the approximation

$$g(t)^n \simeq (1 - t^2/2)^n \simeq e^{-nt^2/2}$$

suggests that the width of the neighborhood might be on the order of $1/\sqrt{n}$. Let $s \leq \pi\sigma$ be an arbitrary positive real number and define

$$I_1 = \frac{1}{2\pi i} \int_{-s}^{s} e^{i\beta t} g(t)^n E(t) \frac{dt}{t}.$$

The parameter s will be chosen later, and its optimal value will actually turn out to be $O(n^{-1/4})$.

Recall from (20.7) that $|g(t)| \leq 1 - rt^2/2$ for t on the interval of integration, where the constant r was defined earlier to be $r = 16m/(\pi C\sigma)^2$, m is the smallest probability of a point on the support of some certificate, and C is the L_1-norm of that certificate.

Theorem 20.4.

$$|I_0 - I_1| \leq \frac{\exp(-nrs^2/2)}{2nrs^2}.$$

Proof. Use $|g(t)| \leq 1 - rt^2/2$ and $1 - x \leq e^{-x}$ to get

$$|g(t)|^n \leq (1 - rt^2/2)^n \leq \exp(-nrt^2/2).$$

The difference $I_0 - I_1$ is the sum of a right tail and a left tail integral that are bounded in exactly the same way. So it suffices to multiply the bound on the upper tail by 2, Recall the upper bound $|E(t)| \leq \pi/2$, and use the bound on $g(t)^n$:

$$|I_1 - I_0| \leq 2 \left| \frac{1}{2\pi i} \int_s^{\pi\sigma} e^{i\beta t} g(t)^n E(t) \frac{dt}{t} \right|$$

$$\leq 2 \cdot \frac{1}{2\pi} \cdot \frac{\pi}{2} \int_s^\infty e^{-nrt^2/2} \frac{dt}{t} \leq \frac{e^{-nrs^2/2}}{2nrs^2}$$

where the last inequality is WGFF3. $\qquad\square$

20.2.4 Bounding the Power Series Tail

In order to analyze the $g(t)^n$ term in the integrand it is convenient to introduce notational shorthand for terms and tails of the power series of $g(t)$:

$$g_j := \frac{v_j(it)^j}{j!}, \quad G_k := \sum_{j \geq k} g_j.$$

Although this lighter notation is pleasant it is important to remember that g_k and G_k depend on t. Note that (20.6) says that $|G_k| \leq |g_k|$ for even k, and $|G_k| \leq \bar{v}_k |t|^k/k!$ for odd k.

The critical term in the integrand of I_1 is $g(t)^n$, and the purpose of this section is to bound the error incurred in the approximations

$$g(t)^n = \exp(n \log(1 + G_2)) \simeq \exp(nG_2) \simeq \exp(n(g_2 + g_3)).$$

It is convenient to introduce further notation. Let

$$q_1 := \frac{1}{5} + \frac{v_4}{24}.$$

Motivated by replacing $g(t) = \exp(\log(1 + G_2))$ by $\exp(g_2 + g_3)$, define a "remainder" R by

$$R = \log(1 + G_2) - g_2 - g_3.$$

This should be small if n is large.

The following obvious bound (OB) on tails of power series with positive coefficients will be used three times in the text that follows; the proof is embedded in the statement of the lemma (!).

Lemma 20.5. *(OB) Let $P(x) = p_k x^k + p_{k+1} x^{k+1} + \ldots$ be a power series power with nonnegative coefficients p_j, and suppose that $P(y)$ converges for some positive real number y. Then if $|x| \leq y$,*

$$|P(x) - p_k x^k| = |x|^{k+1} \cdot \left| \frac{P(x) - p_k x^k}{x^{k+1}} \right| \leq |x|^{k+1} \left| \frac{P(y) - p_k y^k}{y^{k+1}} \right|. \qquad (20.8)$$

Theorem 20.5. *Let s be a positive real number and assume throughout that $|t| \leq s$. If $s \leq 1$ then $|G_2| \leq 1/2$, and the power series for $\log(g(t)) = \log(1 + G_2)$ converges. Moreover,*

$$|R| \leq q_1 t^4, \quad \text{where } q_1 := \frac{1}{5} + \frac{v_4}{24}, \quad R = \log(1 + G_2) - g_2 - g_3.$$

If also $s \leq (q_1 n)^{-1/4}$, i.e., $n q_1 s^4 \leq 1$, then

$$\left| e^{nR} - 1 \right| \leq n p_0 |R| \leq n p_0 q_1 t^4, \quad \text{where } p_0 := e - 1 \simeq 1.71828.$$

Proof. Part 1 follows from $|G_2| \leq g_2 = t^2/2 \leq s^2/2 \leq 1/2$ and the fact that the logarithm series

$$\log(1 + G_2) = G_2 - G_2^2/2 + G_2^3/3 - \ldots$$

converges by comparison with a geometric series of ratio $1/2$.

For the second part, first note that

$$|R| \leq |\log(1 + G_2) - G_2| + |G_2 - g_2 - g_3|.$$

The second term is just $|G_4| \leq g_4 = v_4 t^4/24$. The first term can be bounded by applying the OB to $P(x) = -\log(1 - x) = \sum x^k/k$ with $k = 1$, $x = -G_2$ and $y = 1/2$. Note that $|x| = |G_2| \leq t^2/2 \leq s^2/2 \leq 1/2$. The OB gives

$$|-\log(1 + G_2) + G_2| \leq G_2^2 \left(\frac{-\log(1 - 1/2) - 1/2}{(1/2)^2} \right)$$

$$= \frac{t^4}{4} \left(\frac{\log(2) - 1/2}{1/4} \right) \leq \frac{t^4}{5}.$$

(By choosing an even smaller bound on s, this could be made as close to $t^4/8$ as desired, but no lower.) All in all this gives

$$|\log(1 + G_2) - g_2 - g_3| \leq q_1 t^4, \quad q_1 = \frac{1}{5} + \frac{v_4}{24}$$

as desired.

For the third part, note that the assumed bound on s implies that

$$|nR| \leq nq_1 t^4 \leq nq_1 s^4 \leq 1.$$

Apply OB to $P(x) = e^x$ with $k = 0$, $x = nR$ and $y = 1$ to get

$$\left|e^{nR} - 1\right| = |nR| \left(\frac{e^1 - 1}{1}\right) \leq np_0 q_1 t^4, \quad p_0 := e - 1$$

as desired. $\qquad \square$

Before using this lemma for the central goal of this section — bounding the difference between I_1 and a soon-to-be defined I_2 — we prove a corollary that will be used later to improve the tail integral bound in the previous subsection.

Corollary 20.1. *With the preceding notation, if $s \leq 1$ then*

$$\int_s^1 |g(t)|^n \frac{dt}{t} \leq e^{-ns^2/2} \left(\frac{1}{ns^2} + \frac{2p_0 q_1}{n} + p_0 q_1 s^2\right).$$

Proof. The last part of the theorem says that

$$|g(t)|^n = \left|e^{ng_2 + ng_3} \left(1 + (e^{nR} - 1)\right)\right| \leq e^{-nt^2/2} \left(1 + \theta n p_0 q_1 t^4\right)$$

so that

$$\int_s^1 |g(t)|^n \frac{dt}{t} \leq \int_s^1 e^{-nt^2/2} \frac{dt}{t} + np_0 q_1 \int_s^1 t^4 e^{-nt^2/2} \frac{dt}{t}.$$

The first of the integrals on the right-hand side can be estimated by (extending the interval to infinity and) using the first WGFF3, and the second can be evaluated exactly by taking the difference of two instances of the second WGFF2. The result is

$$\int_s^1 |g(t)|^n \frac{dt}{t} \leq \left(\frac{e^{-ns^2/2}}{ns^2} + \frac{2np_0 q_1\, e^{-ns^2/2}}{n^2} \left(1 + \frac{ns^2}{2}\right)\right)$$

which simplifies to the expression in the corollary. $\qquad \square$

Returning to the problem of approximating I_1, note that the first two factors in the integrand can be written

$$\begin{aligned}
e^{i\beta t} g(t)^n &= \exp(i\beta t + n\log(1 + G_2)) \\
&= \exp(i\beta t + ng_2 + ng_3 + nR) \\
&= \exp(iu)\exp(-nt^2/2)\exp(nR)
\end{aligned}$$

where the quantity

$$u = \beta t - nv_3 t^3/6$$

captures the key imaginary terms. This motivates the definition of the next integral:

$$I_2 := \frac{1}{2\pi i} \int_{-s}^{s} e^{iu} \, e^{-nt^2/2} E(t) \frac{dt}{t}.$$

Theorem 20.6. *If $s \le \min(1, (q_1 n)^{-1/4})$ (so that previous theorem holds) then*

$$|I_1 - I_2| \le \frac{p_0 q_1}{2n},$$

Proof. From everything in the foregoing (e.g., as in the proof of Theorem 20.4)

$$|I_1 - I_2| \le \frac{1}{2\pi} \int_{-s}^{s} \left| e^{nR} - 1 \right| e^{-nt^2/2} E(t) \frac{dt}{t}$$

$$\le 2 \cdot \frac{1}{2\pi} \cdot \frac{\pi}{2} \cdot n p_0 q_1 \int_{0}^{s} t^4 \, e^{-nt^2/2} \frac{dt}{t}.$$

WGFF2 says that the last integral is equal to

$$\frac{1}{n^2} \left(1 - e^{-ns^2/2} \left(1 + ns^2/2 \right) \right).$$

An easy calculus exercise shows that the factor in parentheses is between 0 and 1, and the upshot is that the $|I_1 - I_2|$ is bounded by $p_0 q_1/(2n)$ as claimed. ∎

Remark. The last factor could be retained explicitly, giving a better estimate, especially when n is small. However, we ignore this because of (1) the general philosophy of not worrying too much about small constant factors, and (2) if one is struggling with having to take n large in an unfavorable situation then the factor will be very close to 1 anyway.

20.2.5 Eliminating E

Define I_3 to be the result of erasing $E(t)$ in the integrand of I_2:

$$I_3 = \frac{1}{2\pi i} \int_{-s}^{s} e^{iu} \, e^{-nt^2/2} \frac{dt}{t}. \tag{20.9}$$

Theorem 20.7. *If $s \le \sigma \pi/3$ then*

$$|I_2 - I_3| \le \frac{p_1}{\mu_2 n}, \quad \text{where } p_1 := \frac{3(\pi - 3)}{\pi^3} \simeq .0136997 \ldots.$$

Proof. The power series $t/\sin(t)$ is even and has positive coefficients, so we apply the OB to it and get

$$\frac{t}{\sin(t)} - 1 \le t^2 \left(\frac{s/\sin(s) - 1}{s^2} \right), \quad \text{if } |t| \le s.$$

Taking $s = \pi/6$ (for simplicity) gives $t/\sin(t) - 1 \le 12(\pi - 3)/\pi^2 t^2$ if $|t| \le \pi/6$. Replacing t by $t/(2\sigma)$ and doing a little algebra gives

$$|E(t) - 1| \le \frac{\pi p_1 t^2}{\mu_2}, \quad \text{if } |t| \le \pi\sigma/3.$$

The theorem now follows from WGFF2:

$$|I_2 - I_3| \le \frac{1}{\pi} \int_0^s e^{-nt^2/2} \frac{\pi p_1 t^2}{\mu_2} \frac{dt}{t} \le \frac{p_1}{\mu_2} \int_0^\infty e^{-nt^2/2} t^2 \frac{dt}{t} = \frac{p_1}{\mu_2 n}.$$

$\square$

20.2.6 Eliminating sin

The integral of the odd function $\cos(u)/t$ on the symmetric interval $t \in [-s, s]$ is 0 (using the earlier principal value convention). Then $e^{iu} = \cos(u) + i\sin(u)$ can be replaced $i\sin(u)$ and therefore

$$I_3 = \frac{1}{2\pi} \int_{-s}^s \sin(u) e^{-nt^2/2} \frac{dt}{t}. \tag{20.10}$$

For small t, the quantity $u = \beta t - n v_3 t^3/6$ is small, and the approximation $\sin(u) \simeq u$ motivates defining

$$I_4 := \frac{1}{2\pi} \int_{-s}^s u e^{-nt^2/2} \frac{dt}{t}. \tag{20.11}$$

Theorem 20.8.

$$|I_3 - I_4| \le \frac{q_5}{\sqrt{2\pi} \, n^{3/2}}$$

where

$$q_3 = |\beta|, \quad q_4 = |v_3|/6, \quad q_5 = \frac{q_3^3}{6} + \frac{3 q_3^2 q_4}{2} + \frac{15 q_3 q_4^2}{2} + \frac{35 q_4^3}{2}.$$

Proof. For any real number x, $|\sin(x) - x| \le |x|^3/6$. Applying this to $u = \beta t - n v_3 t^3/6$ gives

$$|\sin(u) - u| \le \frac{|u|^3}{6} \le \frac{1}{6} \left(q_3^3 t^3 + 3 q_3^2 q_4 n t^5 + 3 q_3 q_4^2 n^2 t^6 + q_4^3 n^3 t^9 \right).$$

The claimed inequality follows by integrating and using the second WGFF1, i.e.,

$$\int_0^\infty t^{2k-1} e^{-ct^2/2} \frac{dt}{t} = c^{-k+1} (2k-3)!! \sqrt{\frac{\pi}{2c}}, \tag{20.12}$$

for $k = 2, 3, 4, 5$ to get

$$
\begin{aligned}
|I_4 - I_3| &\leq \frac{1}{2\pi} \int_{-s}^{s} \frac{|\dot{u}|^3}{6} e^{-nt^2/2} \frac{dt}{t} \\
&\leq \frac{1}{6\pi} \int_{0}^{\infty} \left(q_3^3 t^3 + 3q_3^2 q_4 nt^5 + 3q_3 q_4^2 n^2 t^6 + q_4^3 n^3 t^9 \right) e^{-nt^2/2} \, dt \\
&= \frac{1}{6\pi} \sqrt{\frac{\pi}{2n}} \left(\frac{q_3^3 + 9q_3^2 q_4 + 45q_3 q_4^2 + 105q_4^3}{n} \right).
\end{aligned}
$$

$\square$

20.2.7 Extending to R

The next (and final!) integral is obtained by extending the interval of integration to the whole real line, i.e.,

$$
I_5 := \frac{1}{2\pi} \int_{-\infty}^{\infty} u \, e^{-nt^2/2} \frac{dt}{t}.
$$

This integral can be evaluated immediately using WGFF1 given earlier for $k = 1$ and $k = 2$:

$$
\begin{aligned}
I_5 &= \frac{1}{2\pi} \int_{-\infty}^{\infty} (\beta - n\nu_3 t^2/6) \, e^{-nt^2/2} \, dt \\
&= \frac{\beta}{\pi} \int_{0}^{\infty} e^{-nt^2/2} \, dt - \frac{n\nu_3}{6\pi} \int_{0}^{\infty} t^2 e^{-nt^2/2} \, dt \\
&= \frac{1}{\sqrt{2\pi n}} \left(\beta - \frac{\nu_3}{6} \right) = \frac{L_-}{\sqrt{2\pi n}}.
\end{aligned}
$$

Theorem 20.9.

$$
|I_4 - I_5| \leq \left(\frac{q_3 + q_4 + q_4 ns^2}{\pi ns} \right) e^{-ns^2/2}.
$$

Proof. The two tails of $I_5 - I_4$ are equal. Multiplying by 2 and using the last two parts of WGFF3 gives

$$
\begin{aligned}
|I_5 - I_4| &\leq \frac{1}{\pi} \int_{s}^{\infty} (q_3 + q_4 nt^2) \, e^{-nt^2/2} \, dt \\
&\leq \frac{q_3}{\pi} \cdot \frac{e^{-ns^2/2}}{ns} + \frac{q_4 n}{\pi} \cdot \frac{e^{-ns^2/2}}{n^2 s} \cdot (1 + ns^2) \\
&= (q_3 + q_4(1 + ns^2)) \frac{e^{-ns^2/2}}{\pi ns}.
\end{aligned}
$$

$\square$

20.2.8 Finishing the Proof

During the proof the constant s was required to be smaller than $\pi\sigma$, 1, $\pi\sigma/3$, and $1/(q_1 n)^{1/4}$, which can be summarized as

$$'s \le \min(1, \pi\sigma/3, (q_1 n)^{-1/4}). \qquad (20.13)$$

We remind the reader of the notation and then state the main theorem: X is a lattice random variable with mean 0, span 1, shift a, and finitely many values; $X[n]$ denotes the sum of n IID copies of X. The standard deviation of X is σ, and the moments and normalized moments are denoted $\mu_k = \mathbf{E}X^k$, $v_k = \mu_k/\sigma^k$. Moreover, $\beta = \beta_n$ is shorthand for $(1 - \{na\})/\sigma$. Various further constants are defined as follows:

$$p_0 = e - 1 \simeq 1.71828\ldots \qquad\qquad p_1 := 3(\pi - 3)/\pi^3 \simeq .0136997\ldots$$

$$q_1 := \frac{1}{5} + \frac{v_4}{24} \qquad\qquad q_2 := \frac{p_0 q_1}{2} + \frac{p_1}{\mu_2}$$

$$q_3 := |\beta| \qquad\qquad q_4 = |v_3|/6$$

$$q_5 := \frac{q_3^3}{6} + \frac{3 q_3^2 q_4}{2} + \frac{15 q_3 q_4^2}{2} + \frac{35 q_4^3}{2} \qquad r := \frac{16m}{(\pi C\sigma)^2}$$

where, as earlier, C is the L_1 norm of a certificate $\{c_x\}$ for X (i.e., a guarantor that X has span 1), and $m = \min(p_x)$ is the minimum probability of the values of X that are in the support of the certificate.

Theorem 20.10. *Fix a positive integer n and let s be a positive real number that satisfies (20.13). Then*

$$\mathbf{Pr}(X[n] < 0) = \frac{1}{2} - \frac{L_-}{\sqrt{2\pi n}} + E_-$$

where

$$L_- = \beta - \frac{v_3}{6}$$

and

$$|E_-| \le \frac{q_2}{n} + \frac{e^{-nr/2}}{2nr} + \frac{q_5}{\sqrt{2\pi}n^{3/2}} + e^{-ns^2/2}$$

$$\left(p_0 q_1 s^2 + \frac{1}{ns^2} + \frac{2 p_0 q_1}{n} + \frac{q_3 + q_4}{\pi ns} + \frac{q_4 s}{\pi}\right).$$

Proof. The bound (20.13) guarantees that all of the results that required s to be small enough are valid. If $I(j, j+1)$ denotes the bound proved earlier on the difference between I_j and I_{j+1} then

$$|E_-| \le I(0, 1) + I(1, 2) + I(2, 3) + I(3, 4) + I(4, 5).$$

The last four terms were proved in Theorem 20.6, Theorem 20.7, Theorem 20.8, and Theorem 20.9 respectively, giving

$$I(1, 2) + I(2, 3) + I(3, 4) + I(4, 5) = \frac{q_2}{n} + \frac{q_5}{\sqrt{2\pi}\, n^{3/2}}$$
$$+ e^{-ns^2/2} \left(\frac{q_3 + q_4 + q_4 n s^2}{\pi n s} \right).$$

We improve the bound for $|I_0 - I_1|$ given in Theorem 20.4 by writing

$$\int_s^{\pi\sigma} |g(t)|^n \frac{dt}{t} = \int_s^1 |g(t)|^n \frac{dt}{t} + \int_1^{\pi\sigma} |g(t)|^n \frac{dt}{t}$$

and apply Theorem 20.4 to the last integral and Corollary 20.1 to the integral from s to 1. The result is

$$|I_1 - I_0| \le e^{-ns^2/2} \left(\frac{1}{ns^2} + \frac{2 p_0 q_1}{n} + p_0 q_1 s^2 \right) + \frac{e^{-nr/2}}{2nr}.$$

The theorem follows with a little algebra.   □

In the next section this will be used to state a theorem about the tilt, and some simple assumptions will be made that will simplify and clarify this expression.

One possible major improvement to the estimate would come from using the next higher order Edgeworth expansion. The error term would become $O(1/n^2)$. However, our hunch is that this would introduce two further problems: the number of terms in the algebraic expressions would explode, and the upper bound on s would probably be $O(n^{-1/6})$, so it is not immediately clear that the resulting estimate of n_0 would be much better.

20.3 Tilt

The goal of this section is to apply the Main Theorem to the tilt.

The first subsection makes a few observations about the error bounds. The second reverts to the case of dice, giving a theorem about the tilt. Finally, the third subsection considers some numerical examples.

20.3.1 Observations

With the notation in Theorem 20.10, fix n and consider the error bound as a function of s. The only two terms that are not obviously decreasing as s

increases are (constant multiples of) $se^{-ns^2/2}$ and $s^2e^{-ns^2}/2$. Differentiating with respect to s shows that these are both decreasing if $ns^2 \geq 2$. Since this is a very mild assumption, we will just assume it from now on. This implies that the optimal s is just the smallest value in (20.13).

In that bound

$$s \leq \min(1, \pi\sigma/3, (q_1n)^{-1/4}).$$

the last quantity will usually be the smallest. To simplify and focus the notation, assume that this is the case, i.e., that

$$\frac{1}{(q_1n)^{1/4}} \leq \min(1, \pi\sigma/3).$$

Assuming this, requiring $ns^2 \geq 2$ as earlier, setting $s = (q_1n)^{-1/4}$, and adopting the notational shorthand

$$\eta = 2\sqrt{n/q_1}$$

(so that $ns^2/2 = \eta$), allows a restatement of the error bound as follows: If

$$n \geq \max\left(\frac{q_1}{4}, \frac{1}{q_1}, \frac{81}{q_1\pi^4\mu_2^2}\right)$$

then

$$|E_-| \leq \frac{q_2}{n} + \frac{e^{-nr/2}}{2nr} + \frac{q_5}{\sqrt{2\pi}n^{3/2}} + e^{-\eta}$$
$$\times \left(\frac{p_0+1}{2\eta} + \frac{2p_0q_1}{n} + \frac{1}{\pi\sqrt[4]{q_1n}}\left(\frac{q_3+q_4}{2\eta} + q_4\right)\right).$$

For the sake of the next section we note that if X is replaced by $-X$ in the theorem then, as should be expected, the bounds on the error are unchanged. Indeed, the only relevant change is that the sign of a is negated, so that q_3 might change. However, if $x = \{na\}$ then and $y = \{-na\}$ then either $x = y = 0$, or $y = 1 - x$. If $x = y = 0$ then q_3 is obviously unchanged. If $y = 1 - x$ then

$$q_3 = \frac{|1/2 - x|}{\sigma} = \frac{|-1/2 + y|}{\sigma} = \frac{|1/2 - y|}{\sigma}$$

and q_3 is again unchanged.

20.3.2 Tilt for Dice

For the sake of applications it is convenient to return to looking at the tilt in the case of dice. Let X be a bounded integer-valued random variable with span b, shift a, mean 0, and moments $\mu_k = \mathbf{E} X^k$, $\sigma^2 = \mu_2$, and $\nu_k = \mu_k/\sigma^k$.

Apply the Main Theorem (using the reformulation in the preceding section) to X/b. Scale-invariant quantities are unchanged, but $\sigma(X/b) = \sigma/b$. The quantity β is unchanged, but note that

$$\beta = \frac{1/2 - \{na/b\}}{\sigma/b} = \frac{b/2 - b\{na/b\}}{\sigma} = \frac{b/2 - na \bmod b}{\sigma}.$$

The other constants are changed only insofar as σ has to be replaced by σ/b:

$$p_0 = e - 1 \simeq 1.71828\ldots \qquad\qquad p_1 := 3(\pi - 3)/\pi^3 \simeq .0136997\ldots$$

$$q_1 := \frac{1}{5} + \frac{\nu_4}{24} \qquad\qquad q_2 := \frac{p_0 q_1}{2} + \frac{b^2 p_1}{\mu_2}$$

$$q_3 := |\beta| \qquad\qquad q_4 = |\nu_3|/6$$

$$q_5 := \frac{q_3^3}{6} + \frac{3q_3^2 q_4}{2} + \frac{15 q_3 q_4^2}{2} + \frac{35 q_4^3}{2} \qquad r := \frac{16 b^2 m}{(\pi C \sigma)^2}$$

where, as earlier, m is the minimum positive probability p_x of a value, and $C = \sum |c_x|$ is the L_1 norm of a certificate $\{c_x\}$ where $\sum c_x = 0$, $\sum c_x x = b$. Note that the inclusion of b in the formulae does not actually change the values of the constants. For instance, in the span-1 case of X/b the standard deviation is σ/b, where now "σ" refers to the span of X and not that of X/b.

Theorem 20.11. *Fix n with $n \geq \max\left(q_1/4, 1/q_1, 81b^4/(q_1\pi^4\mu_2^2)\right)$. Then*

$$T_n = \frac{1}{\sqrt{2\pi n}} \left(\frac{(-na) \bmod b - (na \bmod b)}{\sigma} - \frac{\nu_3}{3} \right) + E$$

where

$$|E| \leq \frac{2q_2}{n} + \frac{e^{-nr/2}}{nr} + \frac{2q_5}{\sqrt{2\pi}n^{3/2}} + e^{-\eta}$$

$$\times \left(\frac{1 + p_0}{\eta} + \frac{4 p_0 q_1}{n} + \frac{1}{\pi \sqrt[4]{q_1 n}} \left(\frac{q_3 + q_4}{\eta} + 2q_4 \right) \right)$$

and $\eta = 2\sqrt{n/q_1}$ so that $ns^2/2 = \eta$.

Proof. The theorem follows from the fact that

$$T_n = \mathbf{Pr}(X[n] > 0) - \mathbf{Pr}(X[n] < 0) = \mathbf{Pr}((-X)[n] < 0) - \mathbf{Pr}(X[n] < 0)$$

and the earlier remarks on the error bounds.   □

20.3.3 Examples

We apply the foregoing results to several examples, comparing the actual n_0 at which asymptopia arrives to various approximations that emerge from Theorem 20.11.

Let X be the random variable that takes values $-3, 1, 5$ with respective probabilities $1/2, 1/4, 1/4$, so that its PGF is

$$X(z) = (2z^{-3} + z + z^5)/4$$

(it is convenient to identify dice with their PGFs). Then X has mean 0, span 4 and shift 1, so there are really four cases that have to be considered: n going to infinity through integers that are $c \bmod 4$ for $c = 0, 1, 2, 3$; in the table that follows data connected with case c is on the line labeled X_c.

With c fixed, let n_0 be the smallest integer such that the sign of T_n is equal to the sign of L for all $n \geq n_0$, $n \equiv c \bmod b$, i.e., n_0 is the exact point at which asymptopia has arrived in the congruence class c.

The term $2q_2/n$ in the error bound for $T_n - L/\sqrt{2\pi n}$ is unavoidable in any bound obtained by using Edgeworth expansions as earlier, and we will call this the "principal term" of the error, motivated by the fact that in sufficiently favorable circumstances it will be the dominant term. In particular, it is impossible for us to *prove* that asymptopia has arrived unless n is large enough so that

$$\sqrt{2\pi n}\, \frac{2q_2}{n} < |L|, \quad \text{i.e.,} \quad n \geq n_1 := \frac{8\pi q_2^2}{L^2}.$$

Thus n_1 is a lower bound on asymptopia arrival that could be proved by our techniques. Given the nature of some of the error bounds, it is reasonable to expect that n_1 might might actually be larger than n_0 in many cases; however, a later example gives an instance where n_1 is lower than n_0.

Finally, let n_2 be the number, produced by using the error estimates for a fixed congruence class modulo b directly as stated in Theorem 20.11, such that for $n \geq n_2$ (and in the given congruence class), $\sqrt{2\pi n}\, EB(n) < |L|$, where $EB(n)$ is the error bound in the theorem. In other words, Theorem 20.11 can be used to show that asymptopia has arrived by n_2 in the given congruence class.

The values found using (a computer and) Theorem 20.11 are

	L	n_0	n_1	n_2
X_0	-0.16446	4	59	74
X_1	0.43856	5	9	37
X_2	-0.16446	2	59	70
X_3	-0.76748	3	3	27

As expected, smaller values of $|L|$ require larger n. A close examination of the seven terms in the error bound show that the last five rapidly become negligible as n becomes large, whereas the first two terms — the principal term and the tail bound, $e^{-rn/2}/(rn)$, are significant. The tail bound can be decreased, as we will see later, by looking at the tail integral more closely. However, in the X_c cases the amount of computer time required to compute the tilts up to n_2 is negligible, so we will not bother trying to improve the tail bound in these cases, despite the gap between n_0 and n_2.

Let Y be the random variable that takes the value -8 with probability $1/2$, the value 0 with probability $1/18$, and the value 9 with probability $4/9$; its PGF is

$$Y(z) = (9z^{-8} + 1 + 8z^9)/18.$$

The span of Y is 1. One (of the several) indications that this might be a problematic case is that the probability of 0 is small, and the span changes to 17 if this probability is set to 0 (while suitably rebalancing the other two probabilities).

To give a sense of the various components of the error, write

$$\text{EB} = \frac{2q_2}{n} + \frac{e^{-rn/2}}{rn} + \text{TR}$$

where EB is the total error bound and TR ("the rest") is the sum of five other terms. In the following table n is either close to $n_0 = 761$, $n_1 = 682$, or $n_2 = 182024$. The columns are: n, the total error EB, the principal error, and the tail error. (All errors have been multiplied by $\sqrt{2\pi n}$ to make comparison with L easy). In all cases, TR is less than 10^{-5} and it is not tabulated. All decimal expansions are *truncated* rather than rounded, and the constant L is equal to $-0.0404226\ldots$.

n	EB	$2q_2/n$	$e^{-rn/2}/(rn)$
681	1128.163	0.0404310	1128.122
682	1127.289	0.0404013	1127.248
761	1063.690	0.0382468	1063.652
182023	0.040423	0.0024730	0.0379500
182024	0.040421	0.0024729	0.0379483

The delicate nature of this case is illustrated by the graph of $\sqrt{2\pi n}\, T_n$ for n from 1 to 800, and for n from 740 to 800, given in Figs. 20.1 and 20.2.

Clearly Y "nearly" has span 17, so that the graph is the result of applying a damping function to a function that is periodic of period 17. However, at $n_0 = 761$ the damping finally forces to the tilt to become, and forever stay, negative. The actual numerical values in the vicinity of n_0, and the next local

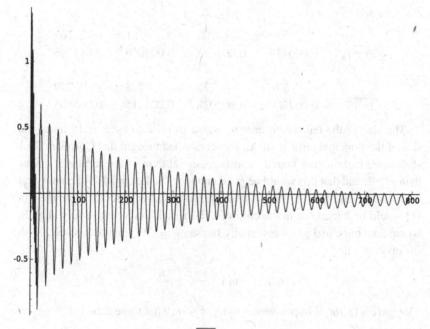

Fig. 20.1. $\sqrt{2\pi n}\, T_n$, $1 \le n \le 800$.

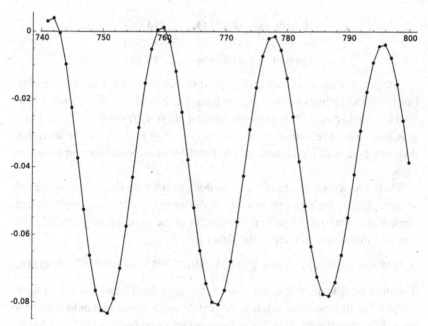

Fig. 20.2. $\sqrt{2\pi n}\, T_n$, $740 \le n \le 800$.

maxima, are

n	759	760	761	762
$\sqrt{2\pi n}\, T_n$	0.000439	0.001195	−0.003066	−0.011796

$$\cdots$$

n	776	777	778	779
$\sqrt{2\pi n}\, T_n$	−0.007300	−0.002028	−0.001415	−0.005505

The size of the tail bound near n_0 was a surprise to us; it is 15 times the size of the principal error term. Moreover, we had thought that Corollary 20.1 shifted the extreme tail bound to an exponential term of the form e^{-cn} rather than $e^{-c\sqrt{n}}$, and that this would be good enough. The key point is of course that the constant r is uncomfortably small. We wondered whether the alternate r in [1] would be better, but in the case of Y Benedick's constant is very slightly worse than ours, and gives essentially the same n_2. We tried to replace r with the optimal value

$$r_{\text{opt}} = \min_{t \in [-\pi, \pi]} \frac{1 - |f(t)|}{t^2}.$$

This gave a factor of improvement to n_2 of somewhat more than 2:

	r	n_2
Benedicks	.0026144...	194,081
ours	.0028144...	182,024
optimal	.0055834...	88,181

The graph of the absolute value $|f(t)|$ of the CF is given in Fig. 20.3; in addition to $|f(t)|$, the lines $x = \pm 1/\sigma$, and the upper bound $1 - rt^2$ are shown. The goal is to find an upper bound for the integral of $|f(t)|^n/t$ outside $[-1/\sigma, 1/\sigma]$, as n gets large. The natural upper bound $(1 - rt^2)^n \le e^{-nrt^2}$ can be used, but does not work well for a characteristic function whose secondary peaks are so high.

There are several things that can be done to improve the bound on the tail integral. The following extremely simple device made a dramatic improvement compared to the n_2 given earlier. The heights of the peaks in the graph of $|f(t)|$ (starting from 0 and moving to the right) are

$$1, 0.88989, 0.99645, 0.89768, 0.98621, 0.91204, 0.97048, 0.93077, 0.95118.$$

The third peak is dominant. It occurs at $t = t_0 = 4\pi/17$ and has height $h_0 = .99645$. The fifth peak has height $h_1 = .98621$ and is the second most dominant peak. Let f^* be the function that is the constant h_1 on $[-\pi, -1/\sigma] \cup [1/\sigma, \pi]$ except for the small section of the parabola $h_0 \cdot (1 - 34(t - t_0)^2)$ that is above

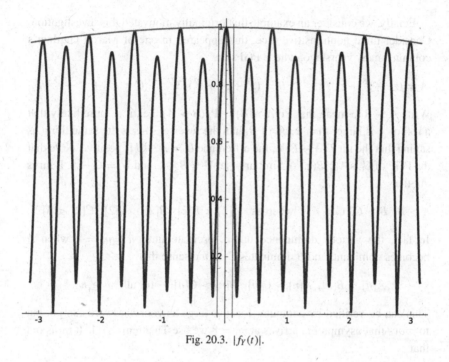

Fig. 20.3. $|f_Y(t)|$.

the line $y = h_1$, and the mirror image of this parabola section at $t = -t_0$. One can check that f^* is an upper bound on $|f|$ outside of $[-1/\sigma, 1/\sigma]$, and it is easy to estimate the integral of $f^*(t)^n/|t|$ on $1/\sigma \le |t| \le \pi$.

Using this tail bound we get an improved $n'_2 = 1455$, which was much smaller than we had expected. The only significant contributions to the error are the $2q_2/n$ term and the tail bound, which are, respectively, 0.02766, and 0.012709 so that the total error is just less than $|L|$:

$$0.0276602867 + 0.01270946 = .040370408 < |L| = .0404226.$$

All other contributions to the error are less than 10^{-6}, and the tail bound has been reduced to under half of the primary contribution.

One can push this further by considering all of the peaks and lowering the "threshold" to, say, the value of the CF at $1/\sigma$, i.e., the value of the normalized CF $g(t)$ at $t = 1$. This seems to move n'_2 further down to an n''_2 just above 1200 (and we think that this is about the limit of what can be done). However, the time required to program this correctly far exceeds the time that a computer takes to compute the tilts between 1200 and 1455, so our general philosophy says that it would be silly to implement this improvement.

Finally, we consider an example that originally motivated this investigation. Consider three nontransitive dice, that appeared in one of Martin Gardner's columns many years ago, whose PGFs are

$$A = (z^2 + z^6 + z^7)/3, \qquad B = (z^1 + z^5 + z^9)/3, \qquad C = (z^3 + z^4 + z^8)/3.$$

Write $A > B$ to mean that $\Pr(A > B) > \Pr(B > A)$, i.e., it is more likely that a roll of A is larger than a roll of B than the reverse. This is the same thing as saying that the tilt $T(A - B)$ of the difference $U = A - B$ is positive. Note that the PGF of U is $A(z)B(z^{-1})$. Similarly, let $V = B - C$ and $W = C - A$. It turns out that

$$A > B, \ B > C, \ C > A, \quad \text{whereas} \quad A[2] < B[2], \ B[2] < C[2], \ C[2] < A[2].$$

In fact, the various dominance orders oscillate until $n = n_0 = 8$ when A becomes dominant, and B dominates C, in the sense that

$$A[n] > B[n], \ A[n] > C[n], \ B[n] > C[n], \quad \text{for all} \quad n \geq n_0.$$

This can be verified by computer for n as large as your hardware can go, but to *prove* that asymptopia arrives at $n_0 = 8$ we use Theorem 20.11. It turns out that

$$U(z) = A(z)B(z^{-1}) = B(z)C(z^{-1}) = V(z)$$

so there are really only two dice to which that theorem has to be applied: U and W. By now we can give a guess as to how large n will have to be, namely, we find the needed error $|L|$ and then choose n large enough so that

$$\sqrt{2\pi n} \, T_n \simeq \sqrt{2\pi n} \, \frac{2q_2}{n} < |L|, \quad \text{i.e.,} \quad n > \frac{8\pi q_2^2}{L^2}.$$

Since $q_2 = 1/5 + v_4/24$, for many random variables it is reasonable to bound q_2 by $1/4$, so that n has to be at least as large as $\pi/(2L^2)$.

For W we find that $L = 0.033310\ldots$, which is unusually small. The approximation $n \simeq 8\pi q_2^2/L^2$ suggests $n_0 \leq 1407$, and in fact in this case L is so small that n is large enough so that all other terms of the error are negligible. In other words, the n required by the principal error term so large that this gives the best possible value.

For U, the limit L is larger, namely $L \simeq -0.14028\ldots$. In this case the bound implied by the principal term is $n = 83$. Although the main tail bound term is then large, we can apply the earlier techniques of piecewise bounding the characteristic function, to show that $n_0 \leq 83$. In other words, this shows that the smallest possible bound is achievable with a bit more work.

All of the aforementioned experiments are summarized in the following table.

	L	n_0	n_1	n_2	n_2'	n_2''
X_0	−0.16446	4	59	74		
X_1	0.43856	5	9	37		
X_2	−0.16446	2	59	70		
X_3	−0.76748	3	3	27		
Y	−0.040422	761	682	182,024	1455	1206
U	−0.14028	9	83	1933	83	
W	0.03333	5	1407	4591	1407	

where n_2' is the result of replacing the tail bound using a little bit of work (e.g., roughly where one could imagine that the necessary estimates could be verified by hand, as in the simple improvement for Y in the foregoing), and n_2'' is the result of replacing the tail bounds by bounds that require a computer to perform all of the verifications, as in the more complicated improved estimate for Y earlier.

We close with some further comments.

1. The primacy of the $1/n$ term is a surprise. If we take the "poor man's approximation" $q_2 \simeq 1/4$, which is a good approximation unless X takes very large values with very small probability, then the error is about $1/2n$, independent of X! And this says that the lower bound n_1, and often the arrival bound n_2, is almost entirely determined by the target error $|L|$. As a first guess, $n_2 \simeq \pi/(2L^2)$, especially if this number is large enough so that the exponential terms are small; if this approximation to n_2 isn't that large then further work might be required to decrease the tail bound.

2. A curious philosophical difficulty is hiding in the weeds. The value of n_0 becomes "obvious" from calculations when the sign of the tilt becomes constant and stays there for as large an n as one cares to compute. However, this gives no hint of how one might prove that this will continue to be the case, and the point of the work here is to be able to actually prove an upper bound on n_0. What computer results are admissible in such a proof? The computation of tilts would seem innocuous to many since any floating point error can be easily bounded, and the programs are short and "easily" proved to be correct — the computer is "just" doing stable, well-understood arithmetic. The error bounds in calculating n_2 straight from Theorem 20.11 can be done by hand, and perhaps the estimates for n_2' could also be done by hand, though few, if any people would do them nowadays without using a computer; the calculations needed to support the determination of n_2'' seem to be intrinsically even more demanding.

3. It would be interesting to apply these ideas here to more general situations (more general RVs, approximation not at the mean, etc.). Extending to higher order Edgeworth approximations seems viable, but would require algebraic stamina. We have thoughts on how this might be automated. However, as noted earlier, we expect that the lower bounds on n will have to increase, so that the utility of this approach isn't entirely clear.

Acknowledgments

We would like to thank Richard Arratia, Steve Butler, Persi Diaconis, Larry Goldstein, Fred Kochman, and Sandy Zabell for interesting and useful feedback.

References

1. Michael Benedicks. An estimate of the modulus of the characteristic function of a lattice distribution with application to remainder term estimates in local limit theorems. *Ann. Probab.* **3** (1975), 162–165.
2. J. P. Buhler, R. L. Graham, and A. W. Hales. Maximally nontransitive dice. *Amer. Math. Monthly*, forthcoming.
3. Rabi N. Bhattacharya and R. Ranga Rao. *Normal Approximation and Asymptotic Expansions*. Society for Industrial and Applied Mathematics (SIAM), Philadelphia, 2010.
4. William Feller. *An Introduction to Probability Theory and Its Applications*. Vol II, 2nd, edn., John Wiley & Sons, New York, 1971.
5. Valentin V. Petrov. *Limit Theorems of Probability Theory*. Oxford Studies in Probability. Clarendon Press, Oxford University Press, New York, 1995.

Printed in the United States
by Baker & Taylor Publisher Services